ELECTRIC MOTORS AND DRIVES

Fundamentals, Types, and Applications

Fourth Edition

AUSTIN HUGHES AND BILL DRURY

Amsterdam • Boston • Heidelberg • London • New York
Oxford • Paris • San Diego • San Francisco • Singapore
Sydney • Tokyo
Newnes is an imprint of Elsevier

Newnes is an imprint of Elsevier
The Boulevard, Langford Lane, Kidlington, Oxford, OX5 1GB
225 Wyman Street, Waltham, MA 02451, USA

First edition 1990
Second edition 1993
Third edition 2006
Reprinted 2006, 2007, 2008 (twice), 2009
Fourth edition 2013

Notices

Knowledge and best practice in this field are constantly changing. As new research and experience broaden our understanding, changes in research methods, professional practices, or medical treatment may become necessary.

Practitioners and researchers must always rely on their own experience and knowledge in evaluating and using any information, methods, compounds, or experiments described herein. In using such information or methods they should be mindful of their own safety and the safety of others, including parties for whom they have a professional responsibility.

To the fullest extent of the law, neither the Publisher nor the authors, contributors, or editors, assume any liability for any injury and/or damage to persons or property as a matter of products liability, negligence or otherwise, or from any use or operation of any methods, products, instructions, or ideas contained in the material herein.

British Library Cataloguing in Publication Data
A catalogue record for this book is available from the British Library

Library of Congress Cataloging-in-Publication Data
A catalog record for this book is available from the Library of Congress

ISBN: 978-0-08-098332-5

Printed and bound in the United Kingdom
13 14 15 16   10 9 8 7 6 5 4 3 2 1

CONTENTS

PREFACE

This fourth edition is again intended primarily for nonspecialist users or students of electric motors and drives. From the outset the aim has been to bridge the gap between specialist textbooks (which are pitched at a level which is too academic for the average user) and the more prosaic handbooks which are full of detailed information but provide little opportunity for the development of any real insight. We intend to continue what has been a successful formula by providing the reader with an understanding of how each motor and drive system works, in the belief that it is only by knowing what should happen (and why) that informed judgements and sound comparisons can be made.

The fact that the book now has joint authors resulted directly from the publisher's successful reviewing process, which canvassed expert opinions about a prospective fourth edition. It identified several new topics needed to bring the work up to date, but these areas were not ones that the original author (AH) was equipped to address, having long since retired. Fortunately, one of the reviewers (WD) turned out to be a willing co-author: he is not only an industrialist (and author) with vast experience in the field, but, at least as importantly, shares the philosophy that guided the first three versions. We enjoy collaborating and hope and believe that our synergy will prove of benefit to our readers.

Given that the book is aimed at readers from a range of disciplines, sections of the book are of necessity devoted to introductory material. The first two chapters therefore provide a gentle introduction to electromagnetic energy conversion and power electronics. Many of the basic ideas introduced here crop up frequently throughout the book (and indeed are deliberately repeated to emphasize their importance), so unless the reader is already well versed in the fundamentals it would be wise to absorb the first two chapters before tackling the later material. At various points later in the book we include more tutorial material, e.g. in Chapter 7 where we prepare the ground for unraveling the mysteries of field-oriented control. A grasp of basic closed-loop principles is also required in order to understand the operation of the various drives, so further introductory material is included in Appendix 1.

The book covers all of the most important types of motor and drive, including conventional and brushless d.c., induction motor, synchronous motors of all types, switched reluctance, and stepping motors (but not highly customized or application-specific systems, e.g. digital hard disk drives). The induction motor and induction motor drives are given most weight, reflecting their dominant market position in terms of numbers. Conventional d.c. machines are deliberately introduced early on, despite their declining importance: this is partly because understanding is relatively easy, but primarily because the fundamental principles that emerge carry forward to other motors. Similarly, d.c. drives are tackled first, because experience shows that readers who manage to grasp the principles of the d.c. drive will find this knowhow invaluable in dealing with other more challenging types.

The third edition has been completely revised and updated. Major additions include an extensive (but largely non-mathematical) treatment of both field-oriented and direct torque control in both induction and synchronous motor drives; a new chapter on permanent magnet brushless machines; new material dealing with self-excited machines, including wind-power generation; and increased emphasis throughout on the inherent ability of electrical machines to act either as a motor or a generator.

Younger readers may be unaware of the radical changes that have taken place over the past 50 years, so a couple of paragraphs are appropriate to put the current scene into perspective. For more than a century, many different types of motor were developed, and each became closely associated with a particular application. Traction, for example, was seen as the exclusive preserve of the series d.c. motor, whereas the shunt d.c. motor, though outwardly indistinguishable, was seen as being quite unsuited to traction applications. The cage induction motor was (and still is) the most numerous type but was judged as being suited only to applications which called for constant speed. The reason for the plethora of motor types was that there was no easy way of varying the supply voltage and/or frequency to obtain speed control, and designers were therefore forced to seek ways of providing for control of speed within the motor itself. All sorts of ingenious arrangements and interconnections of motor windings were invented, but even the best motors had a limited operating range, and they all required bulky electromechanical control gear.

All this changed from the early 1960s, when power electronics began to make an impact. The first major breakthrough came with the thyristor, which provided a relatively cheap, compact, and easily controlled variablespeed drive using the d.c. motor. In the 1970s the second major breakthrough resulted from the development of power electronic inverters, providing a 3phase variable-frequency supply for the cage induction motor and thereby enabling its speed to be controlled. These major developments resulted in the demise of many of the special motors, leaving the majority of applications in the hands of comparatively few types. The switch from analogue to digital control also represented significant progress, but it was the availability of cheap digital processors that sparked the most recent leap forward. Real time modeling and simulation are now incorporated as standard into induction and synchronous motor drives, thereby allowing them to achieve levels of dynamic performance that had long been considered impossible.

The informal style of the book reflects our belief that the difficulty of coming to grips with new ideas should not be disguised. The level at which to pitch the material was based on feedback from previous editions which supported our view that a mainly descriptive approach with physical explanations would be most appropriate, with mathematics kept to a minimum to assist digestion. The most important concepts (such as the inherent e.m.f. feedback in motors, the need for a switching strategy in converters, and the importance of stored energy) are deliberately reiterated to reinforce understanding, but should not prove too tiresome for readers who have already 'got the message'. We have deliberately not included any computed magnetic field plots, nor any results from the excellent motor simulation packages that are now available because experience suggests that simplified diagrams are actually better as learning vehicles.

Finally, we welcome feedback, either via the publisher, or using the e-mail addresses below.

Austin Hughes (a.hughes@leeds.ac.uk)
Bill Drury (w.drury@btinternet.com)

14 October 2012

CHAPTER ONE

Electric Motors – The Basics

1. INTRODUCTION

Electric motors are so much a part of everyday life that we seldom give them a second thought. When we switch on an ancient electric drill, for example, we confidently expect it to run rapidly up to the correct speed, and we don't question how it knows what speed to run at, or how it is that once enough energy has been drawn from the supply to bring it up to speed, the power drawn falls to a very low level. When we put the drill to work it draws more power, and, when we finish, the power drawn from the mains reduces automatically, without intervention on our part.

The humble motor, consisting of nothing more than an arrangement of copper coils and steel laminations, is clearly rather a clever energy converter, which warrants serious consideration. By gaining a basic understanding of how the motor works, we will be able to appreciate its potential and its limitations, and (in later chapters) see how its already remarkable performance is dramatically enhanced by the addition of external electronic controls.

This chapter deals with the basic mechanisms of motor operation, so readers who are already familiar with such matters as magnetic flux, magnetic and electric circuits, torque, and motional e.m.f. can probably afford to skim over much of it. In the course of the discussion, however, several very important general principles and guidelines emerge. These apply to all types of motor and are summarized in section 9. Experience shows that anyone who has a good grasp of these basic principles will be well equipped to weigh the pros and cons of the different types of motor, so all readers are urged to absorb them before tackling other parts of the book.

2. PRODUCING ROTATION

Nearly all motors exploit the force which is exerted on a current-carrying conductor placed in a magnetic field. The force can be demonstrated by placing a bar magnet near a wire carrying current (Figure 1.1), but anyone trying the experiment will probably be disappointed to discover how feeble the force is, and will doubtless be left wondering how such an unpromising effect can be used to make effective motors.

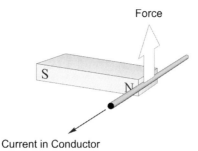

Force

Current in Conductor

Figure 1.1 Mechanical force produced on a current-carrying wire in a magnetic field.

We will see that in order to make the most of the mechanism, we need to arrange for there to be a very strong magnetic field, and for it to interact with many conductors, each carrying as much current as possible. We will also see later that although the magnetic field (or 'excitation') is essential to the working of the motor, it acts only as a catalyst, and all of the mechanical output power comes from the electrical supply to the conductors on which the force is developed.

It will emerge later that in some motors the parts of the machine responsible for the excitation and for the energy-converting functions are distinct and self-evident. In the d.c. motor, for example, the excitation is provided either by permanent magnets or by field coils wrapped around clearly defined projecting field poles on the stationary part, while the conductors on which force is developed are on the rotor and supplied with current via sliding brushes. In many motors, however, there is no such clear-cut physical distinction between the 'excitation' and the 'energy-converting' parts of the machine, and a single stationary winding serves both purposes. Nevertheless, we will find that identifying and separating the excitation and energy-converting functions is always helpful in understanding how motors of all types operate.

Returning to the matter of force on a single conductor, we will look first at what determines the magnitude and direction of the force, before turning to ways in which the mechanism is exploited to produce rotation. The concept of the magnetic circuit will have to be explored, since this is central to understanding why motors have the shapes they do. Before that, a brief introduction to the magnetic field and magnetic flux and flux density is included for those who are not already familiar with the ideas involved.

2.1 Magnetic field and magnetic flux

When a current-carrying conductor is placed in a magnetic field, it experiences a force. Experiment shows that the magnitude of the force depends directly on the current in the wire and the strength of the magnetic field, and that the force is greatest when the magnetic field is perpendicular to the conductor.

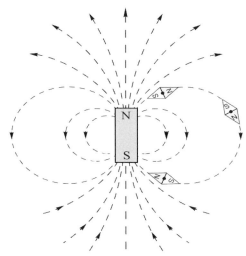

Figure 1.2 Magnetic flux lines produced by a permanent magnet.

In the set-up shown in Figure 1.1, the source of the magnetic field is a bar magnet, which produces a magnetic field as shown in Figure 1.2.

The notion of a 'magnetic field' surrounding a magnet is an abstract idea that helps us to come to grips with the mysterious phenomenon of magnetism: it not only provides us with a convenient pictorial way of visualizing the directional effects, but it also allows us to quantify the 'strength' of the magnetism and hence permits us to predict the various effects produced by it.

The dotted lines in Figure 1.2 are referred to as magnetic flux lines, or simply flux lines. They indicate the direction along which iron filings (or small steel pins) would align themselves when placed in the field of the bar magnet. Steel pins have no initial magnetic field of their own, so there is no reason why one end or the other of the pins should point to a particular pole of the bar magnet.

However, when we put a compass needle (which is itself a permanent magnet) in the field we find that it aligns itself as shown in Figure 1.2. In the upper half of the figure, the S end of the diamond-shaped compass settles closest to the N pole of the magnet, while in the lower half of the figure, the N end of the compass seeks the S of the magnet. This immediately suggests that there is a direction associated with the lines of flux, as shown by the arrows on the flux lines, which conventionally are taken as positively directed from the N to the S pole of the bar magnet.

The sketch in Figure 1.2 might suggest that there is a 'source' near the top of the bar magnet, from which flux lines emanate before making their way to a corresponding 'sink' at the bottom. However, if we were to look at the flux lines inside the magnet, we would find that they were continuous, with no 'start' or 'finish'. (In Figure 1.2 the internal flux lines have been omitted for the sake of clarity, but a very similar field pattern is produced by a circular coil of wire carrying a direct current – see Figure 1.7 where the continuity of the flux lines is clear.) Magnetic flux lines always form closed paths, as we will see when we look at the 'magnetic circuit', and we draw a parallel with the electric circuit, in which the current is also a continuous quantity. (There must be a 'cause' of the magnetic flux, of course, and in a permanent magnet this is usually pictured in terms of atomic-level circulating currents within the magnet material. Fortunately, discussion at this physical level is not necessary for our purposes.)

2.2 Magnetic flux density

As well as showing direction, flux plots convey information about the intensity of the magnetic field. To achieve this, we introduce the idea that between every pair of flux lines (and for a given depth into the paper) there is the same 'quantity' of magnetic flux. Some people have no difficulty with such a concept, while others find that the notion of quantifying something so abstract represents a serious intellectual challenge. But whether the approach seems obvious or not, there is no denying the practical utility of quantifying the mysterious stuff we call magnetic flux, and it leads us next to the very important idea of magnetic flux density (B).

When the flux lines are close together, the 'tube' of flux is squashed into a smaller space, whereas when the lines are further apart the same tube of flux has more breathing space. The flux density (B) is simply the flux in the 'tube' (Φ) divided by the cross-sectional area (A) of the tube, i.e.

$$B = \frac{\Phi}{A} \tag{1.1}$$

The flux density is a vector quantity, and is therefore often written in bold type: its magnitude is given by equation (1.1), and its direction is that of the prevailing flux lines at each point. Near the top of the magnet in Figure 1.2, for example, the flux density will be large (because the flux is squashed into a small area), and pointing upwards, whereas on the equator and far out from the body of the magnet the flux density will be small and directed downwards.

We will see later that in order to create high flux densities in motors, the flux spends most of its life inside well-defined 'magnetic circuits' made of iron or steel, within which the flux lines spread out uniformly to take full advantage of the available area. In the case shown in Figure 1.3, for example, the cross-sectional area of the iron at bb' is twice that at aa', but the flux is constant so the flux density at bb' is half that at aa'.

It remains to specify units for quantity of flux, and flux density. In the SI system, the unit of magnetic flux is the weber (Wb). If one weber of flux is distributed uniformly across an area of one square meter perpendicular to the flux, the flux

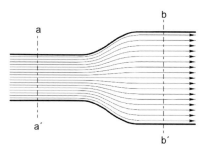

Figure 1.3 Magnetic flux lines inside part of an iron magnetic circuit.

density is clearly one weber per square meter (Wb/m^2). This was the unit of B until about 50 years ago, when it was decided that one weber per square meter would henceforth be known as one tesla (T), in honor of Nikola Tesla, who is generally credited with inventing the induction motor. The widespread use of B (measured in Tesla) in the design stage of all types of electromagnetic apparatus means that we are constantly reminded of the importance of Tesla; but at the same time one has to acknowledge that the outdated unit did have the advantage of conveying directly what flux density is, i.e. flux divided by area.

The flux in a 1 kW motor will be perhaps a few tens of milliwebers, and a small bar magnet would probably only produce a few microwebers. On the other hand, values of flux density are typically around 1 tesla in most motors, which is a reflection of the fact that although the quantity of flux in the 1 kW motor is small, it is also spread over a small area.

2.3 Force on a conductor

We now return to the production of force on a current-carrying wire placed in a magnetic field, as revealed by the set-up shown in Figure 1.1.

The force is shown in Figure 1.1: it is at right angles to both the current and the magnetic flux density, and its direction can be found using Fleming's left-hand rule. If we picture the thumb and the first and middle fingers held mutually perpendicular, then the first finger represents the field or flux density (B), the mIddle finger represents the current (I), and the thumb then indicates the direction of motion, as shown in Figure 1.4.

Clearly, if either the field or the current is reversed, the force acts downwards, and if both are reversed, the direction of the force remains the same.

We find by experiment that if we double either the current or the flux density, we double the force, while doubling both causes the force to increase by a factor of four. But how about quantifying the force? We need to express the force in terms of the product of the current and the magnetic flux density, and this turns out to be very straightforward when we work in SI units.

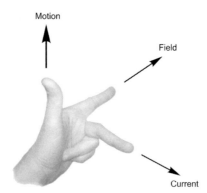

Figure 1.4 Fleming's LH rule for finding direction of force.

The force F on a wire of length l, carrying a current I and exposed to a uniform magnetic flux density B throughout its length is given by the simple expression

$$F = BIl \qquad (1.2)$$

In equation (1.2), F is in newtons when B is in tesla, I in amps, and l in meters.

This is a delightfully simple formula, and it may come as a surprise to some readers that there are no constants of proportionality involved in equation (1.2). The simplicity is not a coincidence, but stems from the fact that the unit of current (the ampere) is actually defined in terms of force.

Equation (1.2) only applies when the current is perpendicular to the field. If this condition is not met, the force on the conductor will be less; and in the extreme case where the current was in the same direction as the field, the force would fall to zero. However, every sensible motor designer knows that to get the best out of the magnetic field it has to be perpendicular to the conductors, and so it is safe to assume in the subsequent discussion that B and I are always perpendicular. In the remainder of this book, it will be assumed that the flux density and current are mutually perpendicular, and this is why, although **B** is a vector quantity (and would usually be denoted by bold type), we can drop the bold notation because the direction is implicit and we are only interested in the magnitude.

The reason for the very low force detected in the experiment with the bar magnet is revealed by equation (1.2). To obtain a high force, we must have a high flux density, and a lot of current. The flux density at the ends of a bar magnet is low, perhaps 0.1 tesla, so a wire carrying 1 amp will experience a force of only 0.1 N/m (approximately 100 gm wt per meter). Since the flux density will be confined to perhaps 1 cm across the end face of the magnet, the total force on the wire will be only 1 gm wt. This would be barely detectable, and is too low to be of any use in a decent motor. So how is more force obtained?

The first step is to obtain the highest possible flux density. This is achieved by designing a 'good' magnetic circuit, and is discussed next. Secondly, as many conductors as possible must be packed in the space where the magnetic field exists, and each conductor must carry as much current as it can without heating up to a dangerous temperature. In this way, impressive forces can be obtained from modestly sized devices, as anyone who has tried to stop an electric drill by grasping the chuck will testify.

3. MAGNETIC CIRCUITS

So far we have assumed that the source of the magnetic field is a permanent magnet. This is a convenient starting point as all of us are familiar with magnets. But in the majority of motors, the magnetic field is produced by coils of wire carrying current, so it is appropriate that we look at how we arrange the coils and their associated 'magnetic circuit' so as to produce high magnetic fields which then interact with other current-carrying conductors to produce force, and hence rotation.

First, we look at the simplest possible case of the magnetic field surrounding an isolated long straight wire carrying a steady current (Figure 1.5). (In the figure, the + sign indicates that current is flowing into the paper, while a dot is used to signify current out of the paper: these symbols can perhaps be remembered by picturing an arrow or dart, with the cross being the rear view of the fletch, and the dot being the approaching point.) The flux lines form circles concentric with the wire, the field strength being greatest close to the wire. As might be expected, the field strength at any point is directly proportional to the current. The convention for determining the direction of the field is that the positive direction is taken to be the direction that a right-handed corkscrew must be rotated to move in the direction of the current.

Figure 1.5 is somewhat artificial as current can only flow in a complete circuit, so there must always be a return path. If we imagine a parallel 'go' and 'return' circuit, for example, the field can be obtained by superimposing the field produced by the positive current in the go side with the field produced by the negative current in the return side, as shown in Figure 1.6.

We note how the field is increased in the region between the conductors, and reduced in the regions outside. Although Figure 1.6 strictly only applies to an infinitely long pair of straight conductors, it will probably not come as a surprise to learn that the field produced by a single turn of wire of rectangular, square or round form is very much the same as that shown in Figure 1.6. This enables us to build up a picture of the field that would be produced – in air – by the sort of coils used in motors, which typically have many turns, as shown, for example, in Figure 1.7.

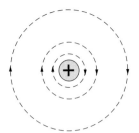

Figure 1.5 Magnetic flux lines produced by a straight, current-carrying wire.

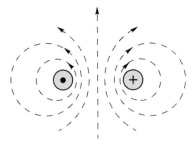

Figure 1.6 Magnetic flux lines produced by current in a parallel go and return circuit.

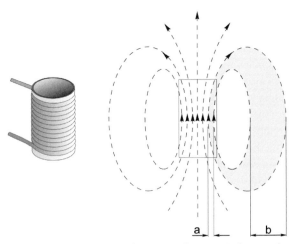

Figure 1.7 Multi-turn cylindrical coil and pattern of magnetic flux produced by current in the coil. (For the sake of clarity, only the outline of the coil is shown on the right.)

The coil itself is shown on the left in Figure 1.7 while the flux pattern produced is shown on the right. Each turn in the coil produces a field pattern, and when all the individual field components are superimposed we see that the field inside the coil is substantially increased and that the closed flux paths closely resemble those of the bar magnet we looked at earlier. The air surrounding the sources of the field offers a homogeneous path for the flux, so once the tubes of flux escape from the concentrating influence of the source, they are free to spread out into the whole of the surrounding space. Recalling that between each pair of flux lines there is an equal amount of flux, we see that because the flux lines spread out as they leave the confines of the coil, the flux density is much lower outside than inside: for example, if the distance 'b' is say four times 'a', the flux density B_b is a quarter of B_a.

Although the flux density inside the coil is higher than outside, we would find that the flux densities which we could achieve are still too low to be of use in a motor. What is needed first is a way of increasing the flux density, and secondly a means for concentrating the flux and preventing it from spreading out into the surrounding space.

3.1 Magnetomotive force (MMF)

One obvious way to increase the flux density is to increase the current in the coil, or to add more turns. We find that if we double the current, or the number of turns, we double the total flux, thereby doubling the flux density everywhere.

We quantify the ability of the coil to produce flux in terms of its magneto-motive force (m.m.f.). The m.m.f. of the coil is simply the product of the number of turns (N) and the current (I), and is thus expressed in ampere-turns. A given m.m.f. can be obtained with a large number of turns of thin wire carrying a low current, or a few turns of thick wire carrying a high current: as long as the product NI is constant, the m.m.f. is the same.

3.2 Electric circuit analogy

We have seen that the magnetic flux which is set up is proportional to the m.m.f. driving it. This points to a parallel with the electric circuit, where the current (amps) which flows is proportional to the electromotive force (e.m.f. volts) driving it.

In the electric circuit, current and e.m.f. are related by Ohm's law, which is

$$\text{Current} = \frac{\text{e.m.f.}}{\text{Resistance}}, \text{ i.e. } I = \frac{V}{R} \qquad (1.3)$$

For a given source e.m.f. (volts), the current depends inversely on the resistance of the circuit, so to obtain more current, the resistance of the circuit has to be reduced.

We can make use of an equivalent 'magnetic Ohm's law' by introducing the idea of reluctance ($\mathscr{R}$). The reluctance gives a measure of how difficult it is for the magnetic flux to complete its circuit, in the same way that resistance indicates how much opposition the current encounters in the electric circuit.

The magnetic Ohm's law is then

$$\text{Flux} = \frac{\text{m.m.f.}}{\text{Reluctance}}, \text{ i.e. } \Phi = \frac{NI}{\mathcal{R}} \tag{1.4}$$

We see from equation (1.4) that, to increase the flux for a given m.m.f., we need to reduce the reluctance of the magnetic circuit. In the case of the example (Figure 1.7), this means we must replace as much as possible of the air path (which is a 'poor' magnetic material, and therefore constitutes a high reluctance) with a 'good' magnetic material, thereby reducing the reluctance and resulting in a higher flux for a given m.m.f.

The material which we choose is good quality magnetic steel, which for historical reasons is often referred to as 'iron'. This brings several very dramatic and desirable benefits, as shown in Figure 1.8.

First, the reluctance of the iron paths is very much less than that of the air paths which they have replaced, so the total flux produced for a given m.m.f. is very much greater. (Strictly speaking therefore, if the m.m.f.s and cross-sections of the coils in Figures 1.7 and 1.8 are the same, many more flux lines should be shown in Figure 1.8 than in Figure 1.7, but for the sake of clarity a similar number are indicated.) Secondly, almost all the flux is confined within the iron, rather than spreading out into the surrounding air. We can therefore shape the iron parts of the magnetic circuit, as shown in Figure 1.8, in order to guide the flux to wherever it is needed. And finally, we see that inside the iron, the flux density remains uniform over the whole cross-section, there being so little reluctance that there is no noticeable tendency for the flux to crowd to one side or another.

Before moving on to the matter of the air-gap, a question which is often asked is whether it is important for the coils to be wound tightly onto the magnetic circuit, and whether, if there is a multi-layer winding, the outer turns are as effective as the inner ones. The answer, happily, is that the total m.m.f. is determined solely by the number of turns and the current, and therefore every complete turn makes the same contribution to the total m.m.f., regardless of whether it happens to be tightly or loosely wound. Of course it does make sense for the coils to be wound as tightly as is

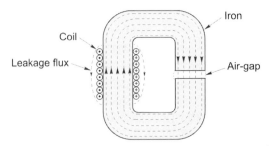

Figure 1.8 Flux lines inside low-reluctance magnetic circuit with air-gap.

practicable, since this not only minimizes the resistance of the coil (and thereby reduces the heat loss) but also makes it easier for the heat generated to be conducted away to the frame of the machine.

3.3 The air-gap

In motors, we intend to use the high flux density to develop force on current-carrying conductors. We have now seen how to create a high flux density in a magnetic circuit, but, of course, it is physically impossible to put current-carrying conductors inside the iron. We therefore arrange for an air-gap in the magnetic circuit, as shown in Figure 1.8. We will see shortly that the conductors on which the force is to be produced will be placed in this air-gap region.

If the air-gap is relatively small, as in motors, we find that the flux jumps across the air-gap as shown in Figure 1.8, with very little tendency to balloon out into the surrounding air. With most of the flux lines going straight across the air-gap, the flux density in the gap region has the same high value as it does inside the iron.

In the majority of magnetic circuits with one or more air-gaps, the reluctance of the iron parts is very much less than the reluctance of the gaps. At first sight this can seem surprising, since the distance across the gap is so much less than the rest of the path through the iron. The fact that the air-gap dominates the reluctance is simply a reflection of how poor air is as a magnetic medium, compared with iron. To put the comparison in perspective, if we calculate the reluctances of two paths of equal length and cross-sectional area, one being in iron and the other in air, the reluctance of the air path will typically be 1000 times greater than the reluctance of the iron path.

Returning to the analogy with the electric circuit, the role of the iron parts of the magnetic circuit can be likened to that of the copper wires in the electric circuit. Both offer little opposition to flow (so that a negligible fraction of the driving force (m.m.f. or e.m.f.) is wasted in conveying the flow to where it is usefully exploited) and both can be shaped to guide the flow to its destination. There is one important difference, however. In the electric circuit, no current will flow until the circuit is completed, after which all the current is confined inside the wires. With an iron magnetic circuit, some flux can flow (in the surrounding air) even before the iron is installed. And although most of the flux will subsequently take the easy route through the iron, some will still leak into the air, as shown in Figure 1.8. We will not pursue leakage flux here, though it is sometimes important, as will be seen later.

3.4 Reluctance and air-gap flux densities

If we neglect the reluctance of the iron parts of a magnetic circuit, it is easy to estimate the flux density in the air-gap. Since the iron parts are then in effect 'perfect conductors' of flux, none of the source m.m.f. (NI) is used in driving the flux through the iron parts, and all of it is available to push the flux across the air-gap.

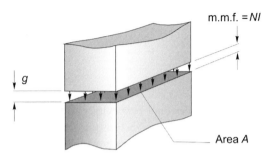

Figure 1.9 Air-gap region, with m.m.f. acting across opposing pole faces.

The situation depicted in Figure 1.8 therefore reduces to that shown in Figure 1.9, where an m.m.f. of NI is applied directly across an air-gap of length g.

To determine how much flux will cross the gap, we need to know its reluctance. As might be expected the reluctance of any part of the magnetic circuit depends on its dimensions and on its magnetic properties, and the reluctance of a rectangular 'prism' of air, of cross-sectional area A and length g, is given by

$$\mathscr{R}_g = \frac{g}{A\mu_0} \tag{1.5}$$

where μ_0 is the so-called 'primary magnetic constant' or 'permeability of free space'. Strictly, as its name implies, μ_0 quantifies the magnetic properties of a vacuum, but for all engineering purposes the permeability of air is also μ_0. The value of the primary magnetic constant (μ_0) in the SI system is $4\pi \times 10^{-7}$ henry/m: rather surprisingly, there is no name for the unit of reluctance.

(In passing, we should note that if we want to include the reluctance of the iron part of the magnetic circuit in our calculation, its reluctance would be given by

$$\mathscr{R}_{\text{iron}} = \frac{l_{\text{iron}}}{A\mu_{\text{iron}}}$$

and we would have to add this to the reluctance of the air-gap to obtain the total reluctance. However, because the permeability of iron (μ_{iron}) is so much higher than μ_0, its reluctance will be very much less than the gap reluctance, despite the path length l_{iron} being considerably longer than the path length (g) in the air.)

Equation (1.5) reveals the expected result that doubling the air-gap would double the reluctance (because the flux has twice as far to go), while doubling the area would halve the reluctance (because the flux has two equally appealing paths in parallel). To calculate the flux, Φ, we use the magnetic Ohm's law (equation (1.4)), which gives

$$\Phi = \frac{\text{m.m.f.}}{\mathscr{R}}, \text{ i.e. } \Phi = \frac{NIA\mu_0}{g} \tag{1.6}$$

We are usually interested in the flux density in the gap, rather than the total flux, so we use equation (1.1) to yield

$$B = \frac{\Phi}{A} = \frac{\mu_0 NI}{g} \tag{1.7}$$

Equation (1.7) is delightfully simple, and from it we can calculate the air-gap flux density once we know the m.m.f. of the coil (NI) and the length of the gap (g). We do not need to know the details of the coil winding as long as we know the product of the turns and the current, and neither do we need to know the cross-sectional area of the magnetic circuit in order to obtain the flux density (though we do if we want to know the total flux; see equation (1.6)).

For example, suppose the magnetizing coil has 250 turns, the current is 2 A, and the gap is 1 mm. The flux density is then given by

$$B = \frac{4\pi \times 10^{-7} \times 250 \times 2}{1 \times 10^{-3}} = 0.63 \text{ tesla}$$

(We could of course create the same flux density with a coil of 50 turns carrying a current of 10 A, or any other combination of turns and current giving an m.m.f. of 500 ampere-turns.)

If the cross-sectional area of the iron was constant at all points, the flux density would be 0.63 T everywhere. Sometimes, as has already been mentioned, the cross-section of the iron reduces at points away from the air-gap, as shown for example in Figure 1.3. Because the flux is compressed in the narrower sections, the flux density is higher, and in Figure 1.3 if the flux density at the air-gap and in the adjacent pole faces is once again taken to be 0.63 T, then at the section aa' (where the area is only half that at the air-gap) the flux density will be $2 \times 0.63 = 1.26$ T.

3.5 Saturation

It would be reasonable to ask whether there is any limit to the flux density at which the iron can be operated. We can anticipate that there must be a limit, or else it would be possible to squash the flux into a vanishingly small cross-section, which we know is not the case. In fact there is a limit, though not a very sharply defined one.

Earlier we noted that the 'iron' has very little reluctance, at least not in comparison with air. Unfortunately this happy state of affairs is only true as long as the flux density remains below about 1.6–1.8 T, depending on the particular magnetic steel in question: if we try to work at higher flux densities, it begins to exhibit significant reluctance, and no longer behaves like an ideal conductor of flux. At these higher flux densities a significant proportion of the source m.m.f. is used in driving the flux through the iron. This situation is obviously undesirable, since less m.m.f. remains to drive the flux across the air-gap. So, just as we would not recommend the use of high-resistance supply leads to the load in an electric circuit, we must avoid overloading the iron parts of the magnetic circuit.

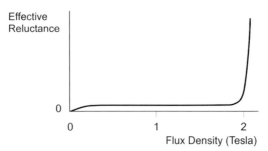

Figure 1.10 Sketch showing how the effective reluctance of iron increases rapidly as the flux density approaches saturation.

The emergence of significant reluctance as the flux density is raised is illustrated qualitatively in Figure 1.10. When the reluctance begins to be appreciable, the iron is said to be beginning to 'saturate'. The term is apt, because if we continue increasing the m.m.f. or reducing the area of the iron, we will eventually reach an almost constant flux density, typically around 2 T. To avoid the undesirable effects of saturation, the sizes of the iron parts of the magnetic circuit are usually chosen so that the flux density does not exceed about 1.5 T. At this level of flux density, the reluctance of the iron parts will be small in comparison with the air-gap.

3.6 Magnetic circuits in motors

The reader may be wondering why so much attention has been focused on the gapped C-core magnetic circuit, when it appears to bear little resemblance to the magnetic circuits found in motors. We will now see that it is actually a short step from the C-core to a typical motor magnetic circuit, and that no fundamentally new ideas are involved.

The evolution from C-core to motor geometry is shown in Figure 1.11, which should be largely self-explanatory, and relates to the field system of a traditional d.c. motor.

We note that the first stage of evolution (Figure 1.11, left) results in the original single gap of length g being split into two gaps of length $g/2$, reflecting the requirement for the rotor to be able to turn. At the same time the single magnetizing coil is split into two to preserve symmetry. (Relocating the magnetizing coil at a different position around the magnetic circuit is of course in order, just as a battery can be placed anywhere in an electric circuit.) Next (Figure 1.11, center), the single magnetic path is split into two parallel paths of half the original cross-section, each of which carries half of the flux; and finally (Figure 1.11, right), the flux paths and pole faces are curved to match the rotor. The coil now has several layers in order to fit the available space, but as discussed earlier this has no adverse effect on the m.m.f. The air-gap is still small, so the flux crosses radially to the rotor.

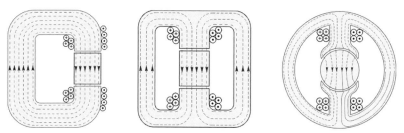

Figure 1.11 Evolution of d.c. motor magnetic circuit from gapped C-core.

4. TORQUE PRODUCTION

Having designed the magnetic circuit to give a high flux density under the poles, we must obtain maximum benefit from it. We therefore need to arrange a set of conductors, fixed to the rotor, as shown in Figure 1.12, and to ensure that conductors under an N-pole (on the left) carry positive current (into the paper), while those under the S-pole carry negative current. The tangential electromagnetic ('BIl') force (see equation (1.2)) on all the positive conductors will be downwards, while the force on the negative ones will be upwards: a torque will therefore be exerted on the rotor, which will be caused to rotate. (The observant reader spotting that some of the conductors appear to have no current in them will find the explanation later, in Chapter 3.)

At this point we should pause and address three questions that often crop up when these ideas are being developed. The first is to ask why we have made no reference to the magnetic field produced by the current-carrying conductors on the rotor. Surely they too will produce a magnetic field, which will presumably interfere with the original field in the air-gap – in which case perhaps the expression used to calculate the force on the conductor will no longer be valid.

The answer to this very perceptive question is that the field produced by the current-carrying conductors on the rotor certainly will modify the original

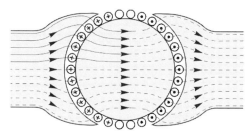

Figure 1.12 Current-carrying conductors on rotor, positioned to maximize torque. (The source of the magnetic flux lines (arrowed) is not shown.)

field (i.e. the field that was present when there was no current in the rotor conductors). But in the majority of motors, the force on the conductor can be calculated correctly from the product of the current and the 'original' field. This is very fortunate from the point of view of calculating the force, but also has a logical feel to it. For example, in Figure 1.1, we would not expect any force on the current-carrying conductor if there was no externally applied field, even though the current in the conductor will produce its own field (upwards on one side of the conductor and downwards on the other). So it seems right that since we only obtain a force when there is an external field, all of the force must be due to that field alone. (In Chapter 3 we will discover that the field produced by the rotor conductors is known as 'armature reaction', and that, especially when the magnetic circuit becomes saturated, its undesirable effects may be combated by fitting additional windings designed to nullify the armature field.)

The second question arises when we think about the action and reaction principle. When there is a torque on the rotor, there is presumably an equal and opposite torque on the stator; and therefore we might wonder if the mechanism of torque production could be pictured using the same ideas as we used for obtaining the rotor torque. The answer is yes, there is always an equal and opposite torque on the stator, which is why it is usually important to bolt a motor down securely. In some machines (e.g. the induction motor) it is easy to see that torque is produced on the stator by the interaction of the air–gap flux density and the stator currents, in exactly the same way that the flux density interacts with the rotor currents to produce torque on the rotor. In other motors (e.g. the d.c. motor we have been looking at), there is no simple physical argument which can be advanced to derive the torque on the stator, but nevertheless it is equal and opposite to the torque on the rotor.

The final question relates to the similarity between the set-up shown in Figure 1.11 and the field patterns produced, for example, by the electromagnets used to lift car bodies in a scrap yard. From what we know of the large force of attraction that lifting magnets can produce, might we not expect there to be a large radial force between the stator pole and the iron body of the rotor? And if there is, what is to prevent the rotor from being pulled across to the stator?

Again the affirmative answer is that there is indeed a radial force due to magnetic attraction, exactly as in a lifting magnet or relay, although the mechanism whereby the magnetic field exerts a pull as it enters iron or steel is entirely different from the 'BIl' force we have been looking at so far.

It turns out that the force of attraction per unit area of pole face is proportional to the square of the radial flux density, and with typical air-gap flux densities of up to 1 T in motors, the force per unit area of rotor surface works out to be about 40 N/cm^2. This indicates that the total radial force can be very large; for example, the force of attraction on a small pole face of only

5 cm × 10 cm is 2000 N, or about 200 kgf. This force contributes nothing to the torque of the motor, and is merely an unwelcome by-product of the '*BIl*' mechanism we employ to produce tangential force on the rotor conductors.

In most machines the radial magnetic force under each pole is actually a good deal bigger than the tangential electromagnetic force on the rotor conductors, and as the question implies, it tends to pull the rotor onto the pole. However, the majority of motors are constructed with an even number of poles equally spaced around the rotor, and the flux density in each pole is the same, so that – in theory at least – the resultant force on the complete rotor is zero. In practice, even a small eccentricity will cause the field to be stronger under the poles where the air-gap is smaller, and this will give rise to an unbalanced pull, resulting in noisy running and rapid bearing wear.

In 99% of motors we can picture how torque is produced via the '*BIl*' approach. The source of the magnetic flux density B may be a winding, as in Figure 1.11, or a permanent magnet. The source (or 'excitation') may be located on the stator (as implied in Figure 1.12) or on the rotor. If the source of B is on the stator, the current carrying conductors on which the force is developed are located on the rotor, whereas if the excitation is on the rotor, the active conductors are on the stator. In all of these '*BIl*' machines, the large radial magnetic forces discussed above are an unwanted by-product.

However, we will see later in the book that in some circumstances the rotor geometry can be arranged so that some of the flux crossing the air-gap to the rotor produces tangential forces (and thus torque) directly on the rotor iron. In these 'reluctance' machines, there are no current-carrying conductors on the rotor, and we have to employ an alternative to the '*BIl*' method to obtain the turning forces.

4.1 Magnitude of torque

Returning to our original discussion, the force on each conductor is given by equation (1.2), and it follows that the total tangential force F depends on the flux density produced by the field winding, the number of conductors on the rotor, the current in each, and the length of the rotor. The resultant torque (T) depends on the radius of the rotor (r), and is given by

$$T = Fr \qquad (1.8)$$

We will return to this after we examine the remarkable benefits gained by putting the rotor conductors into slots.

4.2 The beauty of slotting

If the conductors were mounted on the surface of the rotor iron, as in Figure 1.12, the air-gap would have to be at least equal to the wire diameter, and the conductors would have to be secured to the rotor in order to transmit their turning force to it. The earliest motors were made like this, with string or tape to bind the conductors to the rotor.

Unfortunately, a large air-gap results in an unwelcome high reluctance in the magnetic circuit, and the field winding therefore needs many turns and a high current to produce the desired flux density in the air-gap. This means that the field winding becomes very bulky and consumes a lot of power. The early (nineteenth-century) pioneers soon hit upon the idea of partially sinking the conductors on the rotor into grooves machined parallel to the shaft, the intention being to allow the air-gap to be reduced so that the exciting windings could be smaller. This worked extremely well as it also provided a more positive location for the rotor conductors, and thus allowed the force on them to be transmitted to the body of the rotor. Before long the conductors began to be recessed into ever deeper slots until finally (see Figure 1.13) they no longer stood proud of the rotor surface and the air-gap could be made as small as was consistent with the need for mechanical clearances between the rotor and the stator. The new 'slotted' machines worked very well, and their pragmatic makers were unconcerned by rumblings of discontent from sceptical theorists.

The theorists of the time accepted that sinking conductors into slots allowed the air-gap to be made small, but argued that, as can be seen from Figure 1.13, almost all the flux would now pass down the attractive low-reluctance path through the teeth, leaving the conductors exposed to the very low leakage flux density in the slots. Surely, they argued, little or no '*BIl*' force would be developed on the conductors, since they would only be exposed to a very low flux density.

The sceptics were right in that the flux does indeed flow down the teeth; but there was no denying that motors with slotted rotors produced the same torque as those with the conductors in the air-gap, provided that the average flux densities at the rotor surface were the same. So what could explain this seemingly too good to be true situation?

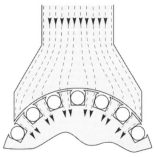

Figure 1.13 Influence on flux paths when the rotor is slotted to accommodate conductors.

The search for an explanation preoccupied some of the leading thinkers long after slotting became the norm, but finally it became possible to show that what happens is that the total force remains the same as it would have been if the conductors were actually in the flux, but almost all of the tangential force now acts on the rotor teeth, rather than on the conductors themselves.

This is remarkably good news. By putting the conductors in slots, we simultaneously enable the reluctance of the magnetic circuit to be reduced, and transfer the force from the conductors themselves to the sides of the iron teeth, which are robust and well able to transfer the resulting torque to the shaft. A further benefit is that the insulation around the conductors no longer has to transmit the tangential forces to the rotor, and its mechanical properties are thus less critical. Seldom can tentative experiments with one aim have yielded rewarding outcomes in almost every other relevant direction.

There are some snags, however. To maximize the torque, we will want as much current as possible in the rotor conductors. Naturally we will work the copper at the highest practicable current density (typically between 2 and 8 A/mm^2), but we will also want to maximize the cross-sectional area of the slots to accommodate as much copper as possible. This will push us in the direction of wide slots, and hence narrow teeth. But we recall that the flux has to pass radially down the teeth, so if we make the teeth too narrow, the iron in the teeth will saturate, and lead to a poor magnetic circuit. There is also the possibility of increasing the depth of the slots, but this cannot be taken too far or the center region of the rotor iron – which has to carry the flux from one pole to another – will become so depleted that it too will saturate. Finally, an unwelcome mechanical effect of slotting is that it increases the frictional drag and acoustic noise, effects which are often minimized by filling the tops of the slot openings so that the rotor becomes smooth.

5. TORQUE AND MOTOR VOLUME

In this section we look at what determines the torque that can be obtained from a rotor of a given size, and see how speed plays a key role in determining the power output.

The universal adoption of slotting to accommodate conductors means that a compromise is inevitable in the crucial air-gap region, and designers constantly have to exercise their skills to achieve the best balance between the conflicting demands on space made by the flux (radial) and the current (axial).

As in most engineering design, guidelines emerge as to what can be achieved in relation to particular sizes and types of machine, and motor designers usually work in terms of two parameters, the specific magnetic loading and the specific electric loading. These parameters will seldom be made available to the user, but, together with the volume of the rotor, they define the torque that can be produced, and are therefore of fundamental importance. An awareness of the existence and signifi-cance of these parameters therefore helps the user to challenge any seemingly extravagant claims that may be encountered.

5.1 Specific loadings

The specific magnetic loading $(\overline{B})$ is the average of the magnitude of the radial flux density over the entire cylindrical surface of the rotor. Because of the slotting, the average flux density is always less than the flux density in the teeth, but in order to calculate the magnetic loading we picture the rotor as being smooth, and calculate the average flux density by dividing the total radial flux from each 'pole' by the surface area under the pole.

The specific electric loading (usually denoted by the symbol $(\overline{A})$, the A standing for amperes) is the axial current per meter of circumference on the rotor. In a slotted rotor, the axial current is concentrated in the conductors within each slot, but to calculate $\overline{A}$ we picture the total current to be spread uniformly over the circumference (in a manner similar to that shown in Figure 1.13, but with the individual conductors under each pole being represented by a uniformly distributed 'current sheet'). For example, if under a pole with a circumferential width of 10 cm we find that there are five slots, each carrying a current of 40 A, the electric loading is

$$\frac{5 \times 40}{0.1} = 2000 \text{ A/m}$$

The discussion in section 4 referred to the conflicting demands of flux and current, so it should be clear that if we seek to increase the electric loading, for example by widening the slots to accommodate more copper, we must be aware that the magnetic loading may have to be reduced because the narrower teeth will mean there is less area for the flux, and therefore a danger of saturating the iron.

Many factors influence the values which can be employed in motor design, but in essence the specific magnetic and electric loadings are limited by the properties of the materials (iron for the flux and copper for the current), and by the cooling system employed to remove heat losses.

The specific magnetic loading does not vary greatly from one machine to another, because the saturation properties of most core steels are similar, so there is an upper limit to the flux density that can be achieved. On the other hand, quite wide variations occur in the specific electric loadings, depending on the type of cooling used.

Despite the low resistivity of the copper conductors, heat is generated by the flow of current, and the current must therefore be limited to a value such that the insulation is not damaged by an excessive temperature rise. The more effective the cooling system, the higher the electric loading can be. For example, if the motor is totally enclosed and has no internal fan, the current density in the copper has to be much lower than in a similar motor which has a fan to provide a continuous flow of ventilating air. Similarly, windings which are fully impregnated with varnish can be worked much harder than those which are surrounded by air, because the solid body of encapsulating varnish provides a much better thermal path along which the heat can flow to the stator body. Overall size also plays a part in determining permissible electric loading, with larger motors generally having higher values than small ones.

In practice, the important point to be borne in mind is that unless an exotic cooling system is employed, most motors (induction, d.c., etc.) of a particular size have more or less the same specific loadings, regardless of type. As we will now see, this in turn means that motors of similar size have similar torque capabilities. This fact is not widely appreciated by users, but is always worth bearing in mind.

5.2 Torque and rotor volume

In the light of the earlier discussion, we can obtain the total tangential force by first considering an area of the rotor surface of width w and length L. The axial current flowing in the width w is given by $I = w\overline{A}$, and on average all of this current is exposed to radial flux density $\overline{B}$ so the tangential force is given (from equation (1.2)) by $\overline{B} \times w\overline{A} \times L$. The area of the surface is wL so the force per unit area is $\overline{B} \times \overline{A}$. We see that the product of the two specific loadings expresses the average tangential stress over the rotor surface.

To obtain the total tangential force we must multiply by the area of the curved surface of the rotor, and to obtain the total torque we multiply the total force by the radius of the rotor. Hence for a rotor of diameter D and length L, the total torque is given by

$$T = (\overline{B}\,\overline{A}) \times (\pi DL) \times \frac{D}{2} = \frac{\pi}{2}(\overline{B}\,\overline{A})D^2 L \qquad (1.9)$$

What this equation tells us is extremely important. The term $D^2 L$ is proportional to the rotor volume, so we see that for given values of the specific magnetic and electric loadings, the torque from any motor is proportional to the rotor volume. We are at liberty to choose a long thin rotor or a short fat one, but once the rotor volume and specific loadings are specified, we have effectively determined the torque.

It is worth stressing that we have not focused on any particular type of motor, but have approached the question of torque production from a completely general viewpoint. In essence our conclusions reflect the fact that all motors are made from iron and copper, and differ only in the way these materials are disposed, and how hard they are worked.

We should also acknowledge that in practice it is the overall volume of the motor which is important, rather than the volume of the rotor. But again we find that, regardless of the type of motor, there is a fairly close relationship between the overall volume and the rotor volume, for motors of similar torque. We can therefore make the bold but generally accurate statement that the overall volume of a motor is determined by the torque it has to produce. There are of course exceptions to this rule, but as a general guideline for motor selection, it is extremely useful.

Having seen that torque depends on rotor volume, we must now turn our attention to the question of power output.

5.3 Output power – importance of speed

Before deriving an expression for power a brief digression may be helpful for those who are more familiar with linear rather than rotary systems.

In the SI system, the unit of work or energy is the joule (J). One joule represents the work done by a force of 1 newton moving 1 meter in its own direction. Hence the work done (W) by a force F which moves a distance d is given by

$$W = F \times d$$

With F in newtons and d in meters, W is clearly in newton–meters (Nm), from which we see that a newton-meter is the same as a joule.

In rotary systems, it is more convenient to work in terms of torque and angular distance, rather than force and linear distance, but these are closely linked as we can see by considering what happens when a tangential force F is applied at a radius r from the center of rotation. The torque is simply given by

$$T = F \times r$$

Now suppose that the arm turns through an angle θ, so that the circumferential distance traveled by the force is $r \times \theta$. The work done by the force is then given by

$$W = F \times (r \times \theta) = (F \times r) \times \theta = T \times \theta \qquad (1.10)$$

We note that whereas in a linear system work is force times distance, in rotary terms work is torque times angle. The units of torque are newton-meters, and the angle is measured in radians (which is dimensionless), so the units of work done are Nm, or joules, as expected. (The fact that torque and work (or energy) are measured in the same units does not seem self-evident to the authors!)

To find the power, or the rate of working, we divide the work done by the time taken. In a linear system, and assuming that the velocity remains constant, power is therefore given by

$$P = \frac{W}{t} = \frac{F \times d}{t} = F \times v, \qquad (1.11)$$

where v is the linear velocity. The angular equivalent of this is

$$P = \frac{W}{t} = \frac{T \times \theta}{t} = T \times \omega, \tag{1.12}$$

where ω is the (constant) angular velocity, in radians per second.

We can now express the power output in terms of the rotor dimensions and the specific loadings using equation (1.9), which yields

$$P = T\omega = \frac{\pi}{2}(\overline{B}\overline{A})D^2L\omega \tag{1.13}$$

Equation (1.13) emphasizes the importance of speed (ω) in determining power output. For given specific and magnetic loadings, if we want a motor of a given power we can choose between a large (and therefore expensive) low-speed motor or a small (and cheaper) high-speed one. The latter choice is preferred for most applications, even if some form of speed reduction (using belts or gears, for example) is needed, because the smaller motor is cheaper. Familiar examples include portable electric tools, where rotor speeds of 12,000 rev/min or more allow powers of hundreds of watts to be obtained, and electric traction: in both the high motor speed is geared down for the final drive. In these examples, where volume and weight are at a premium, a direct drive would be out of the question.

5.4 Power density (specific output power)

By dividing equation (1.13) by the rotor volume, we obtain an expression for the specific power output (power per unit rotor volume), Q, given by

$$Q = \overline{B}\overline{A}\frac{\omega}{2} \tag{1.14}$$

The importance of this simple equation cannot be overemphasized. It is the fundamental design equation that governs the output of any 'BIl' machine, and thus applies to almost all motors.

To obtain the highest possible power from a given volume for given values of the specific magnetic and electric loadings, we must clearly operate the motor at the highest practicable speed. The one obvious disadvantage of a small high-speed motor and gearbox is that the acoustic noise (both from the motor itself and from the power transmission) is higher than it would be from a larger direct drive motor. When noise must be minimized (for example, in ceiling fans), a direct drive motor is therefore preferred, despite its larger size.

In section 5, we began by exploring and quantifying the mechanism of torque production, so not surprisingly it was tacitly assumed that the rotor was at rest, with no work being done. We then moved on to assume that the torque was maintained when the speed was constant and useful power was delivered, i.e. that electrical energy was being converted into mechanical energy. The aim was to establish what factors determine the output of a rotor of given dimensions, and this was possible without reference to any particular type of motor.

In complete contrast, the approach in the next section focuses on a generic 'primitive' motor, and we begin to look in detail at what we have to do at the terminals in order to control the speed and torque.

6. ENERGY CONVERSION – MOTIONAL E.M.F.

We now examine the behavior of a primitive linear machine which, despite its obvious simplicity, encapsulates all the key electromagnetic energy conversion processes that take place in electric motors. We will see how the process of conversion of energy from electrical to mechanical form is elegantly represented in an 'equivalent circuit' from which all the key aspects of motor behavior can be predicted. This circuit will provide answers to such questions as 'How does the motor automatically draw in more power when it is required to work?' and 'What determines the steady speed and current?' Central to such questions is the matter of motional e.m.f., which is explored next.

We have already seen that force (and hence torque) is produced on current-carrying conductors exposed to a magnetic field. The force is given by equation (1.2), which shows that as long as the flux density and current remain constant, the force will be constant. In particular we see that the force does not depend on whether the conductor is stationary or moving. On the other hand, relative movement is an essential requirement in the production of mechanical output power (as distinct from torque), and we have seen that output power is given by the equation $P = T\omega$. We will now see that the presence of relative motion between the conductors and the field always brings 'motional e.m.f.' into play; and we will find that this motional e.m.f. plays a key role in quantifying the energy conversion process.

6.1 Elementary motor – stationary conditions

The primitive linear machine is shown pictorially in Figure 1.14 and in diagrammatic form in Figure 1.15. It consists of a conductor of active[1] length l which can move horizontally perpendicular to a magnetic flux density B.

It is assumed that the conductor has a resistance (R), that it carries a d.c. current (I), and that it moves with a velocity (v) in a direction perpendicular to the field and the current (see Figure 1.15). Attached to the conductor is a string which passes over a pulley and supports a weight, the tension in the string acting as a mechanical 'load' on the rod. Friction is assumed to be zero.

---

[1] The active length is that part of the conductor exposed to the magnetic flux density – in most motors this corresponds to the length of the rotor and stator iron cores.

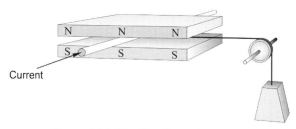

Figure 1.14 Primitive linear d.c. motor.

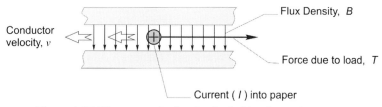

Figure 1.15 Diagrammatic sketch of primitive linear d.c. motor.

We need not worry about the many difficult practicalities of making such a machine, for example how we maintain electrical connections to a moving conductor. The important point is that although this is a hypothetical set-up, it represents what happens in a real motor, and it allows us to gain a clear understanding of how real machines behave before we come to grips with much more complex structures.

We begin by considering the electrical input power with the conductor stationary (i.e. $v = 0$). For the purpose of this discussion we can suppose that the magnetic field (B) is provided by permanent magnets. Once the field has been established (when the magnet was first magnetized and placed in position), no further energy will be needed to sustain the field, which is just as well since it is obvious that an inert magnet is incapable of continuously supplying energy. It follows that when we obtain mechanical output from this primitive 'motor', none of the energy involved comes from the magnet. This is an extremely important point: the field system, whether provided from permanent magnets or 'exciting' windings, acts only as a catalyst in the energy conversion process, and contributes nothing to the mechanical output power.

When the conductor is held stationary the force produced on it (BIl) does no work, so there is no mechanical output power, and the only electrical input power required is that needed to drive the current through the conductor.

The resistance of the conductor is R, the current through it I, so the voltage which must be applied to the ends of the rod from an external source will be given by $V_1 = IR$ and the electrical input power will be V_1I or I^2R. Under these conditions, all the electrical input power will appear as heat inside the conductor, and the power balance can be expressed by the equation

electrical input power $(V_1 I)$ = rate of production of heat in conductor $(I^2 R)$

$$(1.15)$$

Although no work is being done because there is no movement, the stationary condition can only be sustained if there is equilibrium of forces. The tension in the string (T) must equal the gravitational force on the mass (mg), and this in turn must be balanced by the electromagnetic force on the conductor (BIl). Hence under stationary conditions the current must be given by

$$T = mg = BIl, \text{ or } I = \frac{mg}{Bl} \qquad (1.16)$$

This is our first indication of the essential link that always exists (in the steady state) between the mechanical and electric worlds, because we see that in order to maintain the stationary condition, the current in the conductor is determined by the mass of the mechanical load. We will return to this interdependence later.

6.2 Power relationships – conductor moving at constant speed

Now let us imagine the situation where the conductor is moving at a constant velocity (v) in the direction of the electromagnetic driving force that is propelling it. What current must there be in the conductor, and what voltage will have to be applied across its ends?

We start by recognizing that constant velocity of the conductor means that the mass (m) is moving upwards at a constant speed, i.e. it is not accelerating. Hence from Newton's law, there must be no resultant force acting on the mass, so the tension in the string (T) must equal the weight (mg).

Similarly, the conductor is not accelerating, so its net force must also be zero. The string is exerting a braking force (T), so the electromagnetic force (BIl) must be equal to T. Combining these conditions yields

$$T = mg = BIl, \text{ or } I = \frac{mg}{Bl} \qquad (1.17)$$

This is exactly the same equation that we obtained under stationary conditions, and it underlines the fact that the steady-state current is determined by the mechanical load. When we develop the equivalent circuit, we will have to get used to the idea that, in the steady-state, one of the electrical variables (the current) is determined by the mechanical load.

With the mass rising at a constant rate, mechanical work is being done because the potential energy of the mass is increasing. This work is coming from the moving conductor. The mechanical output power is equal to the rate of work, i.e. the force

($T = BIl$) times the velocity (v). The power lost as heat in the conductor is the same as it was when stationary, since it has the same resistance and the same current. The electrical input power supplied to the conductor must continue to furnish this heat loss, but in addition it must now supply the mechanical output power. As yet we do not know what voltage will have to be applied, so we will denote it by V_2. The power balance equation now becomes

$$\text{electrical input power} = \text{rate of production of heat in conductor}$$

$$+ \text{ mechanical output power}$$

i.e.

$$V_2 I = I^2 R + (BIl)v \tag{1.18}$$

We note that the first term on the right-hand side of equation (1.18) represents the heating effect, which is the same as when the conductor was stationary, while the second term corresponds to the additional power that must be supplied to provide the mechanical output. Since the current is the same but the input power is now greater, the new voltage V_2 must be higher than V_1.

By subtracting equation (1.15) from equation (1.18) we obtain

$$V_2 I - V_1 I = (BIl)v,$$

and thus

$$V_2 - V_1 = Blv = E \tag{1.19}$$

Equation (1.19) quantifies the extra voltage to be provided by the source to keep the current constant when the conductor is moving. This increase in source voltage is a reflection of the fact that whenever a conductor moves through a magnetic field, an electromotive force or voltage (E) is induced in it.

We see from equation (1.19) that the e.m.f. is directly proportional to the flux density, to the velocity of the conductor relative to the flux, and to the active length of the conductor. The source voltage has to overcome this additional voltage in order to keep the same current flowing: if the source voltage was not increased, the current would fall as soon as the conductor began to move because of the opposing effect of the induced e.m.f.

We have deduced that there must be an e.m.f. caused by the motion, and have derived an expression for it by using the principle of the conservation of energy, but the result we have obtained, i.e.

$$E = Blv \tag{1.20}$$

is often introduced as the 'flux-cutting' form of Faraday's law, which states that when a conductor moves through a magnetic field an e.m.f. given by equation (1.20) is induced in it. Because motion is an essential part of this mechanism, the e.m.f. induced is referred to as a 'motional e.m.f.'. The 'flux-cutting' terminology arises from attributing the origin of the e.m.f. to the cutting or slicing of the lines of flux by the passage of the conductor. This is a useful mental picture, though it must not be pushed too far: after all, the flux lines are merely inventions which we find helpful in coming to grips with magnet matters.

Before turning to the equivalent circuit of the primitive motor two general points are worth noting. First, whenever energy is being converted from electrical to mechanical form, as here, the induced e.m.f. always acts in opposition to the applied (source) voltage. This is reflected in the use of the term 'back e.m.f.' to describe motional e.m.f. in motors. Secondly, although we have discussed a particular situation in which the conductor carries current, it is certainly not necessary for any current to be flowing in order to produce an e.m.f.: all that is needed is relative motion between the conductor and the magnetic field.

7. EQUIVALENT CIRCUIT

We can represent the electrical relationships in the primitive machine in an equivalent circuit as shown in Figure 1.16.

The resistance of the conductor and the motional e.m.f. together represent in circuit terms what is happening in the conductor (though in reality the e.m.f. and the resistance are distributed, not lumped as separate items). The externally applied source that drives the current is represented by the voltage V on the left (the old-fashioned battery symbol being deliberately used to differentiate the applied voltage V from the induced e.m.f. E). We note that the induced motional e.m.f. is shown as opposing the applied voltage, which applies in the 'motoring' condition we have been discussing. Applying Kirchhoff's law we obtain the voltage equation as

$$V = E + IR, \text{ or } I = \frac{V - E}{R} \tag{1.21}$$

Multiplying equation (1.21) by the current gives the power equation as

electrical input power (VI) = mechanical output power (EI)

$$+ \text{ copper loss } (I^2 R) \tag{1.22}$$

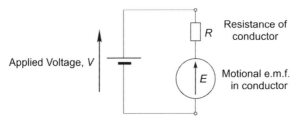

Figure 1.16 Equivalent circuit of primitive d.c. motor.

(Note that the term 'copper loss' used in equation (1.22) refers to the heat generated by the current in the windings: all such losses in electric motors are referred to in this way, even when the conductors are made of aluminium or bronze!)

It is worth seeing what can be learned from these equations because, as noted earlier, this simple elementary 'motor' encapsulates all the essential features of real motors. Lessons which emerge at this stage will be invaluable later, when we look at the way actual motors behave.

7.1 Motoring and generating

If the e.m.f. E is less than the applied voltage V, the current will be positive, and electrical power will flow from the source, resulting in motoring action in which energy is converted from electrical to mechanical form. The first term on the right-hand side of equation (1.22), which is the product of the motional e.m.f. and the current, represents the mechanical output power developed by the primitive linear motor, but the same simple and elegant result applies to real motors. We may sometimes have to be a bit careful if the e.m.f. and the current are not simple d.c. quantities, but the basic idea will always hold good.

Now let us imagine that we push the conductor along at a steady speed that makes the motional e.m.f. greater than the applied voltage. We can see from the equivalent circuit that the current will now be negative (i.e. anticlockwise), flowing back into the supply and thus returning energy to the supply. And if we look at equation (1.22), we see that with a negative current, the first term $(-VI)$ represents the power being returned to the source, the second term $(-EI)$ corresponds to the mechanical power being supplied by us pushing the rod along, and the third term is the heat loss in the conductor.

For readers who prefer to argue from the mechanical standpoint, rather than the equivalent circuit, we can say that when we are generating a negative current $(-I)$, the electromagnetic force on the conductor is $(-BIl)$, i.e. it is directed in the opposite direction to the motion. The mechanical power is given by the product of force and velocity, i.e. $(-BIlv)$, or $-EI$, as above.

The fact that exactly the same kit has the inherent ability to switch from motoring to generating without any interference by the user is an extremely desirable property of all electromagnetic energy converters. Our primitive set-up is simply a machine which is equally at home acting as motor or generator.

Finally, it is obvious that in a motor we want as much as possible of the electrical input power to be converted to mechanical output power, and as little as possible to be converted to heat in the conductor. Since the output power is EI, and the heat loss is I^2R, we see that ideally we want EI to be much greater than I^2R, or in other words E should be much greater than IR. In the equivalent circuit (Figure 1.16) this means that the majority of the applied voltage V is accounted for by the motional e.m.f. (E), and only a little of the applied voltage is used in overcoming the resistance.

8. CONSTANT VOLTAGE OPERATION

Up to now, we have studied behavior under 'steady-state' conditions, which in the context of motors means that the load is constant and conditions have settled to a steady speed. We saw that with a constant load, the current was the same at all steady speeds, the voltage being increased with speed to take account of the rising motional e.m.f. This was a helpful approach to take in order to illuminate the energy conversion process, but is seldom typical of normal operation. We therefore turn to how the moving conductor will behave under conditions where the applied voltage V is constant, since this corresponds more closely with normal operation of a real motor.

Matters inevitably become more complicated because we consider how the motor gets from one speed to another, as well as what happens under steady-state conditions. As in all areas of dynamics, study of the transient behavior of our primitive linear motor brings into play additional parameters, such as the mass of the conductor (equivalent to the inertia of a rotary motor), which are absent from steady-state considerations.

8.1 Behavior with no mechanical load

In this section we assume that the hanging weight has been removed, and that the only force on the conductor is its own electromagnetically generated one. Our primary interest will be in what determines the steady speed of the primitive motor, but we begin by considering what happens when we first apply the voltage.

With the conductor stationary when the voltage V is applied, there is no motional e.m.f. and the current will immediately rise to a value of V/R, since the only thing which limits the current is the resistance. (Strictly we should allow for the effect of inductance in delaying the rise of current, but we choose to ignore it here in the interests of simplicity.) The resistance will be small, so the current will be large, and a high 'BIl' force will therefore be developed on the conductor. The conductor will therefore accelerate at a rate equal to the force on it divided by its mass.

As the speed (v) increases, the motional e.m.f. (equation (1.20)) will grow in proportion to the speed. Since the motional e.m.f. opposes the applied voltage, the current will fall (equation (1.21)), so the force and hence the acceleration will reduce, though the speed will continue to rise. The speed will increase as long as there is an accelerating force, i.e. as long as there is a current in the conductor. We can see from equation (1.21) that the current will finally fall to zero when the speed reaches a level at which the motional e.m.f. is equal to the applied voltage. The speed and current therefore vary as shown in Figure 1.17, both curves having the exponential shape which characterizes the response of systems governed by a first-

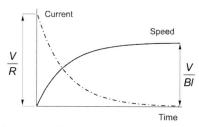

Figure 1.17 Dynamic (run-up) behavior of primitive d.c. motor with no mechanical load.

order differential equation. The fact that the steady-state current is zero is in line with our earlier observation that the mechanical load (in this case zero) determines the steady-state current.

We note that in this idealized situation (in which there is no load applied, and where no friction forces exist), the conductor will continue to travel at a constant speed, because with no net force acting on it there is no acceleration. Of course, no mechanical power is being produced, since we have assumed that there is no opposing force on the conductor, and there is no input power because the current is zero. This hypothetical situation nevertheless corresponds closely to the so-called 'no-load' condition in a motor, the only difference being that a motor will have some friction (and therefore it will draw a small current), whereas we have assumed no friction in order to simplify the discussion.

An elegant self-regulating mechanism is evidently at work here. When the conductor is stationary, it has a high force acting on it, but this force tapers off as the speed rises to its target value, which corresponds to the back e.m.f. being equal to the applied voltage. Looking back at the expression for motional e.m.f. (equation (1.18)), we can obtain an expression for the no-load speed v_0 by equating the applied voltage and the back e.m.f., which gives

$$E = V = Blv_0, \text{ i.e. } v_0 = \frac{V}{Bl} \tag{1.23}$$

Equation (1.23) shows that the steady-state no-load speed is directly proportional to the applied voltage, which indicates that speed control can be achieved by means of the applied voltage. We will see later that one of the main reasons why d.c. motors held sway in the speed-control arena for so long is that their speed can be controlled via the applied voltage.

Rather more surprisingly, equation (1.23) reveals that the speed is inversely proportional to the magnetic flux density, which means that the weaker the field, the higher the steady-state speed. This result can cause raised eyebrows, and with good reason. Surely, it is argued, since the force is produced by the action of the field, the conductor will not go as fast if the field is weaker. This view is wrong, but understandable.

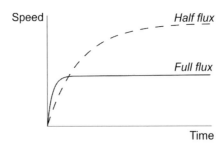

Figure 1.18 Effect of flux density on the acceleration and steady running speed of primitive d.c. motor with no mechanical load.

The flaw in the argument is to equate force with speed. When the voltage is first applied, the force on the conductor certainly will be less if the field is weaker, and the initial acceleration will be lower. But in both cases the acceleration will continue until the current has fallen to zero, and this will only happen when the induced e.m.f. has risen to equal the applied voltage. With a weaker field, the speed needed to generate this e.m.f. will be higher than with a strong field: there is 'less flux', so what there is has to be cut at a higher speed to generate a given e.m.f. The matter is summarized in Figure 1.18, which shows how the speed will rise for a given applied voltage, for 'full' and 'half' fields, respectively. Note that the initial acceleration (i.e. the slope of the speed–time curve) in the half-flux case is half that of the full flux case, but the final steady speed is twice as high. In motors the technique of reducing the flux density in order to increase speed is known as 'field weakening'.

8.2 Behavior with a mechanical load

Suppose that, with the primitive linear motor up to its no-load speed, we suddenly attach the string carrying the weight, so that we now have a steady force T $(= mg)$ opposing the motion of the conductor. At this stage there is no current in the conductor and thus the only force on it will be T. The conductor will therefore begin to decelerate. But as soon as the speed falls, the back e.m.f. will become less than V, and current will begin to flow into the conductor, producing an electromagnetic driving force. The more the speed drops, the bigger the current, and hence the larger the force developed by the conductor. When the force developed by the conductor becomes equal to the load (T), the deceleration will cease, and a new equilibrium condition will be reached. The speed will be lower than at no-load, and the conductor will now be producing continuous mechanical output power, i.e. acting as a motor.

We recall that the electromagnetic force on the conductor is directly proportional to the current, so it follows that the steady-state current is directly proportional to the load which is applied, as we saw earlier. If we were to explore the transient behavior mathematically, we would find that the drop in speed followed the same first-order exponential response that we saw in the run-up period. Once again the self-regulating property is evident, in that when load is applied the speed drops just enough to allow sufficient current to flow to produce the force required to balance the load. We could hardly wish for anything better in terms of performance, yet the conductor does it without any external intervention on our part.

(Readers who are familiar with closed-loop control systems will probably recognize that the reason for this excellent performance is that the primitive motor possesses inherent negative-speed feedback via the motional e.m.f. This matter is explored more fully in Appendix 1.)

Returning to equation (1.21), we note that the current depends directly on the difference between V and E, and inversely on the resistance. Hence for a given resistance, the larger the load (and hence the steady-state current), the greater the required difference between V and E, and hence the lower the steady running speed, as shown in Figure 1.19.

We can also see from equation (1.21) that the higher the resistance of the conductor, the more it slows down when a given load is applied. Conversely, the lower the resistance, the more the conductor is able to hold its no-load speed in the face of applied load, as also shown in Figure 1.19. We can deduce that the only way we could obtain an absolutely constant speed with this type of motor is for the resistance of the conductor to be zero, which is of course not possible. Nevertheless, real d.c. motors generally have resistances which are small, and their speed does not fall much when load is applied – a characteristic which is highly desirable for most applications.

We complete our exploration of the performance when a load is applied by asking how the flux density influences behavior. Recalling that the electromagnetic force is proportional to the flux density as well as the current, we can deduce that to develop a given force, the current required will be higher with a weak flux than with a strong one. Hence in view of the fact that there will always be an upper limit to the current which the conductor can safely carry, the maximum force which can be developed will vary in direct proportion to the flux density, with a weak flux leading to a low maximum force and vice versa. This underlines the importance of operating with maximum flux density whenever possible.

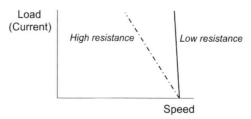

Figure 1.19 Influence of resistance on the ability of the motor to maintain speed when load is applied.

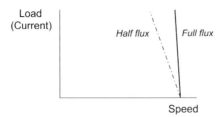

Figure 1.20 Influence of flux on the drop in steady running speed when load is applied.

We can also see another disadvantage of having a low flux density by noting that to achieve a given force, the drop in speed will be disproportionately high when we go to a lower flux density. We can see this by imagining that we want a particular force, and considering how we achieve it first with full flux, and secondly with half flux. With full flux, there will be a certain drop in speed which causes the motional e.m.f. to fall enough to admit the required current. But with half the flux, for example, twice as much current will be needed to develop the same force. Hence the motional e.m.f. must fall by twice as much as it did with full flux. However, since the flux density is now only half, the drop in speed will have to be four times as great as it was with full flux. The half-flux 'motor' therefore has a load characteristic with a load/speed gradient four times more droopy than the full-flux one. This is shown in Figure 1.20, the applied voltages having been adjusted so that in both cases the no-load speed is the same. The half-flux motor is clearly inferior in terms of its ability to hold the set speed when the load is applied.

We may have been tempted to think that the higher speed which we can obtain by reducing the flux somehow makes for better performance, but we can now see that this is not so. By halving the flux, for example, the no-load speed for a given voltage is doubled, but when the load is raised until rated current is flowing in the conductor, the force developed is only half, so the mechanical power is the same. We are in effect trading speed against force, and there is no suggestion of getting something for nothing.

8.3 Relative magnitudes of *V* and *E*, and efficiency

Invariably we want machines which have high efficiency. From equation (1.20), we see that to achieve high efficiency, the copper loss (I^2R) must be small compared with the mechanical power (EI), which means that the resistive volt-drop in the conductor (IR) must be small compared with either the induced e.m.f. (E) or the applied voltage (V). In other words we want most of the applied voltage to be accounted for by the 'useful' motional e.m.f., rather than the wasteful volt-drop in the wire. Since the motional e.m.f. is proportional to speed, and the resistive volt-drop depends on the conductor resistance, we see that a good energy converter requires the conductor resistance to be as low as possible, and the speed to be as high as possible.

To provide a feel for the sorts of numbers likely to be encountered, we can consider a conductor with resistance of 0.5 Ω, capable of carrying a current of 4 A without overheating, and moving at a speed such that the motional e.m.f. is 8 V. From equation (1.19), the supply voltage is given by

$$V = E + IR = 8 + (4 \times 0.5) = 10 \text{ volts}$$

Hence the electrical input power (VI) is 40 watts, the mechanical output power (EI) is 32 watts, and the copper loss (I^2R) is 8 watts, giving an efficiency of 80%.

If the supply voltage was doubled (i.e. $V = 20$ volts), however, and the resisting force is assumed to remain the same (so that the steady-state current is still 4 A), the motional e.m.f. is given by equation (1.21) as

$$E = 20 - (4 \times 0.5) = 18 \text{ volts}$$

which shows that the speed will have rather more than doubled, as expected. The electrical input power is now 80 watts, the mechanical output power is 72 watts, and the copper loss is still 8 watts. The efficiency has now risen to 90%, underlining the fact that the energy conversion process gets better at higher speeds.

When we operate the machine as a generator, we again benefit from the higher speeds. For example, with the battery voltage at a maintained 10 V, and the conductor being propelled by an external force so that its e.m.f. is 12 V, the allowable current of 4 A would now be flowing *into* the battery, with energy being converted from mechanical to electrical form. The power into the battery (VI) is 40 W, the mechanical input power (EI) is 48 W and the heat loss is 8 W. In this case efficiency is defined as the ratio of useful electrical power divided by mechanical input power, i.e. 40/48, or 83.3%.

If we double the battery voltage to 20 V and increase the driven speed so that the motional e.m.f. rises to 22 V, we will again supply the battery with 4 A, but the efficiency will now be 80/88, or 90.9%.

The ideal situation is clearly one where the term IR in equation (1.22) is negligible, so that the back e.m.f. is equal to the applied voltage. We would then have an ideal machine with an efficiency of 100%, in which the steady-state speed would be directly proportional to the applied voltage and independent of the load.

In practice the extent to which we can approach the ideal situation discussed above depends on the size of the machine. Tiny motors, such as those used in wrist-watches, are awful, in that most of the applied voltage is used up in overcoming the resistance of the conductors, and the motional e.m.f. is very small: these motors are much better at producing heat than they are at producing mechanical output power! Small machines, such as those used in hand tools, are a good deal better with the motional e.m.f. accounting for perhaps 70–80% of the applied voltage. Industrial machines are very much better: the largest ones (of many hundreds of kW) use only 1 or 2% of the applied voltage in overcoming resistance, and therefore have very high efficiencies.

8.4 Analysis of primitive machine – conclusions

All of the lessons learned from looking at the primitive machine will find direct parallels in almost all of the motors we look at in the rest of this book, so it is worth reminding ourselves of the key points.

Although this book is primarily about motors, perhaps the most important conclusion so far is that electrical machines are inherently bi-directional energy converters, and any motor can be made to generate, or vice versa. We also saw that the efficiency of the energy-conversion process improves at high speeds, which explains why direct-drive low-speed motors are not widely used.

In terms of the theoretical underpinning, we will make frequent reference to the formula for the force on a conductor in a magnetic field, i.e.

$$\text{Force}, F = BIl \tag{1.24}$$

and to the formula for the motional induced e.m.f., i.e.

$$\text{Motional e.m.f.}, E = Blv \tag{1.25}$$

where B is the magnetic flux density, I is the current, l is the length of conductor and v is the velocity perpendicular to the field.

Specifically in relation to d.c. machines, we have seen that the *speed* at which the primitive motor runs unloaded is determined by the *applied voltage*, while the steady-state *current* that the motor draws is determined by the *mechanical load*. Exactly the same results will hold when we examine real d.c. motors, and very similar relationships will also emerge when we look at the most important type – the induction motor.

9. GENERAL PROPERTIES OF ELECTRIC MOTORS

All electric motors are governed by the laws of electromagnetism, and are subject to essentially the same constraints imposed by the materials (copper, iron and insulation) from which they are made. We should therefore not be surprised to find that at the fundamental level all motors – regardless of type – have a great deal in common.

These common properties, most of which have been touched on in this chapter, are not usually given prominence. Books tend to concentrate on the differences between types of motor, and manufacturers are usually interested in promoting the virtues of their particular motor at the expense of the competition. This divisive emphasis can cause the underlying unity to be obscured, leaving users with little opportunity to absorb the sort of knowledge which will equip them to make informed judgments.

The most useful ideas worth bearing in mind are therefore given below, with brief notes accompanying each. Experience indicates that users who have these basic ideas firmly in mind will find themselves better able to understand why one motor is better than another, and will feel more confident when faced with the difficult task of weighing the pros and cons of competing types.

9.1 Operating temperature and cooling

The cooling arrangement is the single most important factor in determining the permissible output from any given motor.

Any motor will give out more power if its electric circuit is worked harder (i.e. if the current is allowed to increase). The limiting factor is normally the allowable temperature rise of the windings, which depends on the class of insulation.

For class F insulation (the most widely used) the permissible temperature rise is 100 K, whereas for class H it is 125 K. Thus if the cooling remains the same, more output can be obtained simply by using the higher-grade insulation. Alternatively, with a given insulation the output can be increased if the cooling system is improved. A through-ventilated motor, for example, might give perhaps twice the output power of an otherwise identical but totally enclosed machine.

9.2 Torque per unit volume

For motors with similar cooling systems, the rated torque is approximately proportional to the rotor volume, which in turn is roughly proportional to the overall motor volume.

This stems from the fact that for a given cooling arrangement, the specific and magnetic loadings of machines of different types will be more or less the same. The torque per unit length therefore depends first and foremost on the square of the diameter, so motors of roughly the same diameter and length can be expected to produce roughly the same torque.

9.3 Power per unit volume and efficiency – importance of speed

Output power per unit volume is directly proportional to speed.

Low-speed motors are unattractive for most applications because they are large and therefore expensive. It is usually better to use a high-speed motor with a mechanical speed reduction. For example, a direct drive motor for a portable electric screwdriver would be an absurd proposition. On the other hand, the reliability and inefficiency of gearboxes may sometimes outweigh the size argument, especially in high-power applications.

The efficiency of a motor improves with speed.

For a given torque, power output usually rises in direct proportion to speed, while electrical losses tend to rise less rapidly, so that efficiency rises with speed.

9.4 Size effects – specific torque and efficiency

Large motors have a higher specific torque (torque per unit volume) and are more efficient than small ones.

In large motors the specific electric loading is normally much higher than in small ones, and the specific magnetic loading is somewhat higher. These two factors combine to give the higher specific torque.

Very small motors are inherently very inefficient (e.g. 1% in a wrist-watch), whereas motors of over say 100 kW have efficiencies above 96%. The reasons for this scale effect are complex, but stem from the fact that the resistance volt-drop term can be made relatively small in large electromagnetic devices, whereas in small ones the resistance becomes the dominant term.

9.5 Rated voltage

A motor can be provided to suit any voltage.

Within limits it is possible to rewind a motor for a different voltage without affecting its performance. A 200 V, 10 A motor could be rewound for 100 V, 20 A simply by using half as many turns per coil of wire having twice the cross-sectional area. The total amounts of active material, and hence the performance, would be the same. This argument breaks down if pushed too far of course: a very small motor originally wound for 100 V would almost certainly require a larger frame if required to operate at 690 V, because of the additional space required for insulation.

9.6 Short-term overload

Most motors can be overloaded for short periods without damage.

The continuous electric loading (i.e. the current) cannot be exceeded without overheating and damaging the insulation, but if the motor has been running with reduced current for some time, it is permissible for the current (and hence the torque) to be much greater than normal for a short period of time. The principal factors which influence the magnitude and duration of the permissible overload are the thermal time constant (which governs the rate of rise of temperature) and the previous pattern of operation. Thermal time constants range from a few seconds for small motors to many minutes or even hours for large ones. Operating patterns are obviously very variable, so rather than rely on a particular pattern being followed, it is usual for motors to be provided with over-temperature protective devices (e.g. thermistors) which trigger an alarm and/or trip the supply if the safe temperature is exceeded.

CHAPTER TWO

Introduction to Power Electronic Converters for Motor Drives

1. INTRODUCTION

In this chapter we look at examples of the power converter circuits which are widely used with motor drives, providing either d.c. or a.c. outputs, and working from either a d.c. (battery) supply, or from the conventional (50 or 60 Hz) utility supply. The coverage is not intended to be exhaustive, but rather to highlight the most important features and aspects of behavior which recur in many types of drive converter.

Although there are many different types of converter, all except very low-power ones are based on electronic switching. The need to adopt a switching strategy is emphasized in the first example, where the consequences are explored in some depth. We will see that switching is essential in order to achieve high-efficiency power conversion, but that the resulting waveforms are inevitably less than ideal from the point of view of the motor and the power supply.

The examples have been chosen to illustrate typical practice, so for each the most commonly used switching devices (e.g. thyristor, transistor) are shown. In many cases, several different switching devices may be suitable (see later), so we should not identify a particular circuit as being the exclusive preserve of a particular device.

Before discussing particular circuits it will be useful to take an overall look at a typical drive system, so that the role of the converter can be seen in its proper context.

1.1 General arrangement of drive

A complete drive system is shown in block diagram form in Figure 2.1.

The job of the converter is to draw electrical energy from the utility supply (at constant voltage and frequency) and supply electrical energy to the motor at whatever voltage and frequency are necessary to achieve the desired mechanical output. In Figure 2.1, the 'demanded' output is the speed of the motor, but equally it could be the torque, the position of the motor shaft, or some other system variable.

Except in the very simplest converter (such as a basic diode rectifier), there are two distinct parts to the converter. The first is the power stage, through which the

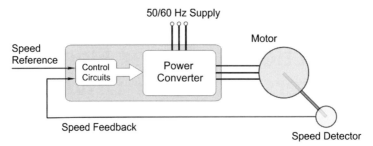

Figure 2.1 General arrangement of speed-controlled drive.

energy flows to the motor, and the second is the control section, which regulates the power flow. Low-power control signals tell the converter what it is supposed to be doing, while other low-power feedback signals are used to measure what is actually happening. By comparing the demand and feedback signals, and adjusting the output accordingly, the target output is maintained.

The basic arrangement shown in Figure 2.1 is clearly a speed control system, because the signal representing the demand or reference quantity is speed, and we note that it is a 'closed-loop' system because the quantity that is to be controlled is measured and fed back to the controller so that action can be taken if the two signals do not correspond. All drives employ some form of closed-loop (feedback) control, so readers who are unfamiliar with the basic principles might find it helpful to read the introduction in Appendix 1.

In later chapters we will explore the internal workings and control arrangements at greater length, but it is worth mentioning that all drives employ current feedback in order to control the motor torque, and that in all except high-performance drives it is unusual to find external transducers, which can account for a significant fraction of the total cost of the drive system. Instead of measuring the actual speed using, for example, a shaft-mounted tachogenerator as shown in Figure 2.1, speed is more likely to be derived from sampled measurements of motor voltages, currents, and frequency, used in conjunction with a stored mathematical model of the motor.

A characteristic of power electronic converters which is shared with most electrical systems is that they have very little capacity for storing energy. This means that any sudden change in the power supplied by the converter to the motor must be reflected in a sudden increase in the power drawn from the supply. In most cases this is not a serious problem, but it does have two drawbacks. First, sudden increases in the current drawn from the supply will cause momentary drops in the supply voltage, because of the effect of the supply impedance. These voltage 'spikes' will appear as unwelcome distortion to other users on the same supply. And secondly, there may be an enforced delay before the supply can furnish extra power. For example, with a single-phase utility supply, there can be no sudden increase in the power supply at the instant where the utility voltage is zero, because instantaneous power is necessarily zero at this point in the cycle since the voltage is itself zero.

It would be ideal if the converter could store at least enough energy to supply the motor for several cycles of the 50/60 Hz supply, so that short-term energy demands could be met instantly, thereby reducing rapid fluctuations in the power drawn from the mains. But unfortunately this is just not economic: most converters do have a small store of energy in their smoothing inductors and capacitors, but the amount is not sufficient to buffer the supply sufficiently to shield it from anything more than very-short-term fluctuations.

2. VOLTAGE CONTROL – D.C. OUTPUT FROM D.C. SUPPLY

In Chapter 1 we saw that control of the basic d.c. machine is achieved by controlling the current in the conductor, which is readily done by variation of the voltage. A controllable voltage source is therefore a key element of a motor drive, as we will see in later chapters.

However, for the sake of simplicity we will begin by exploring the problem of controlling the voltage across a 2 Ω resistive load, fed from a 12 V constant-voltage source such as a battery. Three different methods are shown in Figure 2.2, in which the circle on the left represents an ideal 12 V d.c. source, the tip of the arrow indicating the positive terminal. Although this set-up is not quite the same as if the load was a d.c. motor, the conclusions which we draw are more or less the same.

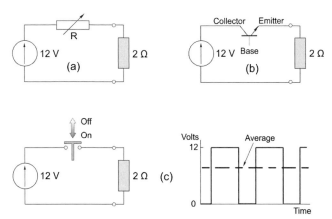

Figure 2.2 Methods of obtaining a variable-voltage output from a constant-voltage source.

Method (a) uses a variable resistor (R) to absorb whatever fraction of the battery voltage is not required at the load. It provides smooth (albeit manual) control over the full range from 0 to 12 V, but the snag is that power is wasted in the control resistor. For example, if the load voltage is to be reduced to 6 V, the resistor R must be set to 2 Ω, so that half of the battery voltage is dropped across R. The current will be 3 A, the load power will be 18 W, and the power dissipated in R will also be 18 W. In terms of overall power conversion efficiency (i.e. useful power delivered to the load divided by total power from the source), the efficiency is a very poor 50%. If R is increased further, the efficiency falls still lower, approaching zero as the load voltage tends to zero. This method of control is therefore unacceptable for motor control, except perhaps in applications such as toy racing cars.

Method (b) is much the same as (a) except that a transistor is used instead of a manually operated variable resistor. The transistor in Figure 2.2(b) is connected with its collector and emitter terminals in series with the voltage source and the load resistor. The transistor is a variable resistor, of course, but a rather special one in which the effective collector–emitter resistance can be controlled over a wide range by means of the base–emitter current. The base–emitter current is usually very small, so it can be varied by means of a low-power electronic circuit (not shown in Figure 2.2) whose losses are negligible in comparison with the power in the main (collector–emitter) circuit.

Method (b) shares the drawback of method (a) above, i.e. the efficiency is very low. But even more seriously, the 'wasted' power (up to a maximum of 18 W in this case) is burned off inside the transistor, which therefore has to be large, well-cooled, and hence expensive. Transistors are never operated in this 'linear' way when used in power electronics, but are widely used as switches, as discussed below.

2.1 Switching control

The basic ideas underlying a switching power regulator are shown by the arrangement in Figure 2.2(c), which uses a mechanical switch. By operating the switch repetitively and varying the ratio of 'on' to 'off' time, the average load voltage can be varied continuously between 0 V (switch off all the time) through 6 V (switch on and off for half of each cycle) to 12 V (switch on all the time).

The circuit shown in Figure 2.2(c) is often referred to as a 'chopper', because the battery supply is 'chopped' on and off. A constant repetition frequency is normally used, and the width of the on pulse is varied to control the mean output voltage (see the waveform in Figure 2.2): this is known as 'pulse-width modulation' (PWM).

The main advantage of the chopper circuit is that no power is wasted, and the efficiency is thus 100%. When the switch is on, current flows through it, but the voltage across it is zero because its resistance is negligible. The power dissipated in the switch is therefore zero. Likewise, when 'off', the current through it is zero, so although the voltage across the switch is 12 V, the power dissipated in it is again zero.

The obvious disadvantage is that by no stretch of the imagination could the load voltage be seen as 'good' d.c.: instead it consists of a mean or 'd.c.' level, with a superimposed 'a.c.' component. Bearing in mind that we really want the load to be a d.c. motor, rather than a resistor, we are bound to ask whether the pulsating voltage will be acceptable. Fortunately, the answer is yes, provided that the chopping frequency is high enough. We will see later that the inductance of the motor causes the current to be much smoother than the voltage, which means that the motor torque fluctuates much less than we might suppose; and the mechanical inertia of the motor filters the torque ripples so that the speed remains almost constant, at a value governed by the mean (or d.c.) level of the chopped waveform.

Obviously a mechanical switch would be unsuitable, and could not be expected to last long when pulsed at high frequency. So an electronic power switch is used instead. The first of many devices to be used for switching was the bipolar junction transistor (BJT), so we will begin by examining how such devices are employed in chopper circuits. If we choose a different device, such as a metal oxide semi-conductor field effect transistor (MOSFET) or an insulated gate bipolar transistor (IGBT), the detailed arrangements for turning the device on and off will be different, but the main conclusions we draw will be much the same.

2.2 Transistor chopper

As noted earlier, a transistor is effectively a controllable resistor, i.e. the resistance between collector and emitter depends on the current in the base–emitter junction. In order to mimic the operation of a mechanical switch, the transistor would have to be able to provide infinite resistance (corresponding to an open switch) or zero resistance (corresponding to a closed switch). Neither of these ideal states can be reached with a real transistor, but both can be approximated closely.

The typical relationship between the collector–emitter voltage and the collector current for a range of base currents rising from zero is shown in Figure 2.3. The bulk of the diagram represents the so-called 'linear' region, where the transistor exhibits the remarkable property of the collector current remaining more or less constant for a wide range of collector–emitter voltages: when the transistor operates in this region there will be significant power loss. For power electronic applications, we want the device to behave like a switch, so we operate on the margins of the diagram, where either the voltage or the current is close to zero, and the heat released inside the device is therefore very low.

The transistor will be 'off' when the base–emitter current (I_b) is zero. Viewed from the main (collector–emitter) circuit, its resistance will be very high, as shown by the region Oa in Figure 2.3.

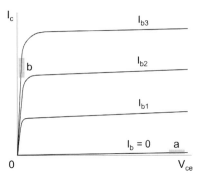

Figure 2.3 Transistor characteristics showing high-resistance (cut-off) region *Oa* and low-resistance (saturation) region *Ob*. Typical 'off' and 'on' operating states are shown by the shaded areas a and b, respectively.

Under this 'cut-off' condition, only a tiny current (I_c) can flow from the collector to the emitter, regardless of the voltage (V_{ce}) between the collector and emitter. The power dissipated in the device will therefore be negligible, giving an excellent approximation to an open switch.

To turn the transistor fully 'on', a base–emitter current must be provided. The base current required will depend on the prospective collector–emitter current, i.e. the current in the load. The aim is to keep the transistor 'saturated' so that it has a very low resistance, corresponding to the region *Ob* in Figure 2.3. In the example shown in Figure 2.2, if the resistance of the transistor is very low, the current in the circuit will be almost 6 A, so we must make sure that the base–emitter current is sufficiently large to ensure that the transistor remains in the saturated condition when $I_c = 6$ A.

Typically in a bipolar transistor (BJT) the base current will need to be around 5–10% of the collector current to keep the transistor in the saturation region: in the example (Figure 2.2), with the full load current of 6 A flowing, the base current might be 400 mA, and the collector–emitter voltage might be say 0.33 V, giving an on-state dissipation of 2 W in the transistor when the load power is 72 W. The power conversion efficiency is not 100%, as it would be with an ideal switch, but it is acceptable.

We should note that the on-state base–emitter voltage is very low, which, coupled with the small base current, means that the power required to drive the transistor is very much less than the power being switched in the collector–emitter circuit. Nevertheless, to switch the transistor in the regular pattern shown in Figure 2.2, we obviously need a base current waveform which goes on and off periodically, and we might wonder how we obtain this 'control' signal. In most modern drives the signal originates from a microprocessor, many of which are designed with PWM auxiliary functions which can be used for this purpose. Depending on the base circuit power requirements of the main switching transistor, it may be possible to feed it directly from the microprocessor, but it is more usual to see additional transistors interposed between the signal source and the main device to provide the required power amplification.

Just as we have to select mechanical switches with regard to their duty, we must be careful to use the right power transistor for the job in hand. In particular, we need to ensure that when the transistor is 'on', we don't exceed the safe current, or else the active semiconductor region of the device will be destroyed by overheating. And we must make sure that the transistor is able to withstand whatever voltage appears across the collector–emitter junction when it is in the 'off' condition. If the safe voltage is exceeded, the transistor will break down, and be permanently 'on'.

A suitable heatsink will be a necessity. We have already seen that some heat is generated when the transistor is on, and at low switching rates this is the main source of unwanted heat. But at high switching rates, 'switching loss' can also be very important.

Switching loss is the heat generated in the finite time it takes for the transistor to go from on to off or vice versa. The base-drive circuitry will be arranged so that the switching takes place as fast as possible, but in practice it will seldom take less than a few microseconds. During the switch-on period, for example, the current will be building up, while the collector–emitter voltage will be falling towards zero. The peak power reached can therefore be large, before falling to the relatively low on-state value. Of course the total energy released as heat each time the device switches is modest because the whole process happens so quickly. Hence if the switching rate is low (say once every second) the switching power loss will be insignificant in comparison with the on-state power. But at high switching rates, when the time taken to complete the switching becomes comparable with the on time, the switching power loss can easily become dominant. In drives, switching rates from hundreds of hertz to tens of kilohertz are used: higher frequencies would be desirable from the point of view of smoothness of supply, but cannot be used because the resultant high switching loss becomes unacceptable.

2.3 Chopper with inductive load – overvoltage protection

So far we have looked at chopper control of a resistive load, but in a drives context the load will usually mean the winding of a machine, which will invariably be inductive.

Chopper control of inductive loads is much the same as for resistive loads, but we have to be careful to prevent the appearance of dangerously high voltages each time the inductive load is switched 'off'. The root of the problem lies with the energy stored in the magnetic field of the inductor. When an inductance L carries a current I, the energy stored in the magnetic field (W) is given by

$$W = \frac{1}{2}LI^2 \qquad (2.1)$$

If the inductor is supplied via a mechanical switch, and we open the switch with the intention of reducing the current to zero instantaneously, we are in effect seeking to destroy the stored energy. This is not possible, and what happens is that the energy is dissipated in the form of a spark across the contacts of the switch.

The appearance of a spark indicates that there is a very high voltage which is sufficient to break down the surrounding air. We can anticipate this by remembering that the voltage and current in an inductance are related by the equation

$$V_L = L \frac{di}{dt} \tag{2.2}$$

which shows that the self-induced voltage is proportional to the rate of change of current, so when we open the switch in order to force the current to zero quickly, a very large voltage is created in the inductance. This voltage appears across the terminals of the switch and, if sufficient to break down the air, the resulting arc allows the current to continue to flow until the stored magnetic energy is dissipated as heat in the arc.

Sparking across a mechanical switch is unlikely to cause immediate destruction, but when a transistor is used sudden death is certain unless steps are taken to tame the stored energy. The usual remedy lies in the use of a 'freewheel diode' (sometimes called a flywheel diode), as shown in Figure 2.4.

A diode is a one-way valve as far as current is concerned: it offers very little resistance to current flowing from anode to cathode (i.e. in the direction of the broad arrow in the symbol for a diode), but blocks current flow from cathode to

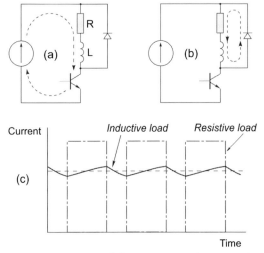

Figure 2.4 Operation of chopper-type voltage regulator.

anode. Actually, when a power diode conducts in the forward direction, the voltage drop across it is usually not all that dependent on the current flowing through it, so the reference above to the diode 'offering little resistance' is not strictly accurate because it does not obey Ohm's law. In practice the volt-drop of power diodes (most of which are made from silicon) is around 0.7 V, regardless of the current rating.

In the circuit of Figure 2.4(a), when the transistor is on, current (I) flows through the load, but not through the diode, which is said to be reverse-biased (i.e. the applied voltage is trying – unsuccessfully – to push current down through the diode). During this period the voltage across the inductance is positive, so the current increases, thereby increasing the stored energy.

When the transistor is turned off, the current through it and the battery drops very quickly to zero. But the stored energy in the inductance means that its current cannot suddenly disappear. So since there is no longer a path through the transistor, the current diverts into the only other route available, and flows upwards through the low-resistance path offered by the diode, as shown in Figure 2.4(b).

Obviously the current no longer has a battery to drive it, so it cannot continue to flow indefinitely. During this period the voltage across the inductance is negative, and the current reduces. If the transistor were left 'off' for a long period, the current would continue to 'freewheel' only until the energy originally stored in the inductance is dissipated as heat, mainly in the load resistance but also in the diode's own (low) resistance. In normal chopping, however, the cycle restarts long before the current has fallen to zero, giving a current waveform as shown in Figure 2.4(c). Note that the current rises and falls exponentially with a time-constant of L/R, though it never reaches anywhere near its steady-state value in Figure 2.4. The sketch corresponds to the case where the time-constant is much longer than one switching period, in which case the current becomes almost smooth, with only a small ripple. In a d.c. motor drive this is just what we want, since any fluctuation in the current gives rise to torque pulsations and consequent mechanical vibrations. (The current waveform that would be obtained with no inductance is also shown in Figure 2.4: the mean current is the same but the rectangular current waveform is clearly much less desirable, given that ideally we would like constant d.c.)

The freewheel (or flywheel) diode was introduced to prevent dangerously high voltages from appearing across the transistor when it switches off an inductive load, so we should check that this has been achieved. When the diode conducts, the volt-drop across it is small – typically 0.7 volts. Hence while the current is free-wheeling, the voltage at the collector of the transistor is only 0.7 volts above the battery voltage. This 'clamping' action therefore limits the voltage across the transistor to a safe value, and allows inductive loads to be switched without damage to the switching element.

We should acknowledge that in this example the discussion has focused on steady-state operation, when the current at the end of every cycle is the same, and it never falls to zero. We have therefore sidestepped the more complex matter of how we get from start-up to the steady state, and we have also ignored the so-called 'discontinuous current' mode. We will touch on the significant consequences of discontinuous operation in drives in later chapters.

We can draw some important conclusions which are valid for all power electronic converters from this simple example. First, efficient control of voltage (and hence power) is only feasible if a switching strategy is adopted. The load is alternately connected and disconnected from the supply by means of an electronic switch, and any average voltage up to the supply voltage can be obtained by varying the mark/space ratio. Secondly, the output voltage is not smooth d.c., but contains unwanted a.c. components which, though undesirable, are tolerable in motor drives. And finally, the load current waveform will be smoother than the voltage waveform if – as is the case with motor windings – the load is inductive.

2.4 Boost converter

The previous section dealt with the so-called step-down or buck converter, which provides an output voltage less than the input. However, if the motor voltage is higher than the supply (for example, in an electric vehicle driven by a 240 V motor from a battery of 48 V), a step-up or boost converter is required. Intuitively this seems a tougher challenge altogether, and we might expect first to have to convert the d.c. to a.c. so that a transformer could be used, but in fact the basic principle of transferring 'packets' of energy to a higher voltage is very simple and elegant, using the circuit shown in Figure 2.5. Operation of this circuit is worth discussing because it again illustrates features common to power-electronic converters.

As is usual, the converter operates repetitively at a rate determined by the frequency of switching on and off the transistor (T). During the 'on' period

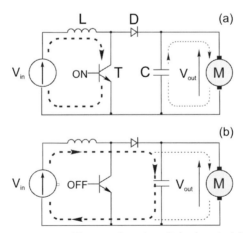

Figure 2.5 Boost converter. The transistor is switched on in (a) and off in (b).

(Figure 2.5(a)), the input voltage (V_{in}) is applied across the inductor (L), causing the current in the inductor to rise linearly, thereby increasing the energy stored in its magnetic field. Meanwhile, the motor current is supplied by the storage capacitor (C), the voltage of which falls only a little during this discharge period. In Figure 2.5(a), the input current is drawn to appear larger than the output (motor) current, for reasons that will soon become apparent. Recalling that the aim is to produce an output voltage greater than the input, it should be clear that the voltage across the diode (D) is negative (i.e. the potential is higher on the right than on the left in Figure 2.5), so the diode does not conduct and the input and output circuits are effectively isolated from each other.

When the transistor turns off, the current through it falls rapidly to zero, and the situation is much the same as in the step-down converter where we saw that, because of the stored energy in the inductor, any attempt to reduce its current results in a self-induced voltage trying to keep the current going. So during the 'off' period, the inductor voltage rises extremely rapidly until the potential at the left side of the diode is slightly greater than V_{out}, the diode then conducts and the inductor current flows into the parallel circuit consisting of the capacitor (C) and the motor, the latter continuing to draw a steady current, while the major share of the inductor current goes to recharge the capacitor. The resulting voltage across the inductor is negative, and of magnitude $V_{out} - V_{in}$, so the inductor current begins to reduce and the extra energy that was stored in the inductor during the 'on' time is transferred into the capacitor. Assuming that we are in the steady state (i.e. power is being supplied to the motor at a constant voltage and current), the capacitor voltage will return to its starting value at the start of the next 'on' time.

If the losses in the transistor and the storage elements are neglected, it is easy to show that the converter functions like an ideal transformer, with output power equal to input power, i.e.

$$V_{in} \times I_{in} = V_{out} \times I_{out}, \text{ or } \frac{V_{out}}{V_{in}} = \frac{I_{in}}{I_{out}}$$

We know that the motor voltage V_{out} is higher than V_{in}, so not surprisingly the quid pro quo is that the motor current is less than the input current, as indicated by line thickness in Figure 2.5.

Also if the capacitor is large enough to hold the output voltage very nearly constant throughout, it is easy to show that the step-up ratio is given by

$$\frac{V_{out}}{V_{in}} = \frac{1}{1 - D}$$

where D is the duty ratio, i.e. the proportion of each cycle for which the transistor is on. Thus for the example at the beginning of this section, where $V_{out}/V_{in} = 240/48 = 5$, the duty ratio is 0.8, i.e. the transistor has to be on for 80% of each cycle.

The advantage of switching at a high frequency becomes clear when we recall that the capacitor has to store enough energy to supply the output during the 'on' period, so if, as is often the case, we want the output voltage to remain almost constant, it should be clear that the capacitor must store a good deal more energy than it gives out each cycle. Given that the size and cost of capacitors depends on the energy they store, it is clearly better to supply small packets of energy at a high rate, rather than use a lower frequency that requires more energy to be stored. Conversely, the switching and other losses increase with frequency, so a compromise is inevitable.

3. D.C. FROM A.C. – CONTROLLED RECTIFICATION

The vast majority of drives of all types draw their power from a constant-voltage 50 or 60 Hz utility supply, and in nearly all converters the first stage consists of a rectifier which converts the a.c. to a crude form of d.c. Where a constant-voltage (i.e. unvarying average) 'd.c.' output is required, a simple (uncontrolled) diode rectifier is sufficient. But where the mean d.c. voltage has to be controllable (as in a d.c. motor drive to obtain varying speeds), a controlled rectifier is used.

Many different converter configurations based on combinations of diodes and thyristors are possible, but we will focus on 'fully controlled' converters in which all the rectifying devices are thyristors, because they are predominant in modern motor drives. Half-controlled converters are used less frequently, so will not be covered here.

From the user's viewpoint, interest centers on the following questions:
- How is the output voltage controlled?
- What does the converter output voltage look like? Will there be any problems if the voltage is not pure d.c.?
- How does the range of the output voltage relate to the utility supply voltage?
- How is the converter and motor drive 'seen' by the supply system? What is the power-factor, and is there waveform distortion and interference to other users?

We can answer these questions without going too thoroughly into the detailed workings of the converter. This is just as well, because understanding all the ins and outs of converter operation is beyond our scope. On the other hand, it is well worth trying to understand the essence of the controlled rectification process, because it assists in understanding the limitations which the converter puts on drive performance (see Chapter 4, etc.). Before tackling the questions posed above, however, it is obviously necessary to introduce the thyristor.

3.1 The thyristor

The thyristor is an electronic switch, with two main terminals (anode and cathode) and a 'switch-on' terminal (gate), as shown in Figure 2.6. Like in a diode, current

Figure 2.6 Circuit diagram of thyristor.

can only flow in the forward direction, from anode to cathode. But unlike a diode, which will conduct in the forward direction as soon as forward voltage is applied, the thyristor will continue to block forward current until a small current pulse is injected into the gate-cathode circuit, to turn it on or 'fire' it. After the gate pulse is applied, the main anode–cathode current builds up rapidly, and as soon as it reaches the 'latching' level, the gate pulse can be removed and the device will remain 'on'.

Once established, the anode–cathode current cannot be interrupted by any gate signal. The non-conducting state can only be restored after the anode–cathode current has reduced to zero, and has remained at zero for the turn-off time (typically 100–200 µs).

When a thyristor is conducting it approximates to a closed switch, with a forward drop of only one or two volts over a wide range of current. Despite the low volt-drop in the 'on' state, heat is dissipated, and heatsinks must usually be provided, perhaps with fan cooling. Devices must be selected with regard to the voltages to be blocked and the r.m.s. and peak currents to be carried. Their overcurrent capability is very limited, and it is usual in drives for devices to have to withstand perhaps twice full-load current for a few seconds only. Special fuses must be fitted to protect against heavy fault currents.

The reader may be wondering why we need the thyristor, since in the previous section we discussed how a transistor could be used as an electronic switch. On the face of it the transistor appears even better than the thyristor because it can be switched off while the current is flowing, whereas the thyristor will remain on until the current through it has been reduced to zero by external means. The primary reason for the use of thyristors is that they are cheaper and their voltage and current ratings extend to higher levels than in power transistors. In addition, the circuit configuration in rectifiers is such that there is no need for the thyristor to be able to interrupt the flow of current, so its inability to do so is no disadvantage. Of course there are other circuits (see, for example, the next section dealing with inverters) where the devices need to be able to switch off on demand, in which case the transistor then has the edge over the thyristor.

3.2 Single pulse rectifier

The simplest phase-controlled rectifier circuit is shown in Figure 2.7. When the supply voltage is positive, the thyristor blocks forward current until the gate pulse arrives, and up to this point the voltage across the resistive load is zero. As soon as a firing pulse is delivered to the gate-cathode circuit (not shown in Figure 2.7) the device turns on, the voltage across it falls to near zero, and the load voltage becomes

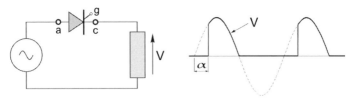

Figure 2.7 Simple single-pulse thyristor-controlled rectifier, with resistive load and firing angle delay α.

equal to the supply voltage. When the supply voltage reaches zero, so does the current. At this point the thyristor regains its blocking ability, and no current flows during the negative half-cycle.

If we neglect the small on-state volt-drop across the thyristor, the load voltage (Figure 2.7) will consist of part of the positive half-cycles of the a.c. supply voltage. It is obviously not smooth, but is 'd.c.' in the sense that it has a positive mean value; and by varying the delay angle (α) of the firing pulses the mean voltage can be controlled. With a purely resistive load, the current waveform will simply be a scaled version of the voltage.

This arrangement gives only one peak in the rectified output for each complete cycle of the supply, and is therefore known as a 'single-pulse' or half-wave circuit. The output voltage (which ideally we would like to be steady d.c.) is so poor that this circuit is never used in drives. Instead, drive converters use four or six thyristors, and produce much superior output waveforms with two or six pulses per cycle, as will be seen in the following sections.

3.3 Single-phase fully controlled converter – output voltage and control

The main elements of the converter circuit are shown in Figure 2.8. It comprises four thyristors, connected in bridge formation. (The term bridge stems from early four-arm measuring circuits which presumably suggested a bridge–like structure to their inventors.)

The conventional way of drawing the circuit is shown in Figure 2.8(a), while in Figure 2.8(b) it has been redrawn to assist understanding. The top of the load can be connected (via T1) to terminal A of the supply, or (via T2) to terminal B of the supply, and likewise the bottom of the load can be connected either to A or to B via T3 or T4, respectively.

We are naturally interested to find what the output voltage waveform on the d.c. side will look like, and in particular to discover how the mean d.c. level can be controlled by varying the firing delay angle α. The angle α is measured from the point on the waveform when a diode in the same circuit position would start to conduct, i.e. when the anode becomes positive with respect to the cathode.

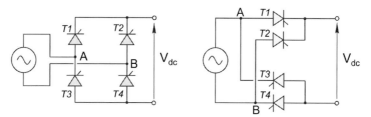

Figure 2.8 Single-phase 2-pulse (full-wave) fully controlled rectifier.

This is not such a simple matter as we might have expected, because it turns out that the mean voltage for a given α depends on the nature of the load. We will therefore look first at the case where the load is resistive, and explore the basic mechanism of phase control. Later, we will see how the converter behaves with a typical motor load.

3.3.1 Resistive load

Thyristors T1 and T4 are fired together when terminal A of the supply is positive, while on the other half-cycle, when B is positive, thyristors T2 and T3 are fired simultaneously. The output voltage and current waveform are shown in Figure 2.9(a) and (b), respectively, the current simply being a replica of the voltage. There are two pulses per mains cycle, hence the description '2-pulse' or full-wave.

At every instant the load is either connected to the mains by the pair of switches T1 and T4, or it is connected the other way up by the pair of switches T2 and T3, or it is disconnected. The load voltage therefore consists of rectified chunks of the incoming supply voltage. It is much smoother than in the single-pulse circuit, though again it is far from pure d.c.

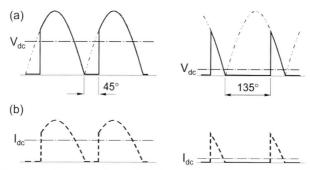

Figure 2.9 Output voltage waveform (a) and current (b) of single-phase fully controlled rectifier with resistive load, for firing angle delays of 45° and 135°.

The waveforms in Figure 2.9 correspond to $\alpha = 45°$ and $\alpha = 135°$, respectively. The mean value, V_{dc}, is shown in each case. It is clear that the larger the delay angle, the lower the output voltage. The maximum output voltage (V_{do}) is obtained with $\alpha = 0°$: this is the same as would be obtained if the thyristors were replaced by diodes, and is given by

$$V_{do} = \frac{2}{\pi}\sqrt{2}V_{rms} \tag{2.3}$$

where V_{rms} is the r.m.s. voltage of the incoming supply. The variation of the mean d.c. voltage with α is given by

$$V_{dc} = \left\{\frac{1 + \cos\alpha}{2}\right\}V_{do} \tag{2.4}$$

from which we see that with a resistive load the d.c. voltage can be varied from a maximum of V_{do} down to zero by varying α from $0°$ to $180°$.

3.3.2 Inductive (motor) load

As mentioned above, motor loads are inductive, and we have seen earlier that the current cannot change instantaneously in an inductive load. We must therefore expect the behavior of the converter with an inductive load to differ from that with a resistive load, in which the current was seen to change instantaneously.

The realization that the mean voltage for a given firing angle might depend on the nature of the load is a most unwelcome prospect. What we would like is to be able to say that, regardless of the load, we can specify the output voltage waveform once we have fixed the delay angle α. We would then know what value of α to select to achieve any desired mean output voltage. What we find in practice is that once we have fixed α, the mean output voltage with a resistive–inductive load is not the same as with a purely resistive load, and therefore we cannot give a simple general formula for the mean output voltage in terms of α. This is of course very undesirable: if, for example, we had set the speed of our unloaded d.c. motor to the target value by adjusting the firing angle of the converter to produce the correct mean voltage, the last thing we would want is for the voltage to fall when the load current drawn by the motor increased, as this would cause the speed to fall below the target.

Fortunately, however, it turns out that the output voltage waveform for a given α does become independent of the load inductance once there is sufficient inductance to prevent the load current from ever falling to zero. This condition is known as 'continuous current', and, happily, many motor circuits do have sufficient self-inductance to ensure that we achieve continuous current. Under continuous current conditions, the output voltage waveform only depends on the firing angle, and not on the actual inductance present. This makes things much more

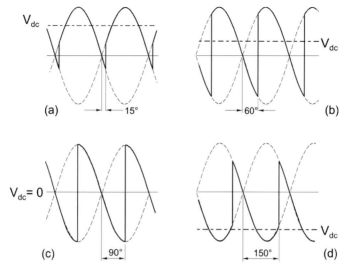

Figure 2.10 Output voltage waveforms of single-phase fully controlled rectifier supplying an inductive (motor) load, for various firing angles.

straightforward, and typical output voltage waveforms for this continuous current condition are shown in Figure 2.10.

The waveforms in Figure 2.10 show that, as with the resistive load, the larger the delay angle the lower the mean output voltage. However, with the resistive load the output voltage was never negative, whereas we see that, although the mean voltage is positive for values of α below $90°$, there are brief periods when the output voltage becomes negative. This is because the inductance smooths out the current (see Figure 4.2, for example) so that at no time does it fall to zero. As a result, one or other pair of thyristors is always conducting, so at every instant the load is connected directly to the supply, and therefore the load voltage always consists of chunks of the supply voltage.

Rather surprisingly, we see that when α is greater than $90°$, the average voltage is negative (though, of course, the current is still positive). The fact that we can obtain a net negative output voltage with an inductive load contrasts sharply with the resistive load case, where the output voltage could never be negative. The combination of negative voltage and positive current means that the power flow is reversed, and energy is fed back to the supply system. We will see later that this facility allows the converter to return energy from the load to the supply, and this is important when we want to use the converter with a d.c. motor in the regenerating mode.

It is not immediately obvious why the current switches over (or 'commutates') from the first pair of thyristors to the second pair when the latter are fired,

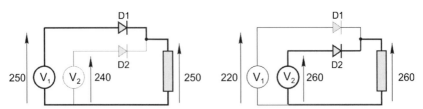

Figure 2.11 Diagram illustrating commutation between diodes: the current flows through the diode with the higher anode potential.

so a brief look at the behavior of diodes in a similar circuit configuration may be helpful at this point. Consider the set-up shown in Figure 2.11, with two voltage sources (each time-varying) supplying a load via two diodes. The question is, what determines which diode conducts, and how does this influence the load voltage?

We can consider two instants as shown in the diagram. On the left, V_1 is 250 V, V_2 is 240 V, and D1 is conducting, as shown by the heavy line. If we ignore the volt-drop across the diode, the load voltage will be 250 V, and the voltage across diode D2 will be $240 - 250 = -10$ V, i.e. it is reverse-biased and hence in the non-conducting state. At some other instant (on the right of the diagram), V_1 has fallen to 220 V while V_2 has increased to 260 V: now D2 is conducting instead of D1, again shown by the heavy line, and D1 is reverse-biased by -40 V. The simple pattern is that the diode with the highest anode potential will conduct, and as soon as it does so it automatically reverse-biases its predecessor. In a single-phase diode bridge, for example, the commutation occurs at the point where the supply voltage passes through zero: at this instant the anode voltage on one pair goes from positive to negative, while on the other pair the anode voltage goes from negative to positive.

The situation in controlled thyristor bridges is very similar, except that before a new device can take over conduction, it must not only have a higher anode potential, but it must also receive a firing pulse. This allows the changeover to be delayed beyond the point of natural (diode) commutation by the angle α, as shown in Figure 2.10. Note that the maximum mean voltage (V_{do}) is again obtained when α is zero, and is the same as for the resistive load (equation (2.3)). It is easy to show that the mean d.c. voltage is now related to α by

$$V_{dc} = V_{d_o}\cos\alpha \qquad (2.5)$$

This equation indicates that we can control the mean output voltage by controlling α, though equation (2.5) shows that the variation of mean voltage with α is different from that for a resistive load (equation (2.4)), not least because when α is greater than 90° the mean output voltage is negative.

It is sometimes suggested (particularly by those with a light-current background) that a capacitor could be used to smooth the output voltage, this being common practice in cheap low-power d.c. supplies. However, the power levels in most drives are such that in order to store enough energy to smooth the voltage waveform over the half-cycle of the utility supply (20 ms at 50 Hz), very bulky and expensive capacitors would be required. Fortunately, as will be seen later, it is not necessary for the voltage to be smooth as it is the current which directly determines the torque, and as already pointed out the current is always much smoother than the voltage because of inductance.

3.4 Three-phase fully controlled converter

The main power elements are shown in Figure 2.12. The 3-phase bridge has only two more thyristors than the single-phase bridge, but the output voltage waveform is vastly better, as shown in Figure 2.13. There are now six pulses of the output voltage per cycle, hence the description '6-pulse'. The thyristors are again fired in

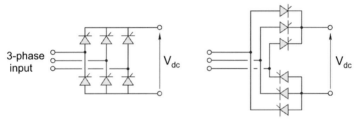

Figure 2.12 Three-phase fully controlled thyristor converter. (The alternative diagram (right) is intended to assist understanding.)

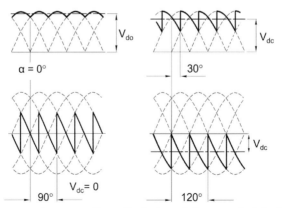

Figure 2.13 Output voltage waveforms for 3-phase fully controlled thyristor converter supplying an inductive (motor) load, for various firing angles from 0° to 120°. The mean d.c. voltage is shown by the horizontal line, except for $\alpha = 90°$ where the mean d.c. voltage is zero.

pairs (one in the top half of the bridge and one – from a different leg – in the bottom half), and each thyristor carries the output current for one-third of the time. As in the single-phase converter, the delay angle controls the output voltage, but now $\alpha = 0$ corresponds to the point at which the phase voltages are equal (see Figure 2.13).

The enormous improvement in the smoothness of the output voltage waveform is clear when we compare Figures 2.13 and 2.10, and it underlines the benefit of choosing a 3-phase converter whenever possible. The very much better voltage waveform also means that the desirable 'continuous current' condition is much more likely to be met, and the waveforms in Figure 2.13 have therefore been drawn with the assumption that the load current is in fact continuous. Occasionally, even a 6-pulse waveform is not sufficiently smooth, and some very large drive converters therefore consist of two 6-pulse converters with their outputs in series. A phase-shifting transformer is used to insert a 30° shift between the a.c. supplies to the two 3-phase bridges. The resultant ripple voltage is then 12-pulse.

Returning to the 6-pulse converter, the mean output voltage can be shown to be given by

$$V_{dc} = V_{do}\cos\alpha = \frac{3}{\pi}\sqrt{2}V_{rms}\cos\alpha \qquad (2.6)$$

We note that we can obtain the full range of output voltages from $+V_{do}$ to close to $-V_{do}$, so that, as with the single-phase converter, regenerative operation will be possible.

It is probably a good idea at this point to remind the reader that, in the context of this book, our first application of the controlled rectifier will be to supply a d.c. motor. When we examine the d.c. motor drive in Chapter 4, we will see that it is the average or mean value of the output voltage from the controlled rectifier that determines the speed, and it is this mean voltage that we refer to when we talk of 'the' voltage from the converter. We must not forget the unwanted a.c. or ripple element, however, as this can be large. For example, we see from Figure 2.13 that to obtain a very low d.c. voltage (to make the motor run very slowly) α will be close to 90°; but if we were to connect an a.c. voltmeter to the output terminals it could register several hundred volts, depending on the incoming supply voltage!

3.5 Output voltage range

In Chapter 4 we will discuss the use of the fully controlled converter to drive a d.c. motor, so it is appropriate at this stage to look briefly at the typical voltages we can expect. Utility supply voltages vary, but single-phase supplies are usually 220–240 V, and we see from equation (2.3) that the maximum mean d.c. voltage available from a single-phase 240 V supply is 216 V. This is suitable for 180–200 V motors. If a higher voltage is needed (for say a 300 V motor), a transformer must be used to step up the incoming supply.

Turning now to typical 3-phase supplies, the lowest 3-phase industrial voltages are usually around 380–440 V. (Higher voltages of up to 11 kV are used for large drives, but these will not be discussed here.) So with $V_{rms} = 400$ V, for example, the maximum d.c. output voltage (equation (2.6)) is 540 volts. After allowances have been made for supply variations and impedance drops, we are not able to rely on obtaining much more than 500–520 V, and it is usual for the motors used with 6-pulse drives fed from 400 V, 3-phase supplies to be rated in the range 430–500 V. (Often the motor's field winding will be supplied from single-phase 230 V, and field voltage ratings are then around 180–200 V, to allow a margin in hand from the theoretical maximum of 207 V referred to earlier.)

3.6 Firing circuits

Since the gate pulses are only of low power, the gate drive circuitry is simple and cheap. Often a single chip contains all the circuitry for generating the gate pulses, and for synchronizing them with the appropriate delay angle α with respect to the supply voltage. To avoid direct electrical connection between the high voltages in the main power circuit and the low voltages used in the control circuits, the gate pulses are coupled to the thyristor either by small pulse transformers or opto-couplers.

---

4. A.C. FROM D.C. – INVERSION

The business of getting a.c. from d.c. is known as inversion, and nine times out of ten we would ideally like to be able to produce sinusoidal output voltages of whatever frequency and amplitude we choose, to achieve speed control of a.c. motors. Unfortunately the constraints imposed by the necessity to use a switching strategy mean that we always have to settle for a voltage waveform which is composed of rectangular chunks, and is thus far from ideal. Nevertheless it turns out that a.c. motors are remarkably tolerant, and will operate satisfactorily despite the inferior waveforms produced by the inverter.

4.1 Single-phase inverter

We can illustrate the basis of inverter operation by considering the single-phase example shown in Figure 2.14. This inverter uses IGBTs (see later) as the switching devices, with diodes to provide the freewheel paths needed when the load is inductive.

The input or d.c. side of the inverter (on the left in Figure 2.14) is usually referred to as the 'd.c. link', reflecting the fact that in the majority of cases the d.c. is

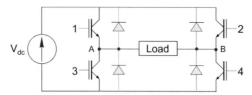

Figure 2.14 Single-phase inverter.

obtained by rectifying the incoming constant-frequency utility supply. The output or a.c. side is taken from terminals A and B in Figure 2.14.

When transistors 1 and 4 are switched on, the load voltage is positive, and equal to the d.c. link voltage, while when 2 and 3 are on it is negative. If no devices are switched on, the output voltage is zero. Typical output voltage waveforms at low and high switching frequencies are shown in Figures 2.15(a) and (b), respectively.

Here each pair of devices is on for one-third of a cycle, and all the devices are off for two periods of one-sixth of a cycle. The output waveform is clearly not a sinewave, but at least it is alternating and symmetrical. The fundamental component is shown dotted in Figure 2.15.

Within each cycle the pattern of switching is regular, and easily generated in software or programmed using appropriate logic circuitry. Frequency variation is obtained by altering the clock frequency controlling the 4-step switching pattern. The effect of varying the switching frequency is shown in Figure 2.15, from which we can see that the amplitude of the fundamental component of voltage remains constant, regardless of frequency. Unfortunately (as explained in Chapter 7) this is not what we want for supplying an induction motor: to prevent the air-gap flux in the motor from falling as the frequency is raised we need to be able to increase the voltage in proportion to the frequency. We will look at voltage control shortly, after a brief digression to discuss the problem of 'shoot-through'.

Inverters with the configurations shown in Figures 2.14 and 2.17 are subject to a potentially damaging condition which can arise if both transistors in one 'leg' of the inverter inadvertently turn on simultaneously. This should never happen if the devices are switched correctly, but if something goes wrong and both devices are on together – even for a very short time – they form a short-circuit across the d.c. link. This fault condition is referred to as 'shoot-through' because a high current is established very rapidly, destroying the devices. A good inverter therefore includes

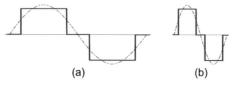

(a)   (b)

Figure 2.15 Inverter output voltage waveforms – resistive load.

provision for protecting against the possibility of shoot-through, usually by imposing a minimum time-delay, or 'dead time' between one device in the leg going off and the other coming on.

4.2 Output voltage control

There are two ways in which the amplitude of the output voltage can be controlled. First, if the d.c. link is provided from the utility supply via a controlled rectifier or from a battery via a chopper, the d.c. link voltage can be varied. We can then set the amplitude of the output voltage to any value within the range of the link. For a.c. motor drives (see Chapter 7) we can arrange for the link voltage to track the output frequency of the inverter, so that at high output frequency we obtain a high output voltage and vice versa. This method of voltage control results in a simple inverter, but requires a controlled (and thus relatively expensive) rectifier for the d.c. link.

The second method, which predominates in all sizes, achieves voltage control by PWM within the inverter itself. A cheaper uncontrolled rectifier can then be used to provide a constant-voltage d.c. link. The principle of voltage control by PWM is illustrated in Figure 2.16.

At low output frequencies, a low output voltage is usually required, so one of each pair of devices is used to chop the voltage, the mark/space ratio being varied to achieve the desired voltage at the output. The low fundamental voltage component at low frequency is shown as a broken line in Figure 2.16(a). At a higher frequency a higher voltage is needed, so the chopping device is allowed to conduct for a longer fraction of each cycle, giving the higher fundamental output shown in Figure 2.16(b). As the frequency is raised still higher, the separate 'on' periods eventually merge, giving the waveform shown in Figure 2.16(c). Any further increase in frequency takes place without further increase in the output voltage, as shown in Figure 2.16(d).

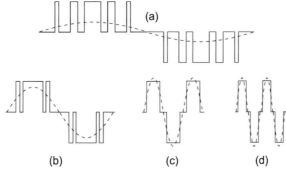

Figure 2.16 Inverter output voltage and frequency control with pulse-width modulation.

When we study a.c. drives later, we will see that the range of frequencies over which the voltage/frequency ratio can be kept constant is sometimes known as the 'constant-torque' region, and the upper limit of the range is usually taken to define the 'base speed' of the motor. Above this frequency, the inverter can no longer match voltage to frequency, the inverter effectively having run out of steam as far as voltage is concerned. The maximum voltage is thus governed by the link voltage, which must therefore be sufficiently high to provide whatever fundamental voltage the motor needs at its base speed.

Beyond the PWM region the voltage waveform is as shown in Figure 2.16(d): this waveform is usually referred to as 'quasi-square', though in the light of the overall object of the exercise (to approximate to a sinewave) a better description might be 'quasi-sine'.

When supplying an inductive motor load, fast recovery freewheel diodes are needed in parallel with each device. These may be discrete devices, or fitted in a common package with the transistor, or integrated to form a single transistor/diode device.

As mentioned earlier, the switching nature of these converter circuits results in waveforms which contain not only the required fundamental component but also unwanted harmonic voltages. It is particularly important to limit the magnitude of the low-order harmonics because these are most likely to provoke an unwanted torque response from the motor, but the high-order harmonics can lead to acoustic noise if they happen to excite a mechanical resonance.

The number, width and spacing of the pulses are therefore optimized to keep the harmonic content as low as possible. There is an obvious advantage in using a high switching frequency since there are more pulses to play with. Ultrasonic frequencies are now widely used, and, as devices improve, frequencies continue to rise. Most manufacturers claim their particular system is better than the competition, but it is not clear which, if any, is best for motor operation. Some early schemes used comparatively few pulses per cycle, and changed the number of pulses in discrete steps rather than smoothly, which earned them the nickname 'gear-changers': their sometimes irritating sound can still be heard in older traction applications.

4.3 Three-phase inverter

Single-phase inverters are seldom used for drives, the vast majority of which use 3-phase motors because of their superior characteristics. A 3-phase output can be obtained by adding only two more switches to the four needed for a single-phase inverter, giving the typical power-circuit configuration shown in Figure 2.17. As usual, a freewheel diode is required in parallel with each transistor to protect against overvoltages caused by an inductive (motor) load. We note that the circuit configuration in Figure 2.17 is the same as for the 3-phase controlled rectifier discussed earlier. We mentioned then that the controlled rectifier could be used to

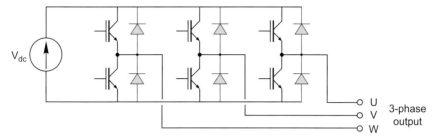

Figure 2.17 Three-phase inverter power circuit.

regenerate, i.e. to convert power from d.c. to a.c., and this is of course 'inversion' as we now understand it.

This circuit forms the basis of the majority of converters for motor drives, and will be explored more fully in Chapter 7. In essence, the output voltage and frequency are controlled in much the same way as for the single-phase inverter discussed in the previous section, but of course the output consists of three identical waveforms displaced by 120° from each other, typically as shown in Figure 2.18.

There are more constraints on switching with this set-up than in the single-phase case. For example, consider the upper and middle waveforms in Figure 2.18: the upper shows the potential of line U with respect to line V (V_{UV}), while the middle shows the potential of line V with respect to line W (V_{VW}). So whenever V_{UV} is positive, the upper right switch and the lower middle switch must be 'on'. If at this instant V_{VW} is also required to be positive we have a problem because this would require the upper middle switch and the lower left switches to be on, which would lead to a short-circuit through the middle leg. Fortunately, these potential problems are limited to well-defined regions of the cycle and switching patterns include built-in delays to prevent such 'shoot-through' faults from occurring.

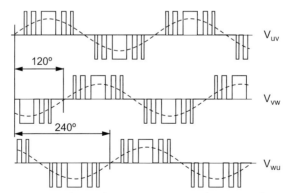

Figure 2.18 Three-phase PWM output voltage waveforms.

5. A.C. FROM A.C.

The converters we have looked at so far have involved a d.c. stage, but the ideal power-electronic converter would allow power conversion in either direction between two systems of any voltage and frequency (including d.c.), and would not involve any intermediate stage, such as a d.c. link. In principle this can be achieved by means of an array of switches that allow any one of a set of input terminals to be connected to any one of a set of output terminals, at any desired instant. The comparatively recent name for such converters is 'matrix converter' (see Chapter 8), but here we will take a brief look at a much older embodiment of the principle – the cycloconverter.

5.1 Cycloconverter

The cycloconverter variable-frequency drive has never become very widespread but is still suitable for very large (e.g. 1 MW and above) low-speed induction motor or synchronous motor drives. The cycloconverter is only capable of producing acceptable output waveforms at frequencies well below the utility frequency, but this, coupled with the fact that it is feasible to make large motors with high pole-numbers (e.g. 20), means that a very low-speed direct (gearless) drive becomes practicable. A 20-pole motor, for example, will have a synchronous speed of only 30 rev/min at 5 Hz, making it suitable for mine winders, kilns, crushers, etc.

The principal advantage of the cycloconverter is that naturally commutated devices (thyristors) can be used instead of self-commutating devices, which means that the cost of each device is lower and higher powers can be achieved.

The power conversion circuit for each of the three output phases is the same, so we can consider the simpler question of how to obtain a single variable-frequency supply from a 3-phase supply of fixed frequency and constant voltage. We will see that in essence, the output voltage is synthesized by switching the load directly across whichever phase of the utility supply gives the best approximation to the desired load voltage at each instant of time.

Assuming that the load is an induction motor, we will discover later in the book that the power-factor varies with load but never reaches unity, i.e. that the current is never in phase with the stator voltage. So during the positive half-cycle of the voltage waveform the current will be positive for some of the time, but negative for the remainder, while during the negative voltage half-cycle the current will be negative for some of the time and positive for the rest of the time. This means that the supply to each phase has to be able to handle any combination of both positive and negative voltage and current.

We have already explored how to achieve a variable-voltage d.c. supply using a thyristor converter, which can handle positive currents, but here we need to

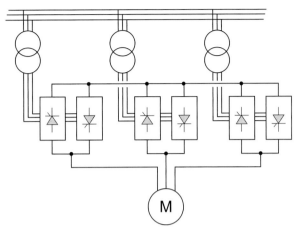

Figure 2.19 Cycloconverter power circuit.

handle negative current as well, so for each of the three motor windings we will need two converters connected in parallel, as shown in Figure 2.19, making a total of 36 thyristors. To avoid short-circuits, isolating transformers are used, again as shown in Figure 2.19.

Previously the discussion focused on the mean or d.c. level of the output voltages, but here we want to provide a low-frequency (preferably sinusoidal) output voltage for an induction motor, and the means for doing this should now be becoming clear. Once we have a double thyristor converter, we can generate a low-frequency sinusoidal output voltage simply by varying the firing angle of the positive-current bridge so that its output voltage increases from zero in a sinusoidal manner with respect to time. Then, when we have completed the positive half-cycle and arrived back at zero voltage, we bring the negative bridge into play and use it to generate the negative half-cycle, and so on.

Hence the output voltage consists of chunks of the incoming supply voltage, and it offers a reasonable approximation to the fundamental-frequency sinewave shown by the dashed line in Figure 2.20.

The output voltage waveform is certainly no worse than the voltage waveform from a d.c. link inverter, and, as we saw in that context, the current waveform in the motor will be much smoother than the voltage waveform, because of the filtering action of the stator leakage inductance. The motor performance will therefore be acceptable, despite the extra losses which arise from the unwanted harmonic components. We should note that, because each phase is supplied from a double converter, the motor can regenerate when required (e.g. to restrain an overhauling load, or to return kinetic energy to the supply when the frequency is lowered to reduce speed).

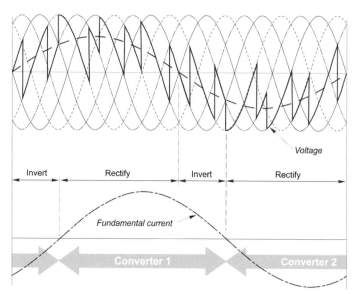

Figure 2.20 *Typical output voltage waveform for one phase of 6-pulse cycloconverter supplying an inductive (motor) load.* (The output frequency shown in the figure is one-third of the utility frequency, and the amplitude of the fundamental component of the output voltage (shown by the dashed line) is about 75% of the maximum that could be obtained. The fundamental component of the load current is shown in order to define the modes of operation of the converters.)

6. INVERTER SWITCHING DEVICES

As far as the user is concerned, it does not really matter what type of switching device is used inside the inverter, but it is probably helpful to mention the three most important[1] families of devices in current use so that the terminology is familiar and the symbols used for each device can be recognized. The common feature of all three devices is that they can be switched on and off by means of a low-power control signal, i.e. they are self-commutating. We have seen earlier that this ability to be turned on or off on demand is essential in any inverter which feeds a passive load, such as an induction motor.

Each device is discussed briefly below, with a broad indication of its most likely range of application. Because there is considerable overlap between competing devices, it is not possible to be dogmatic and specify which device is best, and the reader should not be surprised to find that one manufacturer may offer a 5 kW inverter which uses MOSFETs while another chooses to use IGBTs.

---

[1] The gate turn-off thyristor is now seldom used, so is not discussed here.

Power electronics is still developing, and there are other devices (such as those based on silicon carbide) which have yet to emerge onto the drives scene. Trends which continue include the integration of the drive and protection circuitry in the same package as the switching device (or devices); major reductions in device losses; and improved cooling techniques, all of which have already contributed to impressive reductions in product size.

6.1 Bipolar junction transistor (BJT)

Historically the bipolar junction transistor was the first to be used for power switching. Of the two versions (npn and pnp) only the npn has been widely used in inverters for drives, mainly in applications ranging up to a few kW and several hundred volts.

The npn version is shown in Figure 2.21: the main (load) current flows into the collector (C) and out of the emitter (E), as shown by the arrow on the device symbol. To switch the device on (i.e. to make the resistance of the collector–emitter circuit low, so that load current can flow) a small current must be caused to flow from the base (B) to the emitter. When the base–emitter current is zero, the resistance of the collector–emitter circuit is very high, and the device is switched off.

The advantage of the bipolar transistor is that when it is turned on, the collector–emitter voltage is low (see point b in Figure 2.3) and hence the power dissipation is small in comparison with the load power, i.e. the device is an efficient power switch. The disadvantage is that although the power required in the base–emitter circuit is tiny in comparison with the load power, it is not insignificant and in the largest power transistors can amount to several tens of watts.

6.2 Metal oxide semiconductor field effect transistor (MOSFET)

In the 1980s the power MOSFET superseded the BJT in inverters for drives. Like the BJT, the MOSFET is a three-terminal device and is available in two versions, the n-channel and the p-channel. The n-channel is the most widely used, and is shown in Figure 2.21. The main (load) current flows into the drain (D) and out of the source (S). (Confusingly, the load current in this case flows in the *opposite*

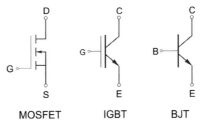

Figure 2.21 Circuit symbols for switching devices.

direction to the arrow on the symbol.) Unlike the BJT, which is controlled by the base *current*, the MOSFET is controlled by the gate-source *voltage*.

To turn the device on, the gate-source voltage must be comfortably above a threshold of a few volts. When the voltage is first applied to the gate, currents flow in the parasitic gate-source and gate-drain capacitances, but once these capacitances have been charged the input current to the gate is negligible, so the steady-state gate drive power is minimal. To turn the device off, the parasitic capacitances must be discharged and the gate-source voltage must be held below the threshold level.

The principal advantage of the MOSFET is that it is a voltage-controlled device which requires negligible power to hold it in the on state. The gate drive circuitry is thus less complex and costly than the base drive circuitry of an equivalent bipolar device. The disadvantage of the MOSFET is that in the 'on' state the effective resistance of the drain-source is higher than for an equivalent bipolar device, so the power dissipation is higher and the device is rather less efficient as a power switch. MOSFETs are used in low and medium power inverters up to a few kW, with voltages generally not exceeding 700 V.

6.3 Insulated gate bipolar transistor (IGBT)

The IGBT (Figure 2.21) is a hybrid device which combines the best features of the MOSFET (i.e. ease of gate turn-on and turn-off from low-power logic circuits) and the BJT (relatively low power dissipation in the main collector–emitter circuit). These obvious advantages give the IGBT the edge over the MOSFET and BJT, and account for its dominance in all but small drives. It is particularly well suited to the medium power, medium voltage range (up to several hundred kW). The path for the main (load) current is from collector to emitter, as in the npn bipolar device.

7. CONVERTER WAVEFORMS, ACOUSTIC NOISE, AND COOLING

In common with most textbooks, the waveforms shown in this chapter (and later in the book) are what we would hope to see under ideal conditions. It makes sense to concentrate on these ideal waveforms from the point of view of gaining a basic understanding, but we ought to be warned that what we see on an oscilloscope may well look rather different, and is often not easy to interpret.

We have seen that the essence of power electronics is the switching process, so it should not come as much of a surprise to learn that in practice the switching is seldom achieved in such a clear-cut fashion as we have assumed. Usually, there will be some sort of high-frequency oscillation or 'ringing' evident, particularly on the voltage waveforms following each transition due to switching. This is due to the effects of stray capacitance and inductance: it will have been anticipated at the design stage, and steps will have been taken to minimize it by fitting 'snubbing' circuits at the appropriate places in the converter. However, complete suppression of all these transient phenomena is seldom economically worthwhile so the user should not be too alarmed to see remnants of the transient phenomena in the output waveforms.

Acoustic noise is also a matter that can worry newcomers. Most power electronic converters emit whining or humming sounds at frequencies corresponding to the fundamental and harmonics of the switching frequency, though when the converter is used to feed a motor, the sound from the motor is usually a good deal louder than the sound from the converter itself. These sounds are very difficult to describe in words, but typically range from a high-pitched hum through a whine to a piercing whistle. In a.c. drives it is often possible to vary the switching frequency, which may allow specific mechanical resonances to be avoided. If taken above the audible range (which is inversely proportional to age) the sound clearly disappears, but at the expense of increased switching losses, so a compromise has to be sought.

7.1 Cooling of switching devices – thermal resistance

We have seen that by adopting a switching strategy the power loss in the switching devices is small in comparison with the power throughput, so the converter has a high efficiency. Nevertheless almost all the heat which is produced in the switching devices is released in the active region of the semiconductor, which is itself very small and will overheat and break down unless it is adequately cooled. It is therefore essential to ensure that even under the most onerous operating conditions, the temperature of the active junction inside the device does not exceed the safe value.

Consider what happens to the temperature of the junction region of the device when we start from cold (i.e. ambient) temperature and operate the device so that its average power dissipation remains constant. At first, the junction temperature begins to rise, so some of the heat generated is conducted to the metal case, which stores some heat as its temperature rises. Heat then flows into the heatsink (if fitted), which begins to warm up, and heat begins to flow to the surrounding air, at ambient temperature. The temperatures of the junction, case and heatsink continue to rise until eventually an equilibrium is reached when the total rate of loss of heat to ambient temperature is equal to the power dissipation inside the device.

The final steady-state junction temperature thus depends on how difficult it is for the power loss to escape down the temperature gradient to ambient, or in other words on the total 'thermal resistance' between the junction inside the device and the surrounding medium (usually air). Thermal resistance is usually expressed in °C/watt, which directly indicates how much temperature rise will occur in the steady state for every watt of dissipated power. It follows that for a given power dissipation, the higher the thermal resistance, the higher the temperature rise, so in order to minimize the temperature rise of the device, the total thermal resistance between it and the surrounding air must be made as small as possible.

The device designer aims to minimize the thermal resistance between the semiconductor junction and the case of the device, and provides a large and flat metal mounting surface to minimize the thermal resistance between the case and the heatsink. The converter designer must ensure good thermal contact between the device and the heatsink, usually by a bolted joint smeared with heat-conducting compound to fill any microscopic voids, and must design the heatsink to minimize the thermal resistance to air (or in some cases oil or water). Heatsink design offers the only real scope for appreciably reducing the total resistance, and involves careful selection of the material, size, shape and orientation of the heatsink, and the associated air-moving system (see below).

One drawback of the good thermal path between the junction and case of the device is that the metal mounting surface (or surfaces in the case of the popular hockey-puck package) can be electrically 'live'. This poses a difficulty for the converter designer, because mounting the device directly on the heatsink causes the latter to be dangerous. In addition, several separate isolated heatsinks may be required in order to avoid short-circuits. The alternative is for the devices to be electrically isolated from the heatsink using thin mica spacers, but then the thermal resistance is increased appreciably.

Increasingly devices come in packaged 'modules' with an electrically isolated metal to get round the 'live' problem. The packages contain combinations of transistors, diodes or thyristors, from which various converter circuits can be built up. Several modules can be mounted on a single heatsink, which does not have to be isolated from the enclosure or cabinet. They are available in ratings suitable for converters up to hundreds of kW, and the range is expanding.

7.2 Arrangement of heatsinks and forced-air cooling

The principal factors which govern the thermal resistance of a heatsink are the total surface area, the condition of the surface and the air flow. Many converters use extruded aluminium heatsinks, with multiple fins to increase the effective cooling surface area and lower the resistance, and with a machined face or faces for mounting the devices. Heatsinks are usually mounted vertically to improve natural air convection. Surface finish is important, with black anodized aluminium being typically 30% better than bright. The cooling performance of heatsinks is, however, a complex technical area, with turbulence being very beneficial in forced air-cooled heatsinks.

A typical layout for a medium-power (say 200 kW) converter is shown in Figure 2.22.

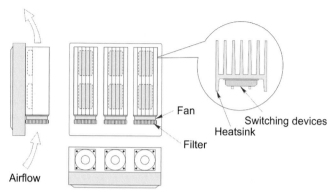

Figure 2.22 Layout of converter showing heatsink and cooling fans.

The fan(s) are positioned at either the top or bottom of the heatsink, and draw external air upwards, assisting natural convection. Even a modest airflow is very beneficial: with an air velocity of only 2 m/s, for example, the thermal resistance is halved as compared with the naturally cooled set-up, which means that for a given temperature rise the heatsink can be half the size of the naturally cooled one. However, large increases in the air velocity bring diminishing returns and also introduce additional noise, which is generally undesirable.

CHAPTER THREE

Conventional D.C. Motors

1. INTRODUCTION

Until the 1980s the conventional (brushed) d.c. machine was the automatic choice where speed or torque control is called for, and large numbers remain in service despite a declining market share that reflects the general move to inverter-fed a.c. motors. D.C. motor drives do remain competitive in some larger ratings (several hundred kW) particularly where drip-proof motors are acceptable, with applications ranging from steel rolling mills, railway traction, through a very wide range of industrial drives.

Given the reduced importance of the d.c. motor, the reader may wonder why a whole chapter is devoted to it. The answer is that, despite its relatively complex construction, the d.c. machine is relatively simple to understand, not least because of the clear physical distinction between its separate 'flux' and 'torque producing' parts. We will find that its performance can be predicted with the aid of a simple equivalent circuit, and that many aspects of its behavior are reflected in other types of motor, where it may be more difficult to identify the sources of flux and torque. The d.c. motor is therefore an ideal learning vehicle, and time spent assimilating the material in this chapter should therefore be richly rewarded later.

Over a very wide power range from several megawatts at the top end down to only a few watts, all d.c. machines have the same basic structure, as shown in Figure 3.1.

The motor has two separate electrical circuits. The smaller pair of terminals (usually designated E1, E2) connect to the field windings, which surround each pole and are normally in series: these windings provide the m.m.f. to set up the flux in the air-gap under the poles. In the steady state all the input power to the field windings is dissipated as heat – none of it is converted to mechanical output power.

The main terminals (usually designated A1, A2) convey the 'torque-producing' or 'work' current to the brushes which make sliding contact to the armature winding on the rotor. The supply to the field (the flux-producing part of the motor) is separate from that for the armature, hence the description 'separately excited'.

As in any electrical machine it is possible to design a d.c. motor for any desired supply voltage, but for several reasons it is unusual to find rated voltages lower than about 6 V or much higher than 700 V. The lower limit arises because the brushes (see below) give rise to an unavoidable volt-drop of perhaps 0.5–1 V, and it is clearly not good practice to let this 'wasted' voltage became a large fraction of the supply voltage. At the other end of the scale it becomes prohibitively expensive to insulate

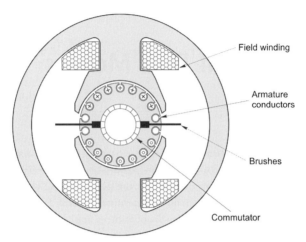

Figure 3.1 Conventional (brushed) d.c. motor.

the commutator segments to withstand higher voltages. The function and operation of the commutator is discussed later, but it is appropriate to mention here that brushes and commutators are troublesome at very high speeds. Small d.c. motors, say up to hundreds of watts output, can run at perhaps 12,000 rev/min, but the majority of medium and large motors are usually designed for speeds below 3000 rev/min.

Motors are usually supplied with power-electronic drives, which draw power from the a.c. utility supply and convert it to d.c. for the motor. Since the utility voltages tend to be standardized (e.g. 110 V, 220–240 V, or 380–440 V, 50 or 60 Hz), motors are made with rated voltages which match the range of d.c. outputs from the converter (see Chapter 2).

As mentioned above, it is quite normal for a motor of a given power, speed and size to be available in a range of different voltages. In principle all that has to be done is to alter the number of turns and the size of wire making up the coils in the machine. A 12 V, 4 A motor, for example, could easily be made to operate from 24 V instead, by winding its coils with twice as many turns of wire having only half the cross-sectional area of the original. The full speed would be the same at 24 V as the original was at 12 V, and the rated current would be 2 A, rather than 4 A. The input power and output power would be unchanged, and externally there would be no change in appearance, except that the terminals might be a bit smaller.

Traditionally, d.c. motors were classified as shunt, series, or separately excited. In addition it was common to see motors referred to as 'compound-wound'. These descriptions date from the period before the advent of power electronics: they reflect the way in which the field and armature circuits are interconnected, which in turn determines the operating characteristics. For example, the series motor has a high starting torque when switched directly on line, so it became the natural choice for traction applications, while applications requiring constant speed would use the shunt connected motor.

However, at the fundamental level there is really no difference between the various types, so we focus attention on the separately excited machine, before taking a brief look at shunt and series motors. Later, in Chapter 4, we will see how the operating characteristics of the separately excited machine with power-electronic supplies equip it to suit any application, and thereby displace the various historic predecessors.

We should make clear at this point that whereas in an a.c. machine the number of poles is of prime importance in determining the speed, the pole-number in a d.c. machine is of little consequence as far as the user is concerned. It turns out to be more economical to use two or four poles (perhaps with a square stator frame) in small or medium size d.c. motors, and more (e.g. ten or 12 or even more) in large ones, but the only difference to the user is that the 2-pole type will have two brushes at 180°, the 4-pole will have four brushes at 90°, and so on. Most of our discussion centers on the 2-pole version in the interests of simplicity, but there is no essential difference as far as operating characteristics are concerned.

2. TORQUE PRODUCTION

Torque is produced by interaction between the axial current-carrying conductors on the rotor and the radial magnetic flux produced by the stator. The flux or 'excitation' can be furnished by permanent magnets (Figure 3.2(a)) or by means of field windings (Figures 3.1 and 3.2(b)).

Permanent magnet versions are available in motors with outputs from a few watts up to a few kilowatts, while wound-field machines begin at about 100 watts and extend to the largest (MW) outputs. The advantages of the permanent magnet type are that no electrical supply is required for the field, and the overall size of the

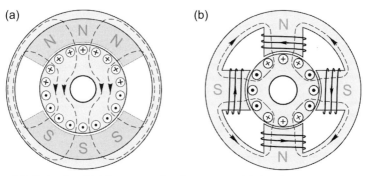

Figure 3.2 Excitation (field) systems for d.c. motors: (a) 2-pole permanent magnet; (b) 4-pole wound field.

motor can be smaller. On the other hand, the strength of the field cannot be varied so one possible option for control is ruled out.

Ferrite magnets have been used for many years. They are relatively cheap and easy to manufacture but their energy product (a measure of their effectiveness as a source of excitation) is poor. Rare earth magnets (e.g. neodymium–iron–boron or samarium–cobalt) provide much higher energy products, and yield high torque/volume ratios: they are used in high-performance servo motors, but are relatively expensive and difficult to manufacture and handle. Nd–Fe–B magnets have the highest energy product but can only be operated at temperatures below about 150°C, which is not sufficient for some high-performance motors.

Although the magnetic field is essential to the operation of the motor, we should recall that in Chapter 1 we saw that none of the mechanical output power actually comes from the field system. The excitation acts like a catalyst in a chemical reaction, making the energy conversion possible but not contributing to the output.

The main (power) circuit consists of a set of identical coils wound in slots on the rotor, and known as the armature. Current is fed into and out of the rotor via carbon 'brushes' which make sliding contact with the 'commutator', which consists of insulated copper segments mounted on a cylindrical former. (The term 'brush' stems from the early attempts to make sliding contacts using bundles of wires bound together in much the same way as the willow twigs in a witch's broomstick. Not surprisingly these primitive brushes soon wore grooves in the commutator.)

The function of the commutator is discussed below, but it is worth stressing here that all the electrical energy which is to be converted into mechanical output has to be fed into the motor through the brushes and commutator. Given that a high-speed sliding electrical contact is involved, it is not surprising that to ensure trouble-free operation the commutator needs to be kept clean, and the brushes and their associated springs need to be regularly serviced. Brushes wear away, of course, though if properly set they can last for thousands of hours. All being well the brush debris (in the form of graphite particles) will be carried out of harm's way by the ventilating air: any build-up of dust on the insulation of the windings of a high-voltage motor risks the danger of short-circuits, while debris on the commutator itself is dangerous and can lead to disastrous 'flashover' faults.

The axial length of the commutator depends on the current it has to handle. Small motors usually have one brush on each side of the commutator, so the commutator is quite short, but larger heavy-current motors may well have many brushes mounted on a common arm, each with its own brushbox (in which it is free to slide) and with all the brushes on one arm connected in parallel via their flexible copper leads or 'pigtails'. The length of the commutator can then be comparable with the 'active' length of the armature (i.e. the part carrying the conductors exposed to the radial flux). (See Plate 3.1.)

2.1 Function of the commutator

Many different winding arrangements are used for d.c. armatures, and it is neither helpful nor necessary for us to delve into the nitty-gritty of winding and commutator design. These are matters which are best left to motor designers and repairers. What we need to do is to focus on what a well-designed commutator-winding actually achieves, and despite the apparent complexity, this can be stated quite simply.

The purpose of the commutator is to ensure that regardless of the position of the rotor, the pattern of current flow in the rotor is always as shown in Figure 3.3.

Current enters the rotor via one brush, flows through all the rotor coils in the directions shown in Figure 3.3, and leaves via the other brush. The first point of contact with the armature is via the commutator segment or segments on which the brush is pressing at the time (the brush is usually wider than a single segment), but since the interconnections between the individual coils are made at each commutator segment, the current actually passes through all the coils via all the commutator segments in its path through the armature.

We can see from Figure 3.3 that all the conductors lying under the N pole carry current in one direction, while all those under the S pole carry current in the opposite direction. All the conductors under the N pole will therefore experience a downward force (which is proportional to the radial flux density B and the armature current I) while all the conductors under the S pole will experience an equal upward force. A torque is thus produced on the rotor, the magnitude of the torque being proportional to the product of the flux density and the current. In practice the flux density will not be completely uniform under the pole, so the force on some of the armature conductors will be greater than on others. However, it is straightforward to show that the total torque developed is given by

$$T = K_T \Phi I \tag{3.1}$$

where Φ is the total flux produced by the field, and K_T is constant for a given motor. In the majority of motors the flux remains constant, so we see that the motor torque is directly proportional to the armature current. This extremely simple result means that if a motor is required to produce constant torque at all speeds, we simply have to arrange to keep the armature current constant. Standard drive packages usually include provision for doing this, as will be seen later. We can also see from equation

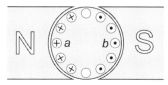

Figure 3.3 Pattern of rotor (armature) currents in 2-pole d.c. motor.

(3.1) that the direction of the torque can be reversed by reversing either the armature current (I) or the flux (Φ). We obviously make use of this when we want the motor to run in reverse, and sometimes when we want regenerative braking.

The alert reader might rightly challenge the claim – made above – that the torque will be constant regardless of rotor position. Looking at Figure 3.3, it should be clear that if the rotor turned just a few degrees, one of the five conductors shown as being under the pole will move out into the region where there is no radial flux, before the next one moves under the pole. Instead of five conductors producing force, there will then be only four, so won't the torque be reduced accordingly?

The answer to this question is yes, and it is to limit this unwelcome variation of torque that most motors have many more coils than are shown in Figure 3.3. Smooth torque is of course desirable in most applications in order to avoid vibrations and resonances in the transmission and load, and is essential in machine tool drives where the quality of finish can be marred by uneven cutting if the torque and speed are not steady.

Broadly speaking the higher the number of coils (and commutator segments) the better, because the ideal armature would be one in which the pattern of current on the rotor corresponded to a 'current sheet', rather than a series of discrete packets of current. If the number of coils was infinite, the rotor would look identical at every position, and the torque would therefore be absolutely smooth. Obviously this is not practicable, but it is closely approximated in most d.c. motors. For practical and economic reasons the number of slots is higher in large motors, which may well have a hundred or more coils and hence very little ripple in their output torque.

2.2 Operation of the commutator – interpoles

Returning now to the operation of the commutator, and focusing on a particular coil (e.g. the one shown as *ab* in Figure 3.3) we note that for half a revolution – while side *a* is under the N pole and side *b* is under the S pole – the current needs to be positive in side *a* and negative in side *b* in order to produce a positive torque. For the other half revolution, while side *a* is under the S pole and side *b* is under the N pole, the current must flow in the opposite direction through the coil for it to continue to produce positive torque. This reversal of current takes place in each coil as it passes through the interpolar axis, the coil being 'switched-round' by the action of the commutator sliding under the brush. Each time a coil reaches this position it is said to be undergoing commutation, and the relevant coil in Figure 3.3 has therefore been shown as having no current to indicate that its current is in the process of changing from positive to negative.

The essence of the current-reversal mechanism is revealed by the simplified sketch shown in Figure 3.4. This diagram shows a single coil fed via the commutator and brushes with current that always flows in at the top brush.

Figure 3.4 Simplified diagram of single-coil motor to illustrate the current-reversing function of the commutator.

In the left-hand sketch, coil-side *a* is under the N pole and carries positive current because it is connected to the shaded commutator segment which in turn is fed from the top brush. Side *a* is therefore exposed to a flux density directed from left (N) to right (S) in the sketch, and will therefore experience a downward force. This force will remain constant while the coil-side remains under the N pole. Conversely, side *b* has negative current but it also lies in a flux density directed from right to left, so it experiences an upward force. There is thus an anticlockwise torque on the rotor.

When the rotor turns to the position shown in the sketch on the right, the current in both sides is reversed, because side *b* is now fed with positive current via the unshaded commutator segment. The direction of force on each coil-side is reversed, which is exactly what we want in order for the torque to remain clockwise. Apart from the short period when the coil is outside the influence of the flux, and undergoing commutation (current-reversal) the torque is constant.

It should be stressed that the discussion above is intended to illustrate the principle involved, and the sketch should not be taken too literally. In a real multi-coil armature, the commutator arc is much smaller than that shown in Figure 3.4 and only one of the many coils is reversed at a time, so the torque remains very nearly constant regardless of the position of the rotor.

The main difficulty in achieving good commutation arises because of the self inductance of the armature coils, and the associated stored energy. As we have seen earlier, inductive circuits tend to resist change of current, and if the current reversal has not been fully completed by the time the brush slides off the commutator segment in question there will be a spark at the trailing edge of the brush.

In small motors some sparking is considered tolerable, but in medium and large wound-field motors small additional stator poles known as interpoles (or compoles) are provided to improve commutation and hence minimize sparking. These extra poles are located midway between the main field poles, as shown in Figure 3.5. Interpoles are not normally required in permanent magnet motors because the absence of stator iron close to the rotor coils results in much lower armature coil inductance.

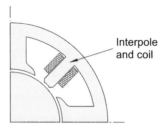

Figure 3.5 Sketch showing location of interpole and interpole winding. (The main field windings have been omitted for the sake of clarity.)

The purpose of the interpoles is to induce a motional e.m.f. in the coil undergoing commutation, in such a direction as to speed up the desired reversal of current, and thereby prevent sparking. The e.m.f. needed is proportional to the current which has to be commutated, i.e. the armature current, and to the speed of rotation. The correct e.m.f. is therefore achieved by passing the armature current through the coils on the interpoles, thereby making the flux from the interpoles proportional to the armature current. The interpole coils therefore consist of a few turns of thick conductor, connected permanently in series with the armature.

3. MOTIONAL E.M.F.

Readers who have skipped Chapter 1 are advised to check that they are familiar with the material covered in section 7 of that chapter before reading the rest of this chapter, as not all of the lessons drawn in Chapter 1 are repeated explicitly here.

When the armature is stationary, no motional e.m.f. is induced in it. But when the rotor turns, the armature conductors cut the radial magnetic flux and an e.m.f. is induced in them.

As far as each individual coil on the armature is concerned, an alternating e.m.f. will be induced in it when the rotor rotates. For the coil *ab* in Figure 3.3, for example, side *a* will be moving upward through the flux if the rotation is clockwise, and an e.m.f. directed out of the plane of the paper will be generated. At the same time the 'return' side of the coil (*b*) will be moving downwards, so the same magnitude of e.m.f. will be generated, but directed into the paper. The resultant e.m.f. in the coil will therefore be twice that in the coil-side, and this e.m.f. will remain constant for almost half a revolution, during which time the coil-sides are cutting a constant flux density. For the comparatively short time when the coil is not cutting any flux the e.m.f. will be zero, and then the coil will begin to cut through the flux again, but now each side is under the other pole, so the e.m.f. is in the opposite direction. The resultant e.m.f. waveform in each coil is therefore a rectangular alternating wave, with magnitude and frequency proportional to the speed of rotation.

The coils on the rotor are connected in series, so if we were to look at the e.m.f. across any given pair of diametrically opposite commutator segments, we would see a large alternating e.m.f. (We would have to station ourselves on the rotor to do this, or else make sliding contacts using slip-rings.)

The fact that the induced voltage in the rotor is alternating may come as a surprise, since we are talking about a d.c. motor rather than an a.c. one. But any worries we may have should be dispelled when we ask what we will see by way of induced e.m.f. when we 'look in' at the brushes. We will see that the brushes and commutator effect a remarkable transformation, bringing us back into the reassuring world of d.c.

The first point to note is that the brushes are stationary. This means that although a particular segment under each brush is continually being replaced by its neighbor, the circuit lying between the two brushes always consists of the same number of coils, with the same orientation with respect to the poles. As a result the e.m.f. at the brushes is direct (i.e. constant), rather than alternating.

The magnitude of the e.m.f. depends on the position of the brushes around the commutator, but they are invariably placed at the point where they continually 'see' the peak value of the alternating e.m.f. induced in the armature. In effect, the commutator and brushes can be regarded as a mechanical rectifier which converts the alternating e.m.f. in the rotating reference frame to a direct e.m.f. in the stationary reference frame. It is a remarkably clever and effective device, its only real drawback being that it is a mechanical system, and therefore subject to wear and tear.

We saw earlier that to obtain smooth torque it was necessary for there to be a large number of coils and commutator segments, and we find that much the same considerations apply to the smoothness of the generated e.m.f. If there are only a few armature coils the e.m.f. will have a noticeable ripple super-imposed on the mean d.c. level. The higher we make the number of coils, the smaller the ripple, and the better the d.c. we produce. The small ripple we inevitably get with a finite number of segments is seldom any problem with motors used in drives, but can sometimes give rise to difficulties when a d.c. machine is used to provide a speed feedback signal in a closed-loop system (see Chapter 4).

In Chapter 1 we saw that when a conductor of length l moves at velocity v through a flux density B, the motional e.m.f. induced is given by $e = Blv$. In the complete machine we have many series-connected conductors; the linear velocity (v) of the primitive machine examined in Chapter 1 is replaced by the tangential velocity of the rotor conductors, which is clearly proportional to the speed of rotation (n); and the average flux density cut by each conductor (B) is directly related to the total flux (Φ). If we roll together the other influential factors (number of conductors, radius, active length of rotor) into a single constant (K_E), it follows that the magnitude of the resultant e.m.f. (E) which is generated at the brushes is given by

$$E = K_E \Phi n \qquad (3.2)$$

This equation reminds us of the key role of the flux, in that until we switch on the field no voltage will be generated, no matter how fast the rotor turns. Once the field is energized, the generated voltage is directly proportional to the speed of rotation, so if we reverse the direction of rotation, we will also reverse the polarity of the generated e.m.f. We should also remember that the e.m.f. depends only on the flux and the speed, and is the same regardless of whether the rotation is provided by some external source (i.e. when the machine is being driven as a generator) or when the rotation is produced by the machine itself (i.e. when it is acting as a motor).

It has already been mentioned that the flux is usually constant at its full value, in which case equations (3.1) and (3.2) can be written in the form

$$T = k_t I \qquad (3.3)$$

$$E = k_e \omega \qquad (3.4)$$

where k_t is the motor torque constant, k_e is the e.m.f. constant, and ω is the angular speed in rad/s.

In this book the international standard (SI) system of units is used throughout. In the SI system, the units for k_t are the units of torque (newton-meter) divided by the unit of current (ampere), i.e. Nm/A; and the units of k_e units are volts/rad/s. (Note, however, that k_e is more often given in volts/1000 rev/min.)

It is not at all clear that the units for the torque constant (Nm/A) and the e.m.f. constant (V/rad/s), which on the face of it measure very different physical phenomena, are in fact the same, i.e. 1 Nm/A = 1 volt/rad/s. Some readers will be content simply to accept it, others may be puzzled, a few may even find it obvious. Those who are surprised and puzzled may feel more comfortable by progressively replacing one set of units by their equivalent, to lead us in the direction of the other, e.g.

$$\frac{(newton)(meter)}{amp} = \frac{joule}{amp} = \frac{(watt)(second)}{amp} = \frac{(volt)(amp)(second)}{amp} = (volt)(second)$$

This still leaves us to ponder what happened to the 'radians' in k_e, but at least the underlying unity is demonstrated, and, after all, a radian is a dimensionless quantity. Delving deeper, we note that 1 volt × 1 second = 1 weber, the unit of magnetic flux. This is hardly surprising because the production of torque and the generation of motional e.m.f. are both brought about by the catalytic action of the magnetic flux.

Returning to more pragmatic issues, we have now discovered the extremely convenient fact that in SI units, the torque and e.m.f. constants are equal, i.e. $k_t = k_e = k$. The torque and e.m.f. equations can thus be further simplified as

$$T = kI \qquad (3.5)$$

$$E = k\omega \qquad (3.6)$$

We will make use of these two delightfully simple equations time and again in the subsequent discussion. Together with the armature voltage equation (see below), they allow us to predict all aspects of behavior of a d.c. motor. There can be few such versatile machines for which the fundamentals can be expressed so simply.

Though attention has been focused on the motional e.m.f. in the conductors, we must not overlook the fact that motional e.m.f.s are also induced in the body of the rotor. If we consider a rotor tooth, for example, it should be clear that it will have an alternating e.m.f. induced in it as the rotor turns, in just the same way as the e.m.f. induced in the adjacent conductor. In the machine shown in Figure 3.1, for example, when the e.m.f. in a tooth under an N pole is positive, the e.m.f. in the diametrically opposite tooth (under an S pole) will be negative. Given that the rotor steel conducts electricity, these e.m.f.s will tend to set up circulating currents in the body of the rotor, so to prevent this happening, the rotor is made not from a solid piece but from thin steel laminations (typically less than 1 mm thick) which have an insulated coating to prevent the flow of unwanted currents. If the rotor was not laminated the induced current would not only produce large quantities of waste heat, but also exert a substantial braking torque.

3.1 Equivalent circuit

The equivalent circuit can now be drawn on the same basis as we used for the primitive machine in Chapter 1, and is shown in Figure 3.6.

The voltage V is the voltage applied to the armature terminals (i.e. across the brushes), and E is the internally developed motional e.m.f. The resistance and

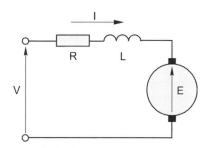

Figure 3.6 Equivalent circuit of a d.c. motor.

inductance of the complete armature are represented by R and L in Figure 3.6. The sign convention adopted is the usual one when the machine is operating as a motor. Under motoring conditions, the motional e.m.f. E always opposes the applied voltage V, and for this reason it is referred to as 'back e.m.f.'. For current to be forced into the motor, V must be greater than E, the armature circuit voltage equation being given by

$$V = E + IR + L\frac{\mathrm{d}I}{\mathrm{d}t} \tag{3.7}$$

The last term in equation (3.7) represents the inductive volt-drop due to the armature self inductance. This voltage is proportional to the rate of change of current, so under steady-state conditions (when the current is constant), the term will be zero and can be ignored. We will see later that the armature inductance has an unwelcome effect under transient conditions, but is also very beneficial in smoothing the current waveform when the motor is supplied by a controlled rectifier.

4. D.C. MOTOR – STEADY-STATE CHARACTERISTICS

From the user's viewpoint the extent to which speed falls when load is applied, and the variation in speed with applied voltage, are usually the first questions which need to be answered in order to assess the suitability of the motor for the job in hand. The information is usually conveyed in the form of the steady-state characteristics, which indicate how the motor behaves when any transient effects (caused, for example, by a sudden change in the load) have died away and conditions have once again become steady. Steady-state characteristics are usually much easier to predict than transient characteristics, and for the d.c. machine they can all be deduced from the simple equivalent circuit in Figure 3.6.

Under steady conditions, the armature current I is constant and equation (3.7) simplifies to

$$V = E + IR, \text{ or } I = \frac{V - E}{R} \tag{3.8}$$

This equation allows us to find the current if we know the applied voltage, the speed (from which we get E via equation (3.6)) and the armature resistance, and we can then obtain the torque from equation (3.5). Alternatively, we may begin with torque and speed, and work out what voltage will be needed.

We will derive the steady-state torque–speed characteristics for any given armature voltage V shortly, but first we begin by establishing the relationship between the no-load speed and the armature voltage, since this is the foundation on which the speed control philosophy is based.

4.1 No-load speed

By 'no-load' we mean that the motor is running light, so that the only mechanical resistance is that due to its own friction. In any sensible motor the frictional torque will be small, and only a small driving torque will therefore be needed to keep the motor running. Since motor torque is proportional to current (equation (3.5)), the no-load current will also be small. If we assume that the no-load current is in fact zero, the calculation of no-load speed becomes very simple. We note from equation (3.8) that zero current implies that the back e.m.f. is equal to the applied voltage, while equation (3.2) shows that the back e.m.f. is proportional to speed. Hence under true no-load (zero torque) conditions, we obtain

$$V = E = K_E \Phi n, \text{ or } n = \frac{V}{K_E \Phi} \tag{3.9}$$

where n is the speed. (We have used equation (3.2) for the e.m.f., rather than the simpler equation (3.4) because the latter only applies when the flux is at its full value, and in the present context it is important for us to see what happens when the flux is reduced.)

At this stage we are concentrating on the steady-state running speeds, but we are bound to wonder how it is that the motor reaches speed from rest. We will return to this when we look at transient behavior, so for the moment it is sufficient to recall that we came across an equation identical to equation (3.9) when we looked at the primitive linear motor in Chapter 1. We saw that if there was no resisting force opposing the motion, the speed would rise until the back e.m.f. equaled the supply voltage. The same result clearly applies to the frictionless and unloaded d.c. motor here.

We see from equation (3.9) that the no-load speed is directly proportional to armature voltage, and inversely proportional to field flux. For the moment we will continue to consider the case where the flux is constant, and demonstrate by means of an example that the approximations used in arriving at equation (3.9) are justified in practice. Later, we can use the same example to study the torque–speed characteristic.

4.2 Performance calculation – example

Consider a 500 V, 9.1 kW, 20 A, permanent-magnet motor with an armature resistance of 1 Ω. (These values tell us that the normal operating voltage is 500 V, the current when the motor is fully loaded is 20 A, and the mechanical output power under these full-load conditions is 9.1 kW.) When supplied at 500 V, the unloaded motor is found to run at 1040 rev/min, drawing a current of 0.8 A.

Whenever the motor is running at a steady speed, the torque it produces must be equal (and opposite) to the total opposing or load torque: if the motor torque was less than the load torque, it would decelerate, and if the motor torque was higher than the load torque it would accelerate. From equation (3.3), we see that the motor torque is determined by its current, so we can make the important statement that, in the steady state, the motor current will be determined by the mechanical load torque. When we make use of the equivalent circuit (Figure 3.6) under steady-state conditions we will need to get used to the idea that the current is determined by the load torque – i.e. one of the principal 'inputs' which will allow us to solve the circuit equations is the mechanical load torque, which is not even shown on the diagram. For those who are not used to electromechanical interactions this can be a source of difficulty.

Returning to our example, we note that because the motor is a real motor, it draws a small current (and therefore produces some torque) even when unloaded. The fact that it needs to produce torque, even though no load torque has been applied and it is not accelerating, is attributable to the inevitable friction in the cooling fan, bearings and brushgear.

If we want to estimate the no-load speed at a different armature voltage (say 250 V), we would ignore the small no-load current and use equation (3.9), giving

$$\text{no-load speed at 250 V} = (250/500) \times 1040 = 520 \text{ rev/min}$$

Since equation (3.9) is based on the assumption that the no-load current is zero, this result is only approximate.

If we insist on being more precise, we must first calculate the original value of the back e.m.f., using equation (3.8), which gives

$$E = 500 - 0.8 \times 1 = 499.2 \text{ volts}$$

As expected the back e.m.f. is almost equal to the applied voltage. The corresponding speed is 1040 rev/min, so the e.m.f. constant must be 499.2/1040, or 480 volts/1000 rev/min. To calculate the no-load speed for $V = 250$ volts, we first need to know the current. We are not told anything about how the friction torque varies with speed so all we can do is to assume that the friction torque is constant, in which case the motor current will be 0.8 A regardless of speed. With this assumption, the back e.m.f. will be given by

$$E = 250 - 0.8 \times 1 = 249.2 \text{ V}$$

And hence the speed will be given by

$$\text{no-load speed at 250 V} = \frac{249.2}{480} \times 1000 = 519.2 \text{ rev/min}$$

The difference between the approximate and true no-load speeds is very small, and is unlikely to be significant. Hence we can safely use equation (3.9) to predict the no-load speed at any armature voltage, and obtain the set of no-load speeds shown in Figure 3.7.

This diagram illustrates the very simple linear relationship between the speed of an unloaded d.c. motor and the armature voltage.

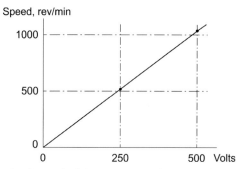

Figure 3.7 No-load speed of d.c. motor as a function of armature voltage.

4.3 Behavior when loaded

Having seen that the no-load speed of the motor is directly proportional to the armature voltage, we need to explore how the speed will vary when we change the load on the shaft.

The usual way we quantify 'load' is to specify the torque needed to drive the load at a particular speed. Some loads, such as a simple drum-type hoist with a constant weight on the hook, require the same torque regardless of speed, but for most loads the torque needed varies with the speed. For a fan, for example, the torque needed varies roughly with the square of the speed. If we know the torque–speed characteristic of the load, and the torque–speed characteristic of the motor (see below), we can find the steady-state speed simply by finding the intersection of the two curves in the torque–speed plane. An example (not specific to a d.c. motor) is shown in Figure 3.8.

At point X the torque produced by the motor is exactly equal to the torque needed to keep the load turning, so the motor and load are in equilibrium and the speed remains steady. At all lower speeds, the motor torque will be higher than the load torque, so the net torque will be positive, leading to an acceleration of the

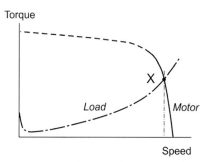

Figure 3.8 Steady-state torque–speed curves for motor and load showing location (*X*) of steady-state operating condition.

motor. As the speed rises towards X the acceleration reduces until the speed stabilizes at X. Conversely, at speeds above X the motor's driving torque is less than the braking torque exerted by the load, so the net torque is negative and the system will decelerate until it reaches equilibrium at X. This example is one which is inherently stable, so that if the speed is disturbed for some reason from the point X, it will always return there when the disturbance is removed.

Turning now to the derivation of the torque–speed characteristics of the d.c. motor, we can profitably extend the previous example to illustrate matters. We can obtain the full-load speed for $V = 500$ volts by first calculating the back e.m.f. at full-load (i.e. when the current is 20 A). From equation (3.8) we obtain

$$E = 500 - 20 \times 1 = 480 \text{ volts}$$

We have already seen that the e.m.f. constant is 480 volts/1000 rev/min, so the full-load speed is clearly 1000 rev/min. From no-load to full-load the speed falls linearly, giving the torque–speed curve for $V = 500$ volts shown in Figure 3.9. Note that from no-load to full-load the speed falls from 1040 rev/min to 1000 rev/min, a drop of only 4%. Over the same range the back e.m.f. falls from very nearly 500 to 480 volts, which of course also represents a drop of 4%.

We can check the power balance using the same approach as in section 7 of Chapter 1. At full-load the electrical input power is given by VI, i.e. $500 \text{ V} \times 20 \text{ A} = 10 \text{ kW}$. The power loss in the armature resistance is $I^2R = 400 \times 1 = 400 \text{ W}$. The power converted from electrical to mechanical form is given by EI, i.e. $480 \text{ V} \times 20 \text{ A} = 9600 \text{ W}$. We can see from the no-load data that the power required to overcome friction and iron losses (eddy currents and hysteresis, mainly in the rotor) at no-load is approximately $500 \text{ V} \times 0.8 \text{ A} = 400 \text{ W}$, so this leaves about 9.2 kW. The rated output power of 9.1 kW indicates that 100 W of additional losses (which we will not attempt to explore here) can be expected under full-load conditions.

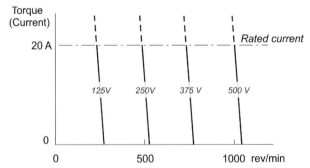

Figure 3.9 Family of steady-state torque–speed curves for a range of armature voltages.

Two important observations follow from these calculations. First, the drop in speed when load is applied (the 'droop') is very small. This is very desirable for most applications, since all we have to do to maintain almost constant speed is to set the appropriate armature voltage and keep it constant. Secondly, a delicate balance between V and E is revealed. The current is in fact proportional to the difference between V and E (equation (3.8)), so that quite small changes in either V or E give rise to disproportionately large changes in the current. In the example, a 4% reduction in E causes the current to rise to its rated value. Hence to avoid excessive currents (which cannot be tolerated in a thyristor supply, for example), the difference between V and E must be limited. This point will be taken up again when transient performance is explored.

A representative family of torque–speed characteristics for the motor discussed above is shown in Figure 3.9. As already explained, the no-load speeds are directly proportional to the applied voltage, while the slope of each curve is the same, being determined by the armature resistance: the smaller the resistance the less the speed falls with load. These operating characteristics are very attractive because the speed can be set simply by applying the correct voltage.

The upper region of each characteristic in Figure 3.9 is shown dashed because in this region the armature current is above its rated value, and therefore the motor cannot be operated continuously without overheating. Motors can and do operate for short periods above rated current, and the fact that the d.c. machine can continue to provide torque in proportion to current well into the overload region makes it particularly well suited to applications requiring the occasional boost of excess torque.

A cooling problem might be expected when motors are run continuously at full current (i.e. full torque) even at very low speed, where the natural ventilation is poor. This operating condition is considered quite normal in converter-fed motor drive systems, and motors are accordingly fitted with a small air-blower motor as standard.

This book is about motors, which convert electrical power into mechanical power. But, in common with all electrical machines, the d.c. motor is inherently capable of operating as a generator, converting mechanical power into electrical power. And although the overwhelming majority of d.c. machines will spend most of their time in motoring mode, there are applications such as rolling mills where frequent reversal is called for, and others where rapid braking is required. In the former, the motor is controlled so that it returns the stored kinetic energy to the supply system each time the rolls have to be reversed, while in the latter case the energy may also be returned to the supply, or dumped as heat in a resistor. These transient modes of operation during which the machine acts as a generator may better be described as 'regeneration' since they only involve recovery of mechanical energy that was originally provided by the motor.

Continuous generation is of course possible using a d.c. machine provided we have a source of mechanical power, such as an internal combustion (IC) engine. In the example discussed above we saw that when connected to a 500 V supply, the unloaded machine ran at 1040 rev/min, at which point the back e.m.f was very nearly 500 V and only a tiny positive current was flowing. As we applied mechanical load to the shaft the steady-state speed fell, thereby reducing the back e.m.f. and increasing the armature current until the motor toque was equal to the opposing load torque and equilibrium returned.

Conversely, if instead of applying an opposing (load) torque, we use the IC engine to supply torque in the opposite direction, i.e. trying to increase the speed of the motor, the increase in speed will cause the motional e.m.f. to be greater than the supply voltage (500 V). This means that the current will flow from the d.c. machine to the supply, resulting in a negative torque and reversal of electrical power flow back into the supply. Stable generating conditions will be achieved when the motor torque (current) is equal and opposite to the torque provided by the IC engine. In the example, the full-load current is 20 A, so in order to drive this current through its own resistance and overcome the supply voltage the e.m.f. must be given by

$$E = IR + V = 20 \times 1 + 500 = 520 \text{ volts}$$

The corresponding speed can be calculated by reference to the no-load e.m.f. (499.2 volts at 1040 rev/min) from which the steady generating speed is given by

$$\frac{N_{gen}}{1040} = \frac{520}{499.2}, \quad \text{i.e. } N_{gen} = 1083 \text{ rev/min}$$

On the torque–speed plot (Figure 3.9) this condition lies on the downward projection of the 500 V characteristic at a current of -20 A. We note that the full range of operation, from full-load motoring to full-load generating, is accomplished with only a modest change in speed from 1000 to 1083 rev/min.

It is worth emphasizing that in order to make the unloaded motor move into the generating mode, all that we had to do was to start supplying mechanical power to the motor shaft. No physical changes had to be made to the motor to make it into a generator – the hardware is equally at home functioning as a motor or as a generator – which is why it is best referred to as a 'machine'. An electric vehicle takes full advantage of this inherent reversibility, recharging the battery when we decelerate or descend a hill. (How pleasant it would be if the IC engine could do the same; whenever we slowed down, we could watch the rising gauge as our kinetic energy was converted back into hydrocarbon fuel in the tank!)

To complete this section we will derive the analytical expression for the steady-state speed as a function of the two variables that we can control, i.e. the applied voltage (V), and the load torque (T_L). Under steady-state conditions the armature current is constant and we can therefore ignore the armature inductance term in equation (3.7); and because there is no acceleration, the motor torque is equal to the load torque. Hence by eliminating the current I between equations (3.5) and (3.7), and substituting for E from equation (3.6), the speed is given by

$$\omega = \frac{V}{k} - \frac{R}{k^2} T_{\mathrm{L}} \qquad (3.10)$$

This equation represents a straight line in the speed–torque plane, as we saw with our previous worked example. The first term shows that the no-load speed is directly proportional to the armature voltage, while the second term gives the drop in speed for a given load torque. The gradient or slope of the torque–speed curve is $-R/k^2$, showing again that the smaller the armature resistance, the smaller the drop in speed when load is applied.

4.4 Base speed and field weakening

Returning to our consideration of motor operating characteristics, when the field flux is at its full value the speed corresponding to full armature voltage and full current (i.e. the rated full-load condition) is known as base speed (see Figure 3.10). The upper part of the figure shows the regions of the torque–speed plane within which the motor can operate without exceeding its maximum or rated current. The lower diagram shows the maximum output power as a function of speed.

The motor can operate at any speed up to base speed, and any torque (current) up to rated value by appropriate choice of armature voltage. This full flux region of operation is indicated by the shaded area *0abc* in Figure 3.10, and is often referred to

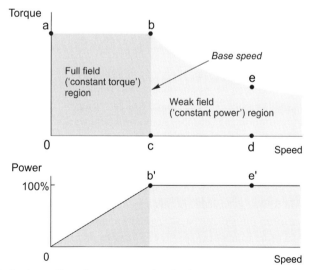

Figure 3.10 Regions of continuous operation in the torque–speed and power–speed planes.

as the 'constant torque' region of the torque–speed characteristic. In this context 'constant torque' signifies that at any speed below base speed the motor is capable of producing its full rated torque. Note that the term constant torque does not mean that the motor *will* produce constant torque, but rather it signifies that the motor *can* produce constant torque if required: as we have already seen, it is the mechanical load we apply to the shaft that determines the steady-state torque produced by the motor.

When the current is at maximum (i.e. along the line *ab* in Figure 3.10), the torque is at its maximum (rated) value. Since mechanical power is given by torque times speed, the power output along *ab* is proportional to the speed, as shown in the lower part of the figure, and the maximum power thus corresponds to the point *b* in Figure 3.10. At point *b* both the voltage and current have their full rated values.

To run faster than base speed the field flux must be reduced, as indicated by equation (3.9). Operation with reduced flux is known as 'field weakening', and we have already discussed this perhaps surprising mode in connection with the primitive linear motor in Chapter 1. For example, by halving the flux (and keeping the armature voltage at its full value), the no-load speed is doubled (point *d* in Figure 3.10). The increase in speed is, however, obtained at the expense of available torque, which is proportional to flux times current (see equation (3.1)). The current is limited to rated value, so if the flux is halved, the speed will double but the maximum torque which can be developed is only half the rated value (point *e* in Figure 3.10). Note that at the point *e* both the armature voltage and the armature current again have their full rated values, so the power is at maximum, as it was at point *b*. The power is constant along the curve through *b* and *e*, and for this reason the shaded area to the right of the line *bc* is referred to as the 'constant power' region. Obviously, field weakening is only satisfactory for applications which do not demand full torque at high speeds, such as electric traction.

The maximum allowable speed under weak field conditions must be limited (to avoid excessive sparking at the commutator), and is usually indicated on the motor rating plate. A marking of 1200/1750 rev/min, for example, would indicate a base speed of 1200 rev/min, and a maximum speed with field weakening of 1750 rev/min. The field weakening range varies widely depending on the motor design, but maximum speed rarely exceeds three or four times base speed.

To sum up, the speed is controlled as follows:
- Below base speed, the flux is maximum, and the speed is set by the armature voltage. Full torque is available at any speed.
- Above base speed the armature voltage is at (or close to) maximum, and the flux is reduced in order to raise the speed. The maximum torque available reduces in proportion to the flux.

To judge the suitability of a motor for a particular application we need to compare the torque–speed characteristic of the prospective load with the operating diagram for the motor: if the load torque calls for operation outside the shaded areas of Figure 3.10, a larger motor is clearly called for.

Finally, we should note that according to equation (3.9), the no-load speed will become infinite if the flux is reduced to zero. This seems an unlikely state of affairs; after all, we have seen that the field is essential for the motor to operate, so it seems unreasonable to imagine that if we removed the field altogether, the speed would rise to infinity. In fact, the explanation lies in the assumption that 'no-load' for a real motor means zero torque. If we could make a motor which had no friction torque whatsoever, the speed would indeed continue to rise as we reduced the field flux towards zero. But as we reduced the flux, the torque per ampere of armature current would become smaller and smaller, so in a real machine with friction, there will come a point where the torque being produced by the motor is equal to the friction torque, and the speed will therefore be limited. Nevertheless, it is quite dangerous to open-circuit the field winding, especially in a large unloaded motor. There may be sufficient 'residual' magnetism left in the poles to produce significant accelerating torque to lead to a run-away situation. Usually, field and armature circuits are interlocked so that if the field is lost, the armature circuit is switched off automatically.

4.5 Armature reaction

In addition to deliberate field weakening, as discussed above, the flux in a d.c. machine can be weakened by an effect known as 'armature reaction'. As its name implies, armature reaction relates to the influence that the armature m.m.f. has on the flux in the machine: in small machines it is negligible, but in large machines the unwelcome field weakening caused by armature reaction is sufficient to warrant extra design features to combat it. A full discussion would be well beyond the needs of most users, but a brief explanation is included for the sake of completeness.

The way armature reaction occurs can best be appreciated by looking at Figure 3.1 and noting that the m.m.f. of the armature conductors acts along the axis defined by the brushes, i.e. the armature m.m.f. acts in quadrature to the main flux axis which lies along the stator poles. The reluctance in the quadrature direction is high because of the large air spaces that the flux has to cross, so despite the fact that the rotor m.m.f. at full current can be very large, the quadrature flux is relatively small; and because it is perpendicular to the main flux, the average value of the latter would not be expected to be affected by the quadrature flux, even though part of the path of the reaction flux is shared with the main flux as it passes (horizontally in Figure 3.1) through the main pole-pieces.

A similar matter was addressed in relation to the primitive machine in Chapter 1. There it was explained that it was not necessary to take account of the flux produced by the conductor itself when calculating the electromagnetic force on it. And if it were not for the non-linear phenomenon of magnetic saturation, the armature reaction flux would have no effect on the average value of the main flux in the machine shown in Figure 3.1: the flux density on one edge of the pole-pieces would be increased by the presence of the reaction flux, but decreased by the same amount on the other edge, leaving the average of the main flux unchanged. However, if the iron in the main magnetic circuit is already some way into saturation, the presence of the rotor m.m.f. will cause less of an increase on the one edge than it causes by way of decrease on the other, and there will be a net reduction in main flux.

We know that reducing the flux leads to an increase in speed, so we can now see that in a machine with pronounced armature reaction, when the load on the shaft is increased and the armature current increases to produce more torque, the field is simultaneously reduced and the motor speeds up. Though this behavior is not a true case of instability, it is not generally regarded as desirable!

Large motors often carry additional windings fitted into slots in the pole-faces and connected in series with the armature. These 'compensating' windings produce an m.m.f. in opposition to the armature m.m.f., thereby reducing or eliminating the armature reaction effect.

4.6 Maximum output power

We have seen that if the mechanical load on the shaft of the motor increases, the speed falls and the armature current automatically increases until equilibrium of torque is reached and the speed again becomes steady. If the armature voltage is at its maximum (rated) value, and we increase the mechanical load until the current reaches its rated value, we are clearly at full-load, i.e. we are operating at the full speed (determined by voltage) and the full torque (determined by current). The maximum current is set at the design stage, and reflects the tolerable level of heating of the armature conductors.

Clearly if we increase the load on the shaft still more, the current will exceed the safe value, and the motor will begin to overheat. But the question which this prompts is: If it were not for the problem of overheating, could the motor deliver more and more power output, or is there a limit?

We can see straightaway that there will be a maximum by looking at the torque–speed curves in Figure 3.9. The mechanical output power is the product of torque and speed, and we see that the power will be zero when either the load torque is zero (i.e. the motor is running light) or the speed is zero (i.e. the motor is stationary). There must be maximum between these two zeroes, and it is easy to show that the peak mechanical power occurs when the speed is half of the no-load speed. However, this operating condition is only practicable in very small motors: in the majority of motors, the supply would simply not be able to provide the very high current required.

Turning to the question of what determines the theoretical maximum power, we can apply the maximum power transfer theorem (from circuit theory) to the equivalent circuit in Figure 3.6. The inductance can be ignored because we assume d.c. conditions. If we regard the armature resistance R as if it were the resistance of the source V, the theorem tells us that in order to transfer maximum power to the load (represented by the motional e.m.f. on the right-hand side of Figure 3.6) we must make the load 'look like' a resistance equal to the source resistance R. This condition is obtained when the applied voltage V divides equally so that half of it is dropped across R and the other half is equal to the e.m.f. E. (We note that the condition $E = V/2$ corresponds to the motor running at half the no-load speed, as stated above.) At the maximum power point, the current is $V/2R$, and the mechanical output power (EI) is given by $V^2/4R$.

The expression for the maximum output power is delightfully simple. We might have expected the maximum power to depend on other motor parameters, but in fact it is determined solely by the armature voltage and the armature resistance. For example, we can say immediately that a 12 V motor with an armature resistance of 1 Ω cannot possibly produce more than 36 W of mechanical output power.

We should of course observe that under maximum power conditions the overall efficiency is only 50% (because an equal power is burned off as heat in the armature resistance); and we emphasize again that only very small motors can ever be operated continuously in this condition. For the vast majority of motors, it is of academic interest only, because the current ($V/2R$) will be far too high for the supply.

---

5. TRANSIENT BEHAVIOR – CURRENT SURGES

It has already been pointed out that the steady-state armature current depends on the small difference between the back e.m.f. E and the applied voltage V. In a converter-fed drive it is vital that the current is kept within safe bounds, otherwise the thyristors or transistors (which have very limited overcurrent capacity) will be destroyed, and it follows from equation (3.8) that in order to prevent the current from exceeding its rated value we cannot afford to let V and E differ by more than IR, where I is the rated current.

It would be unacceptable, for example, to attempt to bring all but the smallest of d.c. motors up to speed simply by switching on rated voltage. In the example studied earlier, rated voltage is 500 V, and the armature resistance is 1 Ω. At standstill the back e.m.f. is zero, and hence the initial current would be $500/1 = 500$ A, or 25 times rated current! This would destroy the thyristors in the supply converter (and/or blow the fuses). Clearly the initial voltage we must apply is much less than 500 V; and if we want to limit the current to rated value (20 A in the example) the voltage needed will be 20×1, i.e. only 20 volts. As the speed picks up, the back e.m.f. rises, and to maintain the full current, V must also be ramped up so that the difference between V and E remains constant at 20 V. Of course, the motor will not accelerate nearly so rapidly when the current is kept in check as it would if we had switched on full voltage, and allowed the current to do as it pleased. But this is the price we must pay in order to protect the converter.

Similar current-surge difficulties occur if the load on the motor is suddenly increased, because this will result in the motor slowing down, with a consequent fall in E. In a sense we welcome the fall in E because this is what brings about the increase in current needed to supply the extra load, but of course we only want the current to rise to its rated value: beyond that point we must be ready to reduce V, to prevent an excessive current.

The solution to the problem of overcurrents lies in providing closed-loop current-limiting as an integral feature of the motor/drive package. The motor current is sensed, and the voltage V is automatically adjusted so that rated current is not exceeded continuously, although typically it is allowed to reach 1.5 times rated current for 60 seconds. We will discuss the current control loop in Chapter 4.

5.1 Dynamic behavior and time-constants

The use of the terms 'surge' and 'sudden' in the discussion above will doubtless have created the impression that changes in the motor current or speed can take place instantaneously, whereas in fact a finite time is always necessary to effect changes in either. (If the current changes, then so does the stored energy in the armature inductance; and if speed changes, so does the rotary kinetic energy stored in the inertia. For either of these changes to take place in zero time it would be necessary for there to be a pulse of infinite power, which is clearly impossible.)

The theoretical treatment of the transient dynamics of the d.c. machine is easier than for any other type of electric motor but is nevertheless beyond our scope. However, it is worth summarizing the principal features of the dynamic behavior, and highlighting the fact that all the transient changes that occur are determined by only two time-constants. The first (and most important from the user's viewpoint) is the electromechanical time-constant, which governs the way the speed settles to a new level following a disturbance such as a change in armature voltage or load torque. The second is the electrical (or armature) time-constant, which is usually much shorter and governs the rate of change of armature current immediately following a change in armature voltage.

When the motor is running, there are two 'inputs' that we can change suddenly, namely the applied voltage and the load torque. When either of these is changed, the motor enters a transient period before settling to its new steady state. It turns out that if we ignore the armature inductance (i.e. we take the armature time-constant to be zero), the transient period is characterized by first-order exponential responses in the speed and current. This assumption is valid for all but the very largest motors.

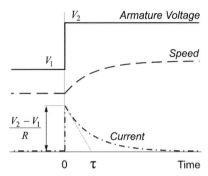

Figure 3.11 Response of d.c. motor to step increase in armature voltage.

We obtained a similar result when we looked at the primitive linear motor in Chapter 1 (see Figure 1.16).

For example, if we suddenly increased the armature voltage of a frictionless and unloaded motor from V_1 to V_2, its speed and current would vary as shown in Figure 3.11.

There is an immediate increase in the current (because we have ignored the inductance), reflecting the fact that the applied voltage is suddenly more than the back e.m.f.; the increased current produces more torque and hence the motor accelerates; the rising speed is accompanied by an increase in back e.m.f., so the current begins to fall; and the process continues until a new steady speed is reached corresponding to the new voltage. In this particular case the steady-state current is zero because we have assumed that there is no friction or load torque, but the shape of the dynamic response would be the same if there had been an initial load, or if we had suddenly changed the load.

The expression describing the current as a function of time (t) is:

$$i = \left\{ \frac{V_2 - V_1}{R} \right\} e^{-t/\tau} \tag{3.11}$$

The expression for the change in speed is similar, the time dependence again featuring the exponential transient term $e^{-t/\tau}$. The significance of the time-constant (τ) is shown in Figure 3.11. If the initial gradient of the current–time graph is projected it intersects the final value after one time-constant. In theory, it takes an infinite time for the response to settle, but in practice the transient is usually regarded as over after about four or five time-constants. We note that the transient response is very satisfactory: as soon as the voltage is increased the current immediately increases to provide more torque and begin the acceleration, but the accelerating torque is reduced progressively to ensure that the new target speed is approached smoothly. Happily, because the system is first-order, there is no suggestion of an oscillatory response with overshoots.

Analysis yields the relationship between the time-constant and the motor/system parameters as

$$\tau = \frac{RJ}{k^2} \tag{3.12}$$

where R is the armature resistance, J is the total rotary inertia of motor plus load, and k is the motor constant (equations (3.3) and (3.4)). The appropriateness of the term 'electromechanical time-constant' should be clear from equation (3.12), because τ depends on the electrical parameters (R and k) and the mechanical parameter J. The fact that if the inertia was doubled, the time-constant would double and transients would take twice as long is perhaps to be expected, but the influence of the motor parameters R and k is probably not so obvious.

The electrical or armature time-constant is defined in the usual way for a series L, R circuit, i.e.

$$\tau_a = \frac{L}{R} \tag{3.13}$$

If we were to hold the rotor of a d.c. motor stationary and apply a step voltage V to the armature, the current would climb exponentially to a final value of V/R with a time-constant τ_a.

If we always applied pure d.c. voltage to the motor we would probably want τ_a to be as short as possible, so that there was no delay in the build-up of current when the voltage is changed.

But given that most motors are fed with voltage waveforms which are far from smooth (see Chapter 2), we are actually rather pleased to find that because of the inductance and associated time-constant, the current waveform (and hence the torque) is smoother than the voltage waveform. So the unavoidable presence of armature inductance turns out (in most cases) to be a blessing in disguise.

So far we have looked at the two time-constants as if they were unrelated in the influence they have on the current. We began with the electromechanical time-constant, assuming that the armature time-constant was zero, and saw that the dominant influence on the current during the transient was the motional e.m.f. We then examined the current when the rotor was stationary (so that the motional e.m.f. is zero), and saw that the growth or decay of current is governed by the armature inductance, manifested via the armature time-constant.

In reality, both time-constants influence the current simultaneously, and the picture is more complicated than we have implied, as the system is in fact a second-order one. However, the good news is that for most motors, and most purposes, we can take advantage of the fact that the armature time-constant is much shorter than the electromechanical time-constant. This allows us to approximate the behavior by decoupling the relatively fast 'electrical transients' in the armature circuit from the much slower 'electromechanical transients' which are apparent to the user. From the latter's point of view, only the electromechanical transient is likely to be of interest.

6. FOUR QUADRANT OPERATION AND REGENERATIVE BRAKING

We have seen that the great beauty of the separately excited d.c. motor is the ease with which it can be controlled. The steady-state speed is determined by the applied voltage, so we can make the motor run at any desired speed in either direction simply by applying the appropriate magnitude and polarity of the armature voltage; and the torque is directly proportional to the armature current, which in turn depends on the difference between the applied voltage V and the back e.m.f. E. We can therefore make the machine develop positive (motoring) or negative (generating) torque simply by controlling the extent to which the applied voltage is greater or less than the back e.m.f. An armature voltage controlled d.c. machine is therefore inherently capable of what is known as 'four-quadrant' operation, by reference to the numbered quadrants of the torque–speed plane shown in Figure 3.12.

Figure 3.12 looks straightforward but experience shows that to draw the diagram correctly calls for a clear head, so it is worth spelling out the key points in detail. A proper understanding of this diagram is invaluable as an aid to understanding how controlled-speed drives operate.

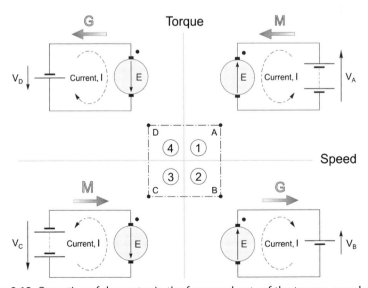

Figure 3.12 Operation of d.c. motor in the four quadrants of the torque–speed plane.

First, one of the motor terminals is shown with a dot, and in all four quadrants the dot is uppermost. The purpose of this convention is to indicate the sign of the torque: if current flows into the dot, the machine produces positive torque, and if current flows out of the dot, the torque is negative.

Secondly, the supply voltage is shown by the old-fashioned battery symbol, as use of the more modern circle symbol for a voltage source would make it more difficult to differentiate between the source and the circle representing the machine armature. The relative magnitudes of applied voltage and motional e.m.f. are emphasized by the use of two battery cells when $V > E$ and one when $V < E$.

We have seen that in a d.c. machine speed is determined by applied voltage and torque is determined by current. Hence on the right-hand side of the diagram the supply voltage is positive (upwards), while on the left-hand side the supply voltage is negative (downwards). And in the upper half of the diagram current is positive (into the dot), while in the lower half it is negative (out of the dot). For the sake of convenience, each of the four operating conditions (A, B, C, D) have the same magnitude of speed and the same magnitude of torque: these translate to equal magnitudes of motional e.m.f. and current for each condition.

When the machine is operating as a motor and running in the forward direction, it is operating in quadrant 1. The applied voltage V_A is positive and greater than the back e.m.f. E, and positive current therefore flows into the motor; in Figure 3.12, the arrow representing V_A has accordingly been drawn larger than E. The power drawn from the supply $V_A I$ is positive in this quadrant, as shown by the shaded arrow labeled M to represent motoring. The power converted to mechanical form is given by EI, and an amount $I^2 R$ is lost as heat in the armature. If E is much greater than IR (which is true in all but small motors), most of the input power is converted to mechanical power, i.e. the conversion process is efficient.

If, with the motor running at position A, we suddenly reduce the supply voltage to a value V_B which is less than the back e.m.f., the current (and hence torque) will reverse direction, shifting the operating point to B in Figure 3.12. There can be no sudden change in speed, so the e.m.f. will remain the same. If the new voltage is chosen so that $E - V_B = V_A - E$ the new current will have the same amplitude as at position A, so the new (negative) torque will be the same as the original positive torque, as shown in Figure 3.12. But now power is supplied from the machine to the supply, i.e. the machine is acting as a generator, as shown by the shaded arrow.

We should be quite clear that all that was necessary to accomplish this remarkable reversal of power flow was a modest reduction of the voltage applied to the machine. At position A, the applied voltage was $E + IR$, while at position B it is $E - IR$. Since IR will be small compared with E, the change ($2IR$) is also small.

Needless to say the motor will not remain at point B if left to its own devices. The combined effect of the load torque and the negative machine torque will cause the speed to fall, so that the back e.m.f. again falls below the applied voltage V_B, the current and torque become positive again, and the motor settles back into quadrant 1, at a lower speed corresponding to the new (lower) supply voltage. During the deceleration phase, kinetic energy from the motor and load inertias is returned to the supply. This is therefore an example of regenerative braking, and it occurs naturally every time we reduce the voltage in order to lower the speed.

If we want to operate continuously at position B, the machine will have to be driven by a mechanical source. We have seen above that the natural tendency of the machine is to run at a lower speed than that corresponding to point B, so we must force it to run faster, and create an e.m.f greater than V_B, if we wish it to generate continuously.

It should be obvious that similar arguments to those set out above apply when the motor is running in reverse (i.e. V is negative). Motoring then takes place in quadrant 3 (point C), with brief excursions into quadrant 4 (point D, accompanied by regenerative braking) whenever the voltage is reduced in order to lower the speed.

6.1 Full-speed regenerative reversal

To illustrate more fully how the voltage has to be varied during sustained regenerative braking, we can consider how to change the speed of an unloaded motor from full speed in one direction to full speed in the other, in the shortest possible time.

At full forward speed the applied armature voltage is taken to be $+V$ (shown as 100% in Figure 3.13), and since the motor is unloaded the no-load current will be very small and the back e.m.f. will be almost equal to V. Ultimately, we will clearly need an armature voltage of $-V$ to make the motor run at full speed in reverse. But we cannot simply reverse the applied voltage: if we did, the armature current immediately afterwards would be given by $(-V - E)/R$, which would be disastrously high. (The motor might tolerate it for the short period for which it would last, but the supply certainly could not!)

What we need to do is adjust the voltage so that the current is always limited to rated value, and in the right direction. Since we want to decelerate as fast as possible, we must aim to keep the current negative, and at rated value (i.e. -100%) throughout the period of deceleration and for the run-up to full speed in reverse. This will give us constant torque throughout, so the deceleration (and subsequent acceleration) will be constant and the speed will change at a uniform rate, as shown in Figure 3.13.

We note that to begin with the applied voltage has to be reduced to less than the back e.m.f., and then ramped down linearly with time so that the difference between V and E is kept constant, thereby keeping the current constant at its rated value. During the reverse run-up, V has to be numerically greater than E, as shown in Figure 3.13. (The difference between V and E has been exaggerated in Figure 3.13 for clarity: in a large motor, the difference may only be 1or 2% at full speed.)

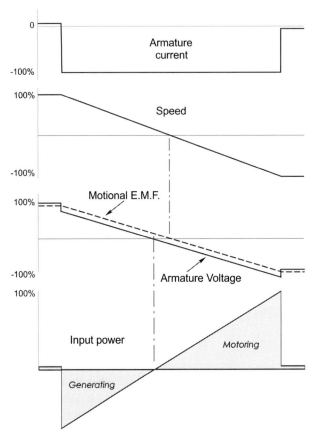

Figure 3.13 Regenerative reversal of d.c. motor from full-speed forward to full-speed reverse, at maximum allowable torque (current).

The power to and from the supply is shown in the bottom plot in Figure 3.13, the energy being represented by the shaded areas. During the deceleration period most of the kinetic energy of the motor (lower shaded area) is progressively returned to the supply, the motor acting as a generator for the whole of this time. The total energy recovered in this way can be appreciable in the case of a large drive such as a steel rolling mill. A similar quantity of energy (upper shaded area) is supplied and stored as kinetic energy as the motor picks up speed in the reverse sense.

Three final points need to be emphasized. First, we have assumed throughout the discussion that the supply can provide positive or negative voltages, and can accept positive or negative currents. A note of caution is therefore appropriate, because many simple power electronic converters do not have this flexibility. Users need to be aware that if full four-quadrant operation (or even two-quadrant regeneration) is called for, a basic converter will probably not be adequate. This point is taken up again in Chapter 4. Secondly, we should not run away with the idea that in order to carry out the reversal in Figure 3.13 we would have to work out in advance how to profile the applied voltage as a function of time. Our drive system will normally have the facility for automatically operating the motor in constant-current mode, and all we will have to do is to tell it the new target speed. This is also taken up in Chapter 4.

6.2 Dynamic braking

A simpler and cheaper but less effective method of braking can be achieved by dissipating the kinetic energy of the motor and load in a resistor, rather than returning it to the supply. A version of this technique is employed in the cheaper power electronic converter drives, which have no facility for returning power to the utility supply.

When the motor is to be stopped, the supply to the armature is removed and a resistor is switched across the armature brushes. The motor e.m.f. drives a (negative) current through the resistor, and the negative torque results in deceleration. As the speed falls, so does the e.m.f., the current, and the braking torque. At low speeds the braking torque is therefore very small. Ultimately, all the kinetic energy is converted to heat in the motor's own armature resistance and the external resistance. Very rapid initial braking is obtained by using a low resistance (or even simply short-circuiting the armature).

7. SHUNT AND SERIES MOTORS

Before variable-voltage supplies became readily available, most d.c. motors were obliged to operate from a single d.c. supply, usually of constant voltage. The armature and field circuits were therefore designed either for connection in parallel (shunt), or in series. The operating characteristics of shunt and series machines differ widely, and hence each type tended to claim its particular niche: shunt motors were judged to be good for constant-speed applications, while series motors were widely used for traction applications.

In a way it is unfortunate that these historical patterns of association became so deep-rooted. The fact is that a converter-fed separately excited motor, freed of any constraint between field and armature, can do everything that a shunt or series motor can, and more; and it is doubtful if shunt and series motors would ever have become widespread if variable-voltage supplies had always been around.

For a given continuous output power rating at a given speed, shunt and series motors are the same size, with the same rotor diameter, the same poles, and the same quantities of copper in the armature and field windings. This is to be expected when we recall that the power output depends on the specific magnetic and electric loadings, so we anticipate that to do a given job, we will need the same amounts of active material. However, differences emerge when we look at the details of the windings, especially the field system, and they can best be illustrated by means of an example which contrasts shunt and series motors for the same output power.

Suppose that for the shunt version the supply voltage is 500 V, the rated armature (work) current is 50 A, and the field coils are required to provide an m.m.f. of 500 ampere-turns (AT). The field might typically consist of say 200 turns of wire with a total resistance of 200 Ω. When connected across the supply (500 V), the field current will be 2.5 A, and the m.m.f. will be 500 AT, as required. The power dissipated as heat in the field will be 500 V × 2.5 A = 1.25 kW, and the total power input at rated load will be 500 V × 52.5 A = 26.25 kW.

To convert the machine into the equivalent series version, the field coils need to be made from much thicker conductor, since they have to carry the armature current of 50 A, rather than the 2.5 A of the shunt motor. So, working at the same current density, the cross-section of each turn of the series field winding needs to be 20 times that of the shunt field wires, but conversely only one-twentieth of the turns (i.e. 10) are required for the same ampere-turns. The resistance of a wire of length l and cross-sectional area A, made from material of resistivity ρ is given by $R = \rho l / A$, so we can use this formula to show that the resistance of the new field winding will be much lower, at 0.5 Ω.

We can now calculate the power dissipated as heat in the series field. The current is 50 A, the resistance is 0.5 Ω, so the volt–drop across the series field is 25 V, and the power wasted as heat is 1.25 kW. This is the same as for the shunt machine, which is to be expected since both sets of field coils are intended to do the same job.

In order to allow for the 25 V dropped across the series field, and still meet the requirement for 500 V at the armature, the supply voltage must now be 525 V. The rated current is 50 A, so the total power input is 525 V × 50 A = 26.25 kW, the same as for the shunt machine.

This example illustrates that in terms of their energy-converting capabilities, shunt and series motors are fundamentally no different. Shunt machines usually have field windings with a large number of turns of fine wire, while series machines have a few turns of thick conductor. But the total amount and disposition of copper is the same, so the energy-converting abilities of both types are identical. In terms of their operating characteristics, however, the two types differ widely, as we will now see.

7.1 Shunt motor – steady-state operating characteristics

A basic shunt-connected motor has its armature and field in parallel across a single d.c. supply, as shown in Figure 3.14(a). Normally, the voltage will be constant and at the rated value for the motor, in which case the steady-state torque–speed curve will be similar to that of a separately excited motor at rated field flux, i.e. the speed will drop slightly with load, as shown by the line *ab* in Figure 3.14(b). Over the normal operating region the torque–speed characteristic is similar to that of the induction motor (see Chapter 6), so shunt motors are suited to the same duties, i.e. what are usually referred to as 'constant speed' applications.

Except for small motors (say less than about 1 kW), it will be necessary to provide an external 'starting resistance' (R_S in Figure 3.14) in series with the armature, to limit the heavy current which would flow if the motor was simply switched directly onto the supply. This starting resistance is progressively reduced as the motor picks up speed, the current falling as the back e.m.f. rises from its initial value of zero.

We should ask what happens if the supply voltage varies for any reason, and as usual the easiest thing to look at is the case where the motor is running light, in which case the back e.m.f. will almost equal the supply voltage. If we reduce the supply voltage, intuition might lead us to anticipate a fall in speed, but in fact two contrary effects occur which leave the speed almost unchanged.

If the voltage is halved, for example, both the field current and the armature voltage will be halved, and if the magnetic circuit is not saturated the flux will also halve. The new steady value of back e.m.f. will have to be half its original value, but since we now have only half as much flux, the speed will be the same. The maximum output power will of course be reduced, since at full-load (i.e. full current) the power available is proportional to the armature voltage. Of course if the magnetic circuit is saturated, a modest reduction in applied voltage may cause very little drop in flux, in which case the speed will fall in proportion to the drop in voltage. We can see from this discussion why, broadly speaking, the shunt motor is not suitable for operation below base speed.

Some measure of speed control is possible by weakening the field (by means of the resistance (R_f) in series with the field winding), and this allows the speed to be raised above base value, but only at the expense of torque. A typical torque–speed characteristic in the field-weakening region is shown by the line *cd* in Figure 3.14(b).

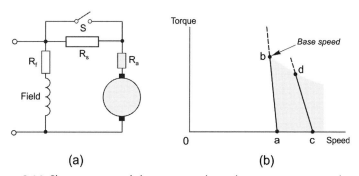

Figure 3.14 Shunt-connected d.c. motor and steady-state torque–speed curve.

Reverse rotation is achieved by reversing the connections to either the field or the armature. The field is usually preferred since the current rating of the switch or contactor will be lower than for the armature.

7.2 Series motor – steady-state operating characteristics

The series connection of armature and field windings (Figure 3.15(a)) means that the field flux is directly proportional to the armature current, and the torque is therefore proportional to the square of the current. Reversing the direction of the applied voltage (and hence current) therefore leaves the direction of torque unchanged. This unusual property is put to good use in the universal motor, but is a handicap when negative (braking) torque is required, since either the field or armature connections must then be reversed.

If the armature and field resistance volt-drops are neglected, and the applied voltage (V) is constant, the current varies inversely with the speed; hence the torque (T) and speed (n) are related by

$$T \propto \left\{ \frac{V}{n} \right\}^2 \tag{3.14}$$

A typical torque–speed characteristic is shown in Figure 3.15(b). The torque at zero speed is not infinite of course, because of the effects of saturation and resistance, both of which are ignored in equation (3.14).

As with the shunt motor, under starting conditions the back e.m.f. is zero, and if the full voltage was applied the current would be excessive, being limited only by the armature and field resistances. Hence for all but small motors a starting resistance is required to limit the current to a safe value.

Returning to Figure 3.15(b), we note that the series motor differs from most other motors in having no clearly defined no-load speed, i.e. no speed (other than infinity) at which the torque produced by the motor falls to zero. This means that when running light, the speed of the motor depends on the windage and friction

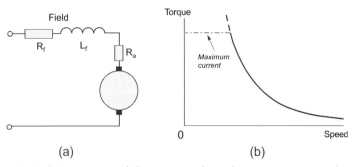

Figure 3.15 Series-connected d.c. motor and steady-state torque–speed curve.

torques, equilibrium being reached when the motor torque equals the total mechanical resisting torque. In large motors, the windage and friction torque is often relatively small, and the no-load speed is then too high for mechanical safety. Large series motors should therefore never be run uncoupled from their loads. As with shunt motors, the connections to either the field or armature must be reversed in order to reverse the direction of rotation.

The reason for the historic use of series motors for traction is that under the simplest possible supply arrangement (i.e. constant voltage) the overall shape of the torque–speed curve fits well with what is needed in traction applications, i.e. high starting torque to provide acceleration from rest and an acceptable reduction in acceleration as the target speed is approached. Early systems achieved coarse speed control via resistors switched in parallel with either the field or the armature in order to divert some of the current and thereby alter the torque–speed curve, but this inherently inefficient approach became obsolete when power electronic converters became available to provide an efficient variable-voltage supply.

7.3 Universal motors

In terms of numbers the main application area for the series commutator motor is in portable power tools, food mixers, vacuum cleaners, etc., where paradoxically the supply is a.c. rather than d.c. Such motors are often referred to as 'universal' motors because they can run from either a d.c. or an a.c. supply.

At first sight the fact that a d.c. machine will work on a.c. is hard to believe. But when we recall that in a series motor the field flux is set up by the current which also flows in the armature, it can be seen that reversal of the current will be accompanied by reversal of the direction of the magnetic flux, thereby ensuring that the torque remains positive. When the motor is connected to a 50 Hz supply, for example, the (sinusoidal) current will change direction every 10 ms, and there will be a peak in the torque 100 times per second. But the torque will always remain unidirectional, and the speed fluctuations will not be noticeable because of the smoothing effect of the armature inertia.

Series motors for use on a.c. supplies are always designed with fully laminated construction (to limit eddy current losses produced by the pulsating flux in the magnetic circuit), and are intended to run at high speeds, say 8–12,000 rev/min at rated voltage. Commutation and sparking are worse than when operating from d.c., and output powers are seldom greater than 1 kW. The advantage of high speed in terms of power output per unit volume was emphasized in Chapter 1, and the universal motor is perhaps the best everyday example which demonstrates how a high power can be obtained with small size by designing for a high speed.

Until recently the universal motor offered the only relatively cheap way of reaping the benefit of high speed from single-phase a.c. supplies. Other small a.c. machines, such as induction motors and synchronous motors, were limited to maximum speeds of 3000 rev/min at 50 Hz (or 3600 rev/min at 60 Hz), and therefore could not compete in terms of power per unit volume. The availability of high-frequency inverters (see Chapter 7) has opened up the prospect of higher specific outputs from induction motors and permanent magnet motors in many appliances. The transition away from the universal motor is under way, but because of the huge investment in high volume manufacturing that has been made over many years, it is still being produced in significant numbers.

Speed control of small universal motors is straightforward using a triac (in effect a pair of thyristors connected back to back) in series with the a.c. supply. By varying the firing angle, and hence the proportion of each cycle for which the triac conducts, the voltage applied to the motor can be varied to provide speed control. This approach is widely used for electric drills, fans, etc. If torque control is required (as in hand power tools, for example), the current is controlled rather than the voltage, and the speed is determined by the load.

8. SELF-EXCITED D.C. MACHINE

At various points in this chapter we have seen that the d.c. machine can change from motoring to generating depending on whether its induced or generated e.m.f. was less than or greater than the voltage applied to the armature terminals. We assumed that the machine was connected to a voltage source, from which it could draw both its excitation (field) and armature currents. But if we want to use the machine as a generator in a location isolated from any voltage source, we have to consider how to set up the flux to enable the machine to work as an energy converter.

When the field is provided by permanent magnets, there is no problem. As soon as the prime-mover (an internal combustion engine, perhaps) starts to turn the rotor, a speed-dependent e.m.f. (voltage) is produced in the armature and we can connect our load to the terminals and start to generate power. To ensure a more or less constant voltage, the engine will be governed to keep the speed constant.

If the machine has a field winding that is intended to be connected in parallel (shunt) with the armature, it is easy to imagine that once the machine is generating an armature voltage, there will be a field current and flux. But before we get a motional e.m.f. we need flux, and we can't have flux without a field current, and we won't get field current until we have generated an e.m.f.

The answer to this apparent difficulty usually lies in taking advantage of the residual magnetic flux remaining in the magnetic circuit from the time the machine was last used. If the residual flux is sufficient we will begin to generate as soon as the rotor is turned, and provided the field is connected the right way round across the armature, the first few milliamps of field current will increase the flux and we will then have positive feedback which we might expect to result in a runaway situation with ever-increasing generated voltage. Fortunately, provided that the driven speed is constant, the growth of flux is stabilized by saturation of the magnetic circuit, as we will discuss next. This procedure is known as self-excitation, for obvious reasons. (Should we be unfortunate enough to connect the field the wrong way round, we will be disappointed: but if the connections were undisturbed from the previous time, we should be OK.)

The stabilizing effect of saturation is shown in Figure 3.16. The solid line in the graph on the right is a typical 'magnetization curve' of the machine. This is a plot of the generated e.m.f. (E) as a function of field current when the rotor is driven at a constant speed. This test is best done with a separate supply to the field, and (usually) at the rated speed. For low and medium values of the field current, the flux is proportional to the current, so the motional e.m.f. is linear, but at higher field currents the magnetic circuit starts to saturate, the flux no longer increases in proportion, and consequently the generated voltage flattens out. The armature circuit is open, so the terminal voltage (V_a) is equal to the induced e.m.f. (E).

When we want the machine to self-excite, we connect the field in parallel with the armature as shown in Figure 3.16. The current in the field is then given by $I_f = V_a/R_f$, where R_f is the resistance of the field winding, and we have neglected the small armature resistance. The two dashed lines relate to field circuit resistances of 100 Ω and 200 Ω, and the intersection of these with the magnetization curve gives the stable operating points. For example, when the field resistance is 100 Ω, and the armature voltage is 200 V, the field current of 2 A is just sufficient to generate the motional e.m.f. of 200 V (point A). At lower values of field current, the generated voltage exceeds the value needed to maintain the field current, so the excess voltage causes the current to increase in the inductive circuit until the steady state is reached.

Increasing the resistance of the field circuit to 200 Ω results in a lower generated voltage (point B), but this is a less stable situation, with small changes in resistance

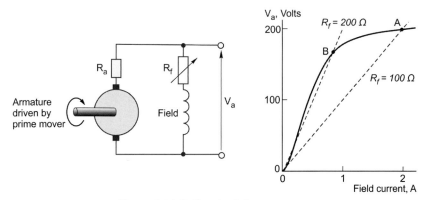

Figure 3.16 Self-excited d.c. generator.

leading to large changes in voltage. It should also be clear that if the resistance is increased above 200 Ω, the resistance line will be above the magnetization curve, and self-excitation will be impossible.

9. TOY MOTORS

The motors used in model cars, trains, etc. are rather different in construction from those discussed so far, primarily because they are designed almost entirely with cost as the primary consideration. They also run at high speeds, so it is not important for the torque to be smooth. A typical arrangement used for rotor diameters from 0.5 cm to perhaps 3 cm is shown in Figure 3.17.

The rotor, made from laminations with a small number (typically three or five) of multi-turn coils in very large 'slots', is simple to manufacture, and because the commutator has few segments, it too is cheap to make. The field system (stator) consists of radially magnetized ceramic magnets with a steel backplate to complete the magnetic circuit.

The rotor clearly has very pronounced saliency, with three very large projections which are in marked contrast to the rotors we looked at earlier where the surface was basically cylindrical. It is easy to imagine that even when there is no current in the rotor coils, there is a strong tendency for the stator magnets to pull one or other of the rotor saliencies into alignment with a stator pole, so that the rotor would tend to lock in any one of six positions. This cyclic 'detent' torque is due to the variation of reluctance with rotor position, an effect which is exploited in a.c. reluctance motors (see Chapter 10), but is unwanted here. To combat the

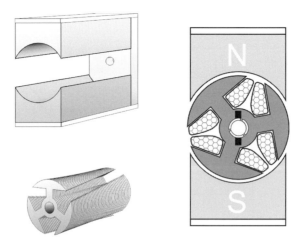

Figure 3.17 Miniature d.c. motor for use in models and toys.

problem the rotor laminations are skewed before the coils are fitted, as shown in the lower sketch on the left.

Each of the three rotor poles carries a multi-turn coil, the start of which is connected to a commutator segment, as shown in the cross-section on the right of Figure 3.17. The three ends are joined together. The brushes are wider than the inter-segment space, so in some rotor positions the current from say the positive brush will divide and flow first through two coils in parallel until it reaches the common point, then through the third coil to the negative brush, while in other positions the current flows through only two coils.

A real stretch of imagination is required to picture the mechanism of torque production using the 'BIl' approach we have followed previously, as the geometry is so different. But we can take a more intuitive approach by considering for each position the polarity and strength of the magnetization of each of the three rotor poles, which depend of course on the direction and magnitude of the respective currents. Following the discussion in the previous paragraph, we can see that in some positions there will be say a strong N and two relatively weak S poles, while at others there will be one strong N, one strong S, and one unexcited pole.

Thus although the rotor appears to be a three-pole device, the magnetization pattern is always two-pole (because two adjacent weak S poles function as a single stronger S pole). When the rotor is rotating, the stator N pole will attract the nearest rotor S pole, pulling it round towards alignment. As the force of attraction diminishes to zero, the commutator reverses the rotor current in that pole so that it now becomes an N and is pushed away towards the S stator pole.

There is a variation of current with angular position as a result of the different resistances seen by the brushes as they alternately make contact with either one or two segments, and the torque is far from uniform, but operating speeds are typically several thousand rev/min and the torque pulsations are smoothed out by the rotor and load inertia.

CHAPTER FOUR

D.C. Motor Drives

1. INTRODUCTION

Until the 1960s, the only satisfactory way of obtaining the variable-voltage d.c. supply needed for speed control of an industrial d.c. motor was to generate it with a d.c. generator. The generator was driven at fixed speed by an induction motor, and the field of the generator was varied in order to vary the generated voltage. For a brief period in the 1950s these 'Ward Leonard' sets were superseded by grid-controlled mercury arc rectifiers, but these were soon replaced by thyristor converters which offered cheaper first cost, higher efficiency (typically over 95%), smaller size, reduced maintenance and faster response to changes in set speed. The disadvantages of rectified supplies are that the waveforms are not pure d.c., that the overload capacity of the converter is very limited, and that a single converter is not inherently capable of regeneration.

Though no longer pre-eminent, study of the d.c. drive is valuable for two reasons:

- The structure and operation of the d.c. drive are reflected in almost all other drives, and lessons learned from the study of the d.c. drive therefore have close parallels in other types.
- Under constant-flux conditions the behavior is governed by a relatively simple set of linear equations, so predicting both steady-state and transient behavior is not difficult. When we turn to the successors of the d.c. drive, notably the induction motor drive, we will find that things are much more complex, and that in order to overcome the poor transient behavior, the control strategies adopted are based on emulating the inherent characteristics of the d.c. drive.

The first and major part of this chapter is devoted to thyristor-fed drives, after which we will look briefly at chopper-fed drives that are used mainly in medium and small sizes, and finally turn our attention to small servo-type drives.

2. THYRISTOR D.C. DRIVES – GENERAL

For motors up to a few kilowatts the armature converter draws power from either a single-phase or 3-phase utility supply. For larger motors 3-phase is preferred because the waveforms are much smoother, although traction uses single-phase with a series inductor to smooth the current. A separate thyristor or diode rectifier is used to supply the field of the motor: the power is much lower than the armature power, the inductance is much higher, and so the supply is often single-phase, as shown in Figure 4.1.

The arrangement shown in Figure 4.1 is typical of most d.c. drives and provides for closed-loop speed control. The function of the two control loops will be explored later, but readers who are not familiar with the basics of feedback and closed-loop systems may find it helpful to read through Appendix 1 at this point.

The main power circuit consists of a six-thyristor bridge circuit (as discussed in Chapter 2) which rectifies the incoming a.c. supply to produce a d.c. supply to the motor armature. By altering the firing angle of the thyristors the mean value of the rectified voltage can be varied, thereby allowing the motor speed to be controlled.

We saw in Chapter 2 that the controlled rectifier produces a crude form of d.c. with a pronounced ripple in the output voltage. This ripple component gives rise to pulsating currents and fluxes in the motor, and in order to avoid excessive eddy-current losses and commutation problems the poles and frame should be of laminated construction. It is accepted practice for motors supplied for use with thyristor drives to have laminated construction, but older motors often have solid poles and/or frames, and these will not always work satisfactorily with a rectifier supply. It is also the norm for d.c. motors for variable-speed operation to be supplied with an attached 'blower' motor as standard. This provides continuous through ventilation and allows the motor to operate continuously at full torque without overheating, even down to the lowest speeds. (See Plate 4.1)

Low-power control circuits are used to monitor the principal variables of interest (usually motor current and speed), and to generate appropriate firing pulses so that the motor maintains constant speed despite variations in the load. The 'speed reference' (Figure 4.1), historically an analogue voltage varying from 0 to 10 V, obtained from a manual speed-setting potentiometer or from elsewhere in the plant, now more typically comes in digital form.

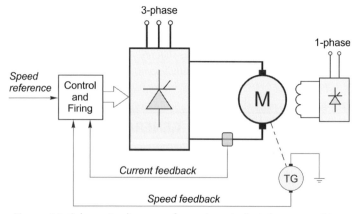

Figure 4.1 Schematic diagram of speed-controlled d.c. motor drive.

The combination of power, control and protective circuits constitutes the converter. Standard modular converters are available as off-the-shelf items in sizes from 100 W up to several hundred kW, while larger drives will be tailored to individual requirements. Individual converters may be mounted in enclosures with isolators, fuses, etc. or groups of converters may be mounted together to form a multi-motor drive.

2.1 Motor operation with converter supply

The basic operation of the rectifying bridge has been discussed in Chapter 2, and we now turn to the matter of how the d.c. motor behaves when supplied with 'd.c.' from a controlled rectifier.

By no stretch of the imagination could the waveforms of armature voltage looked at in Chapter 2 (e.g. Figure 2.13) be thought of as good d.c., and it would not be unreasonable to question the wisdom of feeding such an unpleasant-looking waveform to a d.c. motor. In fact it turns out that the motor works almost as well as it would if fed with pure d.c., for two main reasons. First, the armature inductance of the motor causes the waveform of armature current to be much smoother than the waveform of armature voltage, which in turn means that the torque ripple is much less than might have been feared. And secondly, the inertia of the armature is sufficiently large for the speed to remain almost steady despite the torque ripple. It is indeed fortunate that such a simple arrangement works so well, because any attempt to smooth out the voltage waveform (perhaps by adding smoothing capacitors) would prove to be prohibitively expensive in the power ranges of interest.

2.2 Motor current waveforms

For the sake of simplicity we will look at operation from a single-phase (2-pulse) converter, but similar conclusions apply to the 6-pulse one. The voltage (V_a) applied to the motor armature is typically as shown in Figure 4.2(a): as we saw in Chapter 2, it consists of rectified 'chunks' of the incoming mains voltage, the precise shape and average value depending on the firing angle.

The voltage waveform can be considered to consist of a mean d.c. level (V_{dc}), and a superimposed pulsating or ripple component which we can denote loosely as v_{ac}. If the supply is at 50 Hz, the fundamental frequency of the ripple is 100 Hz. The mean voltage V_{dc} can be altered by varying the firing angle, which also incidentally alters the ripple (i.e. v_{ac}).

The ripple voltage causes a ripple current to flow in the armature, but because of the armature inductance, the amplitude of the ripple current is small. In other words, the armature presents a high impedance to a.c. voltages. This smoothing effect of the armature inductance is shown in Figure 4.2(b), from which it can be seen that the current ripple is relatively small in comparison with the

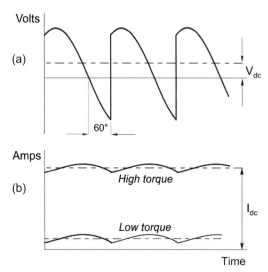

Figure 4.2 Armature voltage (a) and armature current (b) waveforms for continuous-current operation of a d.c. motor supplied from a single-phase fully controlled thyristor converter, with firing angle of 60°.

corresponding voltage ripple. The average value of the ripple current is of course zero, so it has no effect on the average torque of the motor. There is nevertheless a variation in torque every half-cycle of the mains, but because it is of small amplitude and high frequency (100 or 120 Hz for the 1-phase case here, but 300 or 360 Hz for 3-phase) the variation in speed (and hence back e.m.f. E) will not usually be noticeable.

The current at the end of each pulse is the same as at the beginning, so it follows that the average voltage across the armature inductance (L) is zero. We can therefore equate the average applied voltage to the sum of the back e.m.f. (assumed pure d.c. because we are ignoring speed fluctuations) and the average voltage across the armature resistance, to yield

$$V_{dc} = E + I_{dc}R \qquad (4.1)$$

which is exactly the same as for operation from a pure d.c. supply. This is very important, as it underlines the fact that we can control the mean motor voltage, and hence the speed, simply by varying the converter delay angle.

The smoothing effect of the armature inductance is important in achieving successful motor operation: the armature acts as a low-pass filter, blocking most of the ripple, and leading to a more or less constant armature current. For the smoothing to be effective, the armature time-constant needs to be long compared with the pulse duration (half a cycle with a 2-pulse drive, but only one-sixth of a cycle in a 6-pulse drive). This condition is met in all 6-pulse drives, and in many 2-pulse ones. Overall, the motor then behaves much as it would if it was supplied from an ideal d.c. source (though the I^2R loss is higher than it would be if the current was perfectly smooth).

The no-load speed is determined by the applied voltage (which depends on the firing angle of the converter); there is a small drop in speed with load; and, as we have previously noted, the average current is determined by the load. In Figure 4.2, for example, the voltage waveform in (a) applies equally for the two load conditions represented in (b), where the upper current waveform corresponds to a high value of load torque while the lower is for a much lighter load, the speed being almost the same in both cases. (The small difference in speed is due to the IR term, as explained in Chapter 3.) We should note that the current ripple remains the same – only the average current changes with load. Broadly speaking, therefore, we can say that the speed is determined by the converter firing angle, which represents a very satisfactory state of affairs because we can control the firing angle by low-power control circuits and thereby regulate the speed of the drive.

The current waveforms in Figure 4.2(b) are referred to as 'continuous', because there is never any time during which the current is not flowing. This 'continuous current' condition is the norm in most drives, and it is highly desirable because it is only under continuous current conditions that the average voltage from the converter is determined solely by the firing angle, and is independent of the load current. We can see why this is so with the aid of Figure 2.8, imagining that the motor is connected to the output terminals and that it is drawing a continuous current. For half of a complete cycle, the current will flow into the motor from T1 and return to the supply via T4, so the armature is effectively switched across the supply and the armature voltage is equal to the supply voltage, which is assumed to be ideal; i.e. it is independent of the current drawn. For the other half of the time, the motor current flows from T2 and returns to the supply via T3, so the motor is again hooked up to the supply, but this time the connections are reversed. Hence the average armature voltage (and thus, to a first approximation the speed) is defined once the firing angle is set.

2.3 Discontinuous current

We can see from Figure 4.2(b) that as the load torque is reduced, there will come a point where the minima of the current ripple touch the zero-current line, i.e. the current reaches the boundary between continuous and discontinuous current. The load at which this occurs will also depend on the armature inductance, because the higher the inductance the smoother the current (i.e. the less the ripple). Discontinuous current mode is therefore most likely to be encountered in small machines with low inductance (particularly when fed from 2-pulse converters) and under light-load or no-load conditions.

Typical armature voltage and current waveforms in the discontinuous mode are shown in Figure 4.3, the armature current consisting of discrete pulses of current

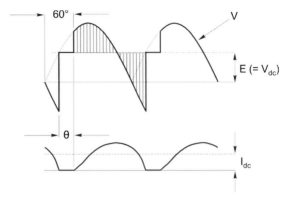

Figure 4.3 Armature voltage and current waveforms for discontinuous-current operation of a d.c. motor supplied from a single-phase fully controlled thyristor converter, with firing angle of 60°.

that occur only while the armature is connected to the supply, with zero current for the period (represented by θ in Figure 4.3) when none of the thyristors is conducting and the motor is coasting free from the supply.

The shape of the current waveform can be understood by noting that with resistance neglected, equation (3.7) can be rearranged as

$$\frac{di}{dt} = \frac{1}{L}(V - E) \tag{4.2}$$

which shows that the rate of change of current (i.e. the gradient of the lower graph in Figure 4.3) is determined by the instantaneous difference between the applied voltage V and the motional e.m.f. E. Values of $(V - E)$ are shown by the vertical hatchings in Figure 4.3, from which it can be seen that if $V > E$, the current is increasing, while if $V < E$, the current is falling. The peak current is thus determined by the area of the upper or lower shaded areas of the upper graph.

The firing angle in Figures 4.2 and 4.3 is the same, at 60°, but the load is less in Figure 4.3 and hence the average current is lower (though, for the sake of the explanation offered below, the current axis in Figure 4.3 is expanded as compared with that in Figure 4.2). It should be clear by comparing these figures that the armature voltage waveforms (solid lines) differ because, in Figure 4.3, the current falls to zero before the next firing pulse arrives and during the period shown as θ the motor floats free, its terminal voltage during this time being simply the motional e.m.f. (E). To simplify Figure 4.3 it has been assumed that the armature resistance is small and that the corresponding volt-drop ($I_a R_a$) can be ignored. In this case, the average armature voltage (V_{dc}) must be equal to the motional e.m.f., because there can be no average voltage across the armature inductance when there is no net change in the current over one pulse: the hatched areas – representing the volt-seconds in the inductor – are therefore equal.

The most important difference between Figures 4.2 and 4.3 is that the average voltage is higher when the current is discontinuous, and hence the speed corresponding to the conditions in Figure 4.3 is higher than in Figure 4.2 despite both having the same firing angle. And whereas in continuous mode a load increase can be met by an increased armature current without affecting the voltage (and hence speed), the situation is very different when the current is discontinuous. In the latter case, the only way that the average current can increase is for the speed (and hence E) to fall so that the shaded areas in Figure 4.3 become larger.

This means that the behavior of the motor in discontinuous mode is much worse than in the continuous current mode, because, as the load torque is increased, there is a serious drop in speed. The resulting torque–speed curve therefore has a very unwelcome 'droopy' characteristic in the discontinuous current region, as shown in Figure 4.4, and in addition the I^2R loss is much higher than it would be with pure d.c.

Under very light or no-load conditions, the pulses of current become virtually non-existent, the shaded areas in Figure 4.3 become very small and the motor speed approaches that at which the back e.m.f. is equal to the peak of the supply voltage (point C in Figure 4.4).

It is easy to see that inherent torque–speed curves with sudden discontinuities of the form shown in Figure 4.4 are very undesirable. If, for example, the firing angle is set to zero and the motor is fully loaded, its speed will settle at point A, its average armature voltage and current having their full (rated) values. As the load is reduced, current remaining continuous, and there is the expected slight rise in speed, until point B is reached. This is the point at which the current is about to enter the discontinuous phase. Any further reduction in the load torque then produces a wholly disproportionate – not to say frightening – increase in speed, especially if the load is reduced to zero when the speed reaches point C.

There are two ways by which we can improve these inherently poor characteristics. First, we can add extra inductance in series with the armature to further smooth the current waveform and lessen the likelihood of discontinuous current.

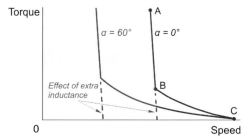

Figure 4.4 Torque–speed curves illustrating the undesirable 'droopy' characteristic associated with discontinuous current. The improved characteristic (shown dashed) corresponds to operation with continuous current.

The effect of adding inductance is shown by the dashed lines in Figure 4.4. And secondly, we can switch from a single-phase converter to a 3-phase one which produces smoother voltage and current waveforms, as discussed in Chapter 2.

When the converter and motor are incorporated in a closed-loop drive system the user should be unaware of any shortcomings in the inherent motor/converter characteristics because the control system automatically alters the firing angle to achieve the target speed at all loads. In relation to Figure 4.4, for example, as far as the user is concerned the control system will confine operation to the shaded region, and the fact that the motor is theoretically capable of running unloaded at the high speed corresponding to point C is only of academic interest.

It is worth mentioning that discontinuous current operation is not restricted to the thyristor converter, but occurs in many other types of power-electronic system. Broadly speaking, converter operation is more easily understood and analyzed when in continuous current mode, and the operating characteristics are more desirable, as we have seen here. We will not dwell on the discontinuous mode in the rest of the book, as it is beyond our scope, and unlikely to be of concern to the drive user.

2.4 Converter output impedance: overlap

So far we have tacitly assumed that the output voltage from the converter was independent of the current drawn by the motor, and depended only on the delay angle α. In other words we have treated the converter as an ideal voltage source.

In practice the a.c. supply has a finite impedance, and we must therefore expect a volt-drop which depends on the current being drawn by the motor. Perhaps surprisingly, the supply impedance (which is mainly due to inductive leakage reactances in transformers) manifests itself at the output stage of the converter as a supply resistance, so the supply volt-drop (or regulation) is directly proportional to the motor armature current.

It is not appropriate to go into more detail here, but we should note that the effect of the inductive reactance of the supply is to delay the transfer (or commutation) of the current between thyristors, a phenomenon known as overlap. The consequence of overlap is that instead of the output voltage making an abrupt jump at the start of each pulse, there is a short period where two thyristors are conducting simultaneously. During this interval the output voltage is the mean of the voltages of the incoming and outgoing voltages, as shown typically in Figure 4.5. When the drive is connected to a 'stiff' (i.e. low impedance) industrial supply the overlap will only last for perhaps a few microseconds, so the 'notch' shown in Figure 4.5 would be barely visible on an oscilloscope. Books always exaggerate the width of the overlap for the sake of clarity, as in Figure 4.5: with a 50 or 60 Hz supply, if the overlap lasts for more than say 1 millisecond, the implication is that the supply system impedance is too high for the size of converter in question, or conversely, the converter is too big for the supply.

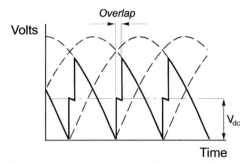

Figure 4.5 Distortion of converter output voltage waveform caused by rectifier overlap.

Returning to the practical consequences of supply impedance, we simply have to allow for the presence of an extra 'source resistance' in series with the output voltage of the converter. This source resistance is in series with the motor armature resistance, and hence the motor torque–speed curves for each value of α have a somewhat steeper droop than they would if the supply impedance was zero. However, as part of a closed-loop control system, the drive would automatically compensate for any speed droop resulting from overlap.

2.5 Four-quadrant operation and inversion

So far we have looked at the converter as a rectifier, supplying power from the a.c. utility supply to a d.c. machine running in the positive direction and acting as a motor. As explained in Chapter 3, this is known as one-quadrant operation, by reference to quadrant 1 of the complete torque–speed plane shown in Figure 3.12.

But suppose we want to run a motor in the opposite direction, with negative speed and torque, i.e. in quadrant 3. How do we do it? And what about operating the machine as a generator, so that power is returned to the a.c. supply, the converter then 'inverting' power rather than rectifying, and the system operating in quadrant 2 or quadrant 4? We need to do this if we want to achieve regenerative braking. Is it possible, and if so how?

The good news is that as we saw in Chapter 3 the d.c. machine is inherently a bi-directional energy converter. If we apply a positive voltage V greater than E, a current flows into the armature and the machine runs as a motor. If we reduce V so that it is less than E, the current, torque and power automatically reverse direction, and the machine acts as a generator, converting mechanical energy (its own kinetic energy in the case of regenerative braking) into electrical energy. And if we want to motor or generate with the reverse direction of rotation, all we have to do is to reverse the polarity of the armature supply. The d.c. machine is inherently a four-quadrant device, but needs a supply which can provide positive or negative voltage, and simultaneously handle either positive or negative current.

This is where we meet a snag: a single thyristor converter can only handle current in one direction, because the thyristors are unidirectional devices. This does not mean that the converter is incapable of returning power to the supply, however. The d.c. current can only be positive, but (provided it is a fully controlled converter) the d.c. output voltage can be either positive or negative (see Chapter 2). The power flow can therefore be positive (rectification) or negative (inversion).

For normal motoring where the output voltage is positive (and assuming a fully controlled converter), the delay angle (α) will be up to 90°. (It is common practice for the firing angle corresponding to rated d.c. voltage to be around 20° when the incoming a.c. voltage is normal: then if the a.c. voltage falls for any reason, the firing angle can be further reduced to compensate and allow full d.c. voltage to be maintained.)

When α is greater than 90°, however, the average d.c. output voltage is negative, as indicated by equation (2.5), and shown in Figure 4.6. A single fully controlled converter therefore has the potential for two-quadrant operation, though it has to be admitted that this capability is not easily exploited unless we are prepared to employ reversing switches in the armature or field circuits. This is discussed next.

2.6 Single-converter reversing drives

We will consider a fully controlled converter supplying a permanent-magnet motor, and see how the motor can be regeneratively braked from full speed in one direction, and then accelerated up to full speed in reverse. We looked at this procedure in principle at the end of Chapter 3, but here we explore the practicalities of achieving it with a converter-fed drive. We should be clear from the outset that in practice all the user has to do is to change the speed reference signal from full forward to full reverse; the control system in the drive converter takes care of matters from then on. What it does, and how, is discussed below.

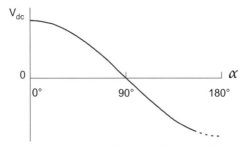

Figure 4.6 Average d.c. output voltage from a fully controlled thyristor converter as a function of the firing angle delay α.

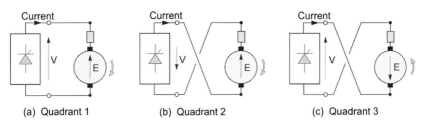

(a) Quadrant 1   (b) Quadrant 2   (c) Quadrant 3

Figure 4.7 Stages in motor reversal using a single-converter drive and mechanical reversing switch.

When the motor is running at full speed forward, the converter delay angle will be small, and the converter output voltage V and current I will both be positive. This condition is shown in Figure 4.7(a), and corresponds to operation in quadrant 1.

In order to brake the motor, the torque has to be reversed. The only way this can be done is by reversing the direction of armature current. The converter can only supply positive current, however, so to reverse the motor torque we have to reverse the armature connections, using a mechanical switch or contactor, as shown in Figure 4.7(b). (Before operating the contactor, the armature current would be reduced to zero by lowering the converter voltage, so that the contactor is not required to interrupt current.) Note that because the motor is still rotating in the positive direction, the back e.m.f. remains in its original sense; but now the motional e.m.f. is seen to be assisting the current and so to keep the current within bounds the converter must produce a negative voltage V which is just a little less than E. This is achieved by setting the delay angle at the appropriate point between $90°$ and $180°$. (The dashed line in Figure 4.6 indicates that the maximum acceptable negative voltage will generally be somewhat less than the maximum positive voltage: this restriction arises because of the need to preserve a margin for commutation of current between thyristors.) Note that the converter current is still positive (i.e. upwards in Figure 4.7(b)), but the converter voltage is negative, and power is thus flowing back to the supply system. In this condition the system is operating in quadrant 2, and the motor is decelerating because of the negative torque. As the speed falls, E reduces, and so V must be reduced progressively to keep the current at full value. This is achieved automatically by the action of the current-control loop, which is discussed later.

The current (i.e. torque) needs to be kept negative in order to run up to speed in the reverse direction, but after the back e.m.f. changes sign (as the motor reverses) the converter voltage again becomes positive and greater than E, as shown in Figure 4.7(c). The converter is then rectifying, with power being fed into the motor, and the system is operating in quadrant 3.

Schemes using reversing contactors are not suitable where the reversing time is critical or where periods of zero torque are unacceptable, because of the delay caused by the mechanical reversing switch, which may easily amount to 200–400 ms. Field reversal schemes operate in a similar way, but reverse the field current instead of the armature current. They are even slower, up to 5 s, because of the relatively long time-constant of the field winding.

2.7 Double converter reversing drives

Where full four-quadrant operation and rapid reversal is called for, two converters connected in anti-parallel are used, as shown in Figure 4.8. One converter supplies positive current to the motor, while the other supplies negative current.

The bridges are operated so that their d.c. voltages are almost equal, thereby ensuring that any d.c. circulating current is small, and a reactor may be placed between the bridges to limit the flow of ripple currents which result from the unequal ripple voltages of the two converters.

In most applications, the reactor can be dispensed with and the converters operated one at a time. The changeover from one converter to the other can only take place after the firing pulses have been removed from one converter and the armature current has decayed to zero. Appropriate zero-current detection circuitry is provided as an integral part of the drive, so that as far as the user is concerned, the two converters behave as if they were a single ideal bi-directional d.c. source. There is a dead (torque-free) time typically of only 10 ms or so during the changeover period from one bridge to the other.

Prospective users need to be aware of the fact that a basic single converter can only provide for operation in one quadrant. If regenerative braking is required, either field or armature reversing contactors will be needed; and if rapid reversal is essential, a double converter has to be used. All these extras naturally push up the purchase price.

2.8 Power factor and supply effects

One of the drawbacks of a converter-fed d.c. drive is that the supply power-factor is very low when the motor is operating at low speed (i.e. low armature voltage), and is less than unity even at base speed and full-load. This is because the supply current waveform lags the supply voltage waveform by the delay angle α, as shown (for

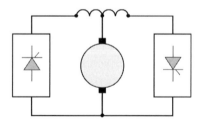

Figure 4.8 Double-converter reversing drive.

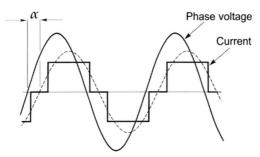

Figure 4.9 Supply voltage and current waveforms for single-phase converter-fed d.c. motor drive.

a 3-phase converter) in Figure 4.9, and also the supply current is approximately rectangular (rather than sinusoidal).

It is important to emphasize that the supply power-factor is always lagging, even when the converter is inverting. There is no way of avoiding the low power-factor, so users of large drives need to be prepared to augment their existing power-factor, correcting equipment if necessary.

The harmonics in the mains current waveform can give rise to a variety of disturbance problems, and supply authorities generally impose statutory limits. For large drives (say hundreds of kW), filters may have to be provided to prevent these limits from being exceeded.

Since the supply impedance is never zero, there is also inevitably some distortion of the mains voltage waveform, as shown in Figure 4.10, which indicates the effect of a 6-pulse converter on the supply line-to-line voltage waveform. The spikes and notches arise because the utility supply is momentarily short-circuited each time the current commutates from one thyristor to the next, i.e. during the overlap period discussed earlier. For the majority of small and medium drives, connected to stiff industrial supplies, these notches are too small to be noticed (they are greatly exaggerated for the sake of clarity in Figure 4.10); but they can cause

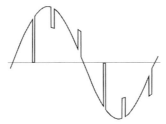

Figure 4.10 Distortion of line voltage waveform caused by overlap in 3-phase fully controlled converter. (The width of the notches has been exaggerated for the sake of clarity.)

serious disturbance to other consumers when a large drive is connected to a weak supply.

3. CONTROL ARRANGEMENTS FOR D.C. DRIVES

The most common arrangement, which is used with only minor variations from small drives of say 0.5 kW up to the largest industrial drives of several MW, is the so-called two-loop control. This has an inner feedback loop to control the current (and hence torque) and an outer loop to control speed. When position control is called for, a further outer position loop is added. A two-loop scheme for a thyristor d.c. drive is discussed first, but the essential features are the same in a chopper-fed drive. Later the simpler arrangements used in low-cost small drives are mentioned.

In order to simplify the discussion we will assume that the control signals are analogue, although in all modern versions the implementation will be digital, and we will limit consideration to those aspects which will be beneficial for the user to know something about. In practice, once a drive has been commissioned, there are only a few adjustments to which the user has access. While most of them are self-explanatory (e.g. max. speed, min. speed, accel. and decel. rates), some are less obvious (e.g. 'current stability', 'speed stability', 'IR comp'.) so these are explained.

To appreciate the overall operation of a two-loop scheme we can consider what we would do if we were controlling the motor manually. For example, if we found by observing the tachogenerator that the speed was below target, we would want to provide more current (and hence torque) in order to produce acceleration, so we would raise the armature voltage. We would have to do this gingerly, however, being mindful of the danger of creating an excessive current because of the delicate balance that exists between the back e.m.f. E and applied voltage V. We would doubtless wish to keep our eye on the ammeter at all times to avoid blowing up the thyristors; and as the speed approached the target, we would trim back the current (by lowering the applied voltage) so as to avoid overshooting the set speed. Actions of this sort are carried out automatically by the drive system, which we will now explore.

A standard d.c. drive system with speed and current control is shown in Figure 4.11. The primary purpose of the control system is to provide speed control, so the 'input' to the system is the speed reference signal on the left, and the output is the speed of the motor (as measured by a tachogenerator) on the right. As with any closed-loop system, the overall performance is heavily dependent on the quality of the feedback signal, in this case the speed-proportional voltage provided by the tachogenerator. It is therefore important to ensure that the tacho is of high quality (so that, for example, its output voltage does not vary with ambient temperature, and is ripple-free) and as a result the cost of the tacho often represents a significant fraction of the total cost.

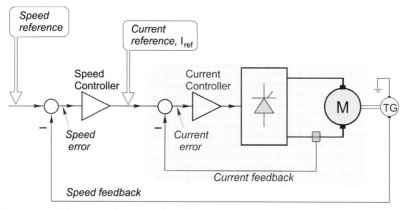

Figure 4.11 Schematic diagram of analogue controlled-speed drive with current and speed feedback control loops.

We will take an overview of how the scheme operates first, and then examine the function of the two loops in more detail.

To get an idea of the operation of the system we will consider what will happen if, with the motor running light at a set speed, the speed reference signal is suddenly increased. Because the set (reference) speed is now greater than the actual speed there will be a speed-error signal (see also Figure 4.12), represented by the output of the left-hand summing junction in Figure 4.11. A speed error indicates that acceleration is required, which in turn means torque, i.e. more current. The speed error is amplified by the speed controller (which is more accurately described as a speed-error amplifier) and the output serves as the reference or input signal to the inner control system. The inner feedback loop is a current-control loop, so when the current reference increases, so does the motor armature current, thereby providing extra torque and initiating acceleration. As the speed rises the speed error reduces and the current and torque therefore reduce to obtain a smooth approach to the target speed.

We will now look in more detail at the inner (current-control) loop, as its correct operation is vital to ensure that the thryistors are protected against excessive current.

3.1 Current limits and protection

The closed-loop current controller, or current loop, is at the heart of the drive system and is indicated by the shaded region in Figure 4.11. The purpose of the current loop is to make the actual motor current follow the current reference signal (I_{ref}) shown in Figure 4.11. It does this by comparing a feedback signal of actual motor current with the current reference signal, amplifying the difference (or current error), and using the resulting amplified current-error signal to control the firing angle α, and hence the output voltage, of the converter. The current feedback signal is obtained

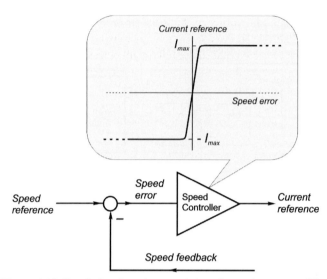

Figure 4.12 Detail showing characteristics of speed-error amplifier.

either from a d.c. current transformer (which gives an isolated analogue voltage output), or from a.c. current transformer/rectifiers in the mains supply lines.

The job of comparing the reference (demand) and actual current signals and amplifying the error signal is carried out by the current-error amplifier. By giving the current-error amplifier a high gain, the actual motor current will always correspond closely to the current reference signal, i.e. the current error will be small, regardless of motor speed. In other words, we can expect the actual motor current to follow the 'current reference' signal at all times, the armature voltage being automatically adjusted by the controller so that, regardless of the speed of the motor, the current has the correct value.

Of course no control system can be perfect, but it is usual for the current-error amplifier to be of the proportional plus integral (PI) type (see below), in which case the actual and demanded currents will be exactly equal under steady-state conditions.

The importance of preventing excessive converter currents from flowing has been emphasized previously, and the current control loop provides the means to this end. As long as the current control loop functions properly, the motor current can never exceed the reference value. Hence by limiting the magnitude of the current reference signal (by means of a clamping circuit), the motor current can never exceed the specified value. This is shown in Figure 4.12, which represents a small portion of Figure 4.11. The characteristics of the speed controller are shown in the shaded panel, from which we can see that for small errors in speed, the current reference increases in proportion to the speed, thereby ensuring 'linear system' behavior with a smooth approach to the target speed. However, once the speed error exceeds a limit, the output of the speed-error amplifier saturates and there is thus no further increase in the current reference. By arranging for this maximum current reference to correspond to the full (rated) current of the system there is no possibility of the current in the motor and converter exceeding its rated value, no matter how large the speed error becomes.

This 'electronic current limiting' is by far the most important protective feature of any drive. It means that if, for example, the motor suddenly stalls because the load seizes (so that the back e.m.f. falls dramatically), the armature voltage will automatically reduce to a very low value, thereby limiting the current to its maximum allowable level.

The inner loop is critical in a two-loop control system, so the current loop must guarantee that the steady-state motor current corresponds exactly with the reference, and the transient response to step changes in the current reference should be fast and well damped. The first of these requirements is satisfied by the integral term in the current-error amplifier, while the second is obtained by judicious choice of the amplifier proportional gain and time-constant. As far as the user is concerned, a 'current stability' adjustment may be provided to allow him/her to optimize the transient response of the current loop.

On a point of jargon, it should perhaps be mentioned that the current-error amplifier is more often than not called either the 'current controller' (as in Figure 4.11) or the 'current amplifier'. The first of these terms is quite sensible, but the second can be very misleading: there is after all no question of the motor current itself being amplified.

3.2 Torque control

For applications requiring the motor to operate with a specified torque regardless of speed (e.g. in line tensioning), we can dispense with the outer (speed) loop, and simply feed a current reference signal directly to the current controller (typically via a 'torque ref' terminal on the control board). This is because torque is directly proportional to current, so the current controller is in effect also a torque controller. We may have to make an allowance for accelerating torque, by means of a transient 'inertia compensating' signal, which could simply be added to the torque demand.

In the current-control mode the current remains constant at the set value, and the steady running speed is determined by the load. If the torque reference signal was set at 50%, for example, and the motor was initially at rest, it would accelerate with a constant current of half rated value until the load torque was equal to the motor torque. Of course, if the motor was running without any load, it would accelerate quickly, the applied voltage ramping up so that it always remained higher than the back e.m.f. by the amount needed to drive the specified current into the armature. Eventually the motor would reach a speed (a little above normal 'full' speed) at which the converter output voltage had reached its upper limit, and it was therefore no longer possible to maintain the set current: thereafter, the motor speed would remain steady.

This discussion assumes that torque is proportional to armature current, which is true only if the flux is held constant, which in turn requires the field current to be constant. Hence in all but small drives the field will be supplied from a thyristor converter with current feedback. Variations in field circuit resistance due to temperature changes, and/or changes in the utility supply voltage, are thereby automatically compensated and the flux is maintained at its rated value.

3.3 Speed control

The outer loop in Figure 4.11 provides speed control. Speed feedback is typically provided by a d.c. tachogenerator, and the actual and required speeds are fed into the speed-error amplifier (often known simply as the speed amplifier or the speed controller).

Any difference between the actual and desired speed is amplified, and the output serves as the input to the current loop. Hence if, for example, the actual motor speed is less than the desired speed, the speed amplifier will demand current in proportion to the speed error, and the motor will therefore accelerate in an attempt to minimize the speed error.

When the load increases, there is an immediate deceleration and the speed-error signal increases, thereby calling on the inner loop for more current. The increased torque results in acceleration and a progressive reduction of the speed error until equilibrium is reached at the point where the current reference (I_{ref}) produces a motor current that gives a torque equal and opposite to the load torque. Looking at Figure 4.12, where the speed controller is shown as a simple proportional amplifier (P control), it will be readily appreciated that in order for there to be a steady-state value of I_{ref}, there would have to be a finite speed error; i.e. a P controller would not allow us to reach exactly the target speed. (We could approach the ideal by increasing the gain of the amplifier, but that might lead us to instability.)

To eliminate the steady-state speed error we can easily arrange for the speed controller to have an integral (I) term as well as a proportional (P) term (see Appendix 1). A PI controller can have a finite output even when the input is zero, which means that we can achieve zero steady-state error if we employ PI control.

The speed will be held at the value set by the speed reference signal for all loads up to the point where full armature current is needed. If the load torque increases any more the speed will drop because the current-loop will not allow any more armature current to flow. Conversely, if the load attempted to force the speed above the set value, the motor current would be reversed automatically, so that the motor acts as a brake and regenerates power to the mains.

To emphasize further the vitally important protective role of the inner loop, we can see what happens when, with the motor at rest (and unloaded for the sake of simplicity), we suddenly increase the speed reference from zero to full value, i.e. we apply a step demand for full speed. The speed error will be 100%, so the output (I_{ref}) from the speed-error amplifier will immediately saturate at its maximum value (I_{max}, as shown in Figure 4.12), this value corresponding to the maximum (rated) current of the motor. The motor current will therefore be at rated value, and the motor will accelerate at full torque. Speed and back e.m.f. (E) will therefore rise at a constant rate, the applied voltage (V) increasing steadily so that the difference ($V - E$) is sufficient to drive rated current (I) through the armature resistance. A very similar sequence of events was discussed in Chapter 3, and is illustrated by the second half of Figure 3.13. (In some drives the current reference is allowed to reach 150% or even 200% of rated value for a few seconds, in order to provide a short torque boost. This is particularly valuable in starting loads with high static friction, and is known as 'two-stage current limit'.)

The output of the speed amplifier will remain saturated until the actual speed is quite close to the target speed, and for all this time the motor current will therefore be held at full value. Only when the speed is within a few per cent of target will the speed-error amplifier come out of saturation. Thereafter, as the speed continues to rise, and the speed error falls, the output of the speed-error amplifier falls below the clamped level. Speed control then enters a linear regime, in which the correcting current (and hence the torque) is proportional to speed error, giving a smooth approach to final speed.

A 'good' speed controller will result in zero steady-state error, and have a well-damped response to step changes in the demanded speed. The integral term in the PI control caters for the requirement of zero steady-state error, while the transient response depends on the setting of the proportional gain and time-constant. The 'speed stability' setting (traditionally a potentiometer) is provided to allow fine tuning of the transient response. (It should be mentioned that in some high-performance drives the controller will be of the PID form, i.e. it will also include a differential (D) term. The D term gives a bit of a kick to the controllers when a step change is called for – in effect advanced warning that we need a bit of smart action to change the current in the inductive circuit.)

It is important to remember that it is much easier to obtain a good transient response with a regenerative drive, which has the ability to supply negative current (i.e. braking torque) should the motor overshoot the desired speed. A non-regenerative drive cannot furnish negative current (unless fitted with reversing contactors), so if the speed overshoots the target the best that can be done is to reduce the armature current to zero and wait for the motor to decelerate naturally. This is not satisfactory, and every effort therefore has to be made to avoid controller settings which lead to an overshoot of the target speed.

As with any closed-loop scheme, problems occur if the feedback signal is lost when the system is in operation. If the tacho feedback became disconnected, the speed amplifier would immediately saturate, causing full torque to be applied. The speed would then rise until the converter output reached its maximum output voltage. To guard against this, many drives incorporate tacho-loss detection circuitry, and in some cases armature voltage feedback (see later section) automatically takes over in the event of tacho failure.

Drives which use field-weakening to extend the speed range include automatic provision for controlling both armature voltage and field current when running above base speed. Typically, the field current is kept at full value until the armature voltage reaches about 95% of rated value. When a higher speed is demanded, the extra armature voltage applied is accompanied by a simultaneous reduction in the field current, in such a way that when the armature voltage reaches 100% the field current is at the minimum safe value.

3.4 Overall operating region

A standard drive with field-weakening provides armature voltage control of speed up to base speed, and field-weakening control of speed thereafter. Any torque up to the rated value can be obtained at any speed below base speed, and as explained in Chapter 3 this region is known as the 'constant torque' region. Above base speed, the maximum available torque reduces inversely with speed, so this is known as the 'constant power' region. For a converter-fed drive the operating region in quadrant 1 of the torque–speed plane is therefore as shown in Figure 3.10. (If the drive is equipped for regenerative and reversing operation, the operating area is mirrored in the other three quadrants, of course.)

3.5 Armature voltage feedback and IR compensation

In low-power drives where precision speed-holding is not essential, and cost must be kept to a minimum, the tachogenerator may be dispensed with and the armature voltage used as a 'speed feedback' instead. Performance is clearly not as good as with tacho feedback, since while the steady-state no-load speed is proportional to armature voltage, the speed falls as the load (and hence armature current) increases.

We saw in Chapter 3 that the drop in speed with load was attributable to the armature resistance volt-drop (IR), and the steady-state drop in speed can therefore be compensated by boosting the applied voltage in proportion to the current. 'IR compensation' would in such cases be provided on the drive circuit for the user to adjust to suit the particular motor. The compensation is far from perfect, since it cannot cope with temperature variation of resistance, or with load transients.

3.6 Drives without current control

Very low-cost, low-power drives may dispense with a full current control loop, and incorporate a crude 'current-limit' which only operates when the maximum set current would otherwise be exceeded. These drives usually have an in-built ramp circuit which limits the rate of rise of the set speed signal so that under normal conditions the current limit is not activated. They are, however, prone to tripping in all but the most controlled of applications and environments.

4. CHOPPER-FED D.C. MOTOR DRIVES

If the source of supply is d.c. (for example, in a battery vehicle or a rapid transit system) a chopper-type converter is usually employed. The basic operation of a single-switch chopper was discussed in Chapter 2, where it was shown that the average output voltage could be varied by periodically switching the battery voltage on and off for varying intervals. The principal difference between the thyristor-controlled rectifier and the chopper is that in the former the motor current always flows through the supply, whereas in the latter the motor current only flows from the supply terminals for part of each cycle.

A single-switch chopper using a transistor, MOSFET or IGBT can only supply positive voltage and current to a d.c. motor, and is therefore restricted to one-quadrant motoring operation. When regenerative and/or rapid speed reversal is called for, more complex circuitry is required, involving two or more power switches, and consequently leading to increased cost. Many different circuits are used and it is not possible to go into detail here, but it will be recalled that the two most important types were described in section 2 of Chapter 2: the simplest or 'buck' converter provides an output voltage in the range $0 < E$, where E is the battery voltage, while the slightly more complex 'boost' converter provides output voltages greater than that of the supply.

4.1 Performance of chopper-fed d.c. motor drives

We saw earlier that the d.c. motor performed almost as well when fed from a phase-controlled rectifier as it does when supplied with pure d.c. The chopper-fed motor is, if anything, rather better than the phase-controlled, because the armature current ripple can be less if a high chopping frequency is used.

A typical circuit and waveforms of armature voltage and current are shown in Figure 4.13: these are drawn with the assumption that the switch is ideal. A chopping frequency of around 100 Hz, as shown in Figure 4.13, is typical of medium and large chopper drives, while small drives often use a much higher chopping frequency, and thus have lower ripple current. As usual, we have assumed that the speed remains constant despite the slightly pulsating torque, and that the armature current is continuous.

The shape of the armature voltage waveform reminds us that when the transistor is switched on, the battery voltage V is applied directly to the armature, and during this period the path of the armature current is indicated by the dotted line in Figure 4.13(a). For the remainder of the cycle the transistor is turned 'off' and the current freewheels through the diode, as shown by the dashed line in

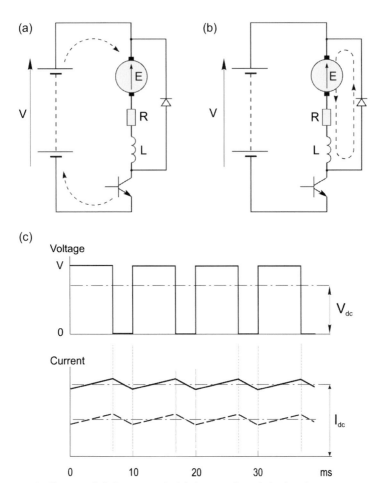

Figure 4.13 Chopper-fed d.c. motor. In (a) the transistor is 'on' and armature current is flowing through the voltage source; in (b) the transistor is 'off' and the armature current freewheels through the diode. Typical armature voltage and current waveforms are shown at (c), with the dotted line representing the current waveform when the load torque is reduced by half.

Figure 4.13(b). When the current is freewheeling through the diode, the armature voltage is clamped at (almost) zero.

The speed of the motor is determined by the average armature voltage (V_{dc}), which in turn depends on the proportion of the total cycle time (T) for which the transistor is 'on'. If the on and off times are defined as $T_{on} = kT$ and $T_{off} = (1 - kT)$, where $0 < k < 1$, then the average voltage is simply given by

$$V_{dc} = kV \tag{4.3}$$

from which we see that speed control is effected via the on time ratio k.

Turning now to the current waveforms shown in Figure 4.13(c), the upper waveform corresponds to full-load, i.e. the average current (I_{dc}) produces the full rated torque of the motor. If now the load torque on the motor shaft is reduced to half rated torque, and assuming that the resistance is negligible, the steady-state speed will remain the same but the new mean steady-state current will be halved, as shown by the lower dotted curve. We note, however, that although, as expected, the mean current is determined by the load, the ripple current is unchanged, and this is explained below.

If we ignore resistance, the equation governing the current during the 'on' period is

$$V = E + L\frac{di}{dt}, \text{ or } \frac{di}{dt} = \frac{1}{L}(V - E) \tag{4.4}$$

Since V is greater than E, the gradient of the current (di/dt) is positive, as can be seen in Figure 4.13(c). During this 'on' period the battery is supplying power to the motor. Some of the energy is converted to mechanical output power, but some is also stored in the magnetic field associated with the inductance. The latter is given by $\frac{1}{2}Li^2$, and so as the current (i) rises, more energy is stored.

During the 'off' period, the equation governing the current is

$$0 = E + L\frac{di}{dt}, \text{ or } \frac{di}{dt} = \frac{-E}{L} \tag{4.5}$$

We note that during the 'off' time the gradient of the current is negative (as shown in Figure 4.13(c)) and it is determined by the motional e.m.f. E. During this period, the motor is producing mechanical output power which is supplied from the energy stored in the inductance, so not surprisingly the current falls as the energy previously stored in the 'on' period is now given up.

We note that the rise and fall of the current (i.e. the current ripple) is inversely proportional to the inductance, but is independent of the mean d.c. current, i.e. the ripple does not depend on the load.

To study the input/output power relationship, we note that the battery current only flows during the on period, and its average value is therefore kI_{dc}. Since the battery voltage is constant, the power supplied is simply given by $V(kI_{dc}) = kVI_{dc}$. Looking at the motor side, the average voltage is given by $V_{dc} = kV$, and the average current (assumed constant) is I_{dc}, so the power input to the motor is again kVI_{dc}, i.e. there is no loss of power in the ideal chopper. Given that k is less than 1, we see that the input (battery) voltage is higher than the output (motor) voltage, but conversely the input current is less than the output current, and in this respect we see that the chopper behaves in much the same way for d.c. as a conventional transformer does for a.c.

4.2 Torque–speed characteristics and control arrangements

Under open-loop conditions (i.e. where the mark/space ratio of the chopper is fixed at a particular value) the behavior of the chopper-fed motor is similar to that of the converter-fed motor discussed earlier (see Figure 4.3). When the armature current is continuous the speed falls only slightly with load, because the mean armature voltage remains constant. But when the armature current is discontinuous (which is most likely at high speeds and light load) the speed falls off rapidly when the load increases, because the mean armature voltage falls as the load increases. Discontinuous current can be avoided by adding an inductor in series with the armature, or by raising the chopping frequency, but when closed-loop speed control is employed, the undesirable effects of discontinuous current are masked by the control loop.

The control philosophy and arrangements for a chopper-fed motor are the same as for the converter-fed motor, with the obvious exception that the mark/space ratio of the chopper is used to vary the output voltage, rather than the firing angle.

5. D.C. SERVO DRIVES

The precise meaning of the term 'servo' in the context of motors and drives is difficult to pin down. Broadly speaking, if a drive incorporates 'servo' in its description, the implication is that it is intended specifically for closed-loop or feedback control, usually of shaft torque, speed or position. Early servomechanisms were developed primarily for military applications, and it quickly became apparent that standard d.c. motors were not always suited to precision control. In particular, high torque to inertia ratios were needed, together with smooth ripple-free torque. Motors were therefore developed to meet these exacting requirements, and not surprisingly they were, and still are, much more expensive than their industrial counterparts. Whether the extra expense of a servo motor can be justified depends on the specification, but prospective users should always be on their guard to ensure they are not pressed into an expensive purchase when a conventional industrial drive could cope perfectly well.

The majority of servo drives are sold in modular form, consisting of a high-performance permanent magnet motor, often with an integral tachogenerator, and a chopper-type power amplifier module. The drive amplifier normally requires a separate regulated d.c. power supply, if, as is normally the case, the power is to be drawn from the utility supply. Continuous output powers range from a few watts up to perhaps 2–5 kW, with voltages of 12 V, 24 V, 48 V and multiples of 50 V being standard.

There has been an even more pronounced movement in the servo market than in industrial drives, away from d.c. in favor of the a.c. permanent magnet or induction motor, although d.c. servos do retain some niche applications.

5.1 Servo motors

Although there is no sharp dividing line between servo motors and ordinary motors, the servo type will be intended for use in applications which require rapid acceleration and deceleration. The design of the motor will reflect this by catering for intermittent currents (and hence torques) of many times the continuously rated value. Because most servo motors are small, their armature resistances are relatively high: the short-circuit (locked-rotor) current at full armature voltage is therefore perhaps only a few times the continuously rated current, and the drive amplifier will normally be selected so that it can cope with this condition, giving the motor a very rapid acceleration from rest. The even more arduous condition in which the full armature voltage is suddenly reversed with the motor running at full speed is also quite normal. (Both of these modes of operation would of course be quite unthinkable with a large d.c. motor, because of the huge currents which would flow as a result of the much lower per-unit armature resistance.)

In order to maximize acceleration, the rotor inertia must be minimized, and one obvious way to achieve this is to construct a motor in which only the electric circuit (conductors) on the rotor moves, the magnetic part (either iron or permanent magnet) remaining stationary. This principle is adopted in 'ironless rotor' and 'printed armature' motors.

In the ironless rotor or moving-coil type (Figure 4.14) the armature conductors are formed as a thin-walled cylinder consisting essentially of nothing more than varnished wires wound in skewed form together with the disc-type commutator (not shown). Inside the armature sits a two-pole (upper N, lower S) permanent magnet which provides the radial flux, and outside it is a steel cylindrical shell which completes the magnetic circuit.

Needless to say the absence of any slots to support the armature winding results in a relatively fragile structure, which is therefore limited to diameters of not much over 1 cm. Because of their small size they are often known as micromotors, and are very widely used in cameras, video systems, card readers, etc.

The printed armature type is altogether more robust, and is made in sizes up to a few kW. They are generally made in disc or pancake form, with the direction of flux axial and the armature current radial. The armature conductors resemble spokes on a wheel, the conductors themselves being formed on a lightweight disc. Early versions were made by using printed-circuit techniques, but pressed fabrication is now more common. Since there are usually at least a hundred armature conductors,

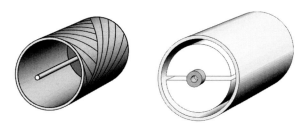

Figure 4.14 Ironless rotor d.c. motor. The commutator (not shown) is usually of the disc type.

the torque remains almost constant as the rotor turns, which allows them to produce very smooth rotation at low speed. Inertia and armature inductance are low, giving a good dynamic response, and the short and fat shape makes them suitable for applications such as machine tools and disc drives where axial space is at a premium.

5.2 Position control

As mentioned earlier many servo motors are used in closed-loop position control applications, so it is appropriate to look briefly at how this is achieved. Later (in Chapter 10) we will see that the stepping motor provides an alternative open-loop method of position control, which can be cheaper for some less demanding applications.

In the example shown in Figure 4.15, the angular position of the output shaft is intended to follow the reference voltage (θ_{ref}), but it should be clear that if the motor drives a toothed belt, linear outputs can also be obtained. The potentiometer mounted on the output shaft provides a feedback voltage proportional to the actual position of the output shaft. The voltage from this potentiometer must be a linear function of angle, and must not vary with temperature, otherwise the accuracy of the system will be in doubt.

The feedback voltage (representing the actual angle of the shaft) is subtracted from the reference voltage (representing the desired position) and the resulting position error signal is amplified and used to drive the motor so as to rotate the output shaft in the desired direction. When the output shaft reaches the target position, the position error becomes zero, no voltage is applied to the motor, and the output shaft remains at rest. Any attempt to physically move the output shaft from its target position immediately creates a position error and a restoring torque is applied by the motor.

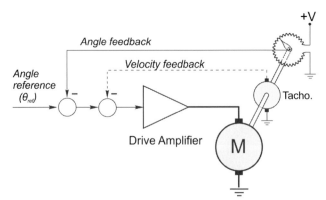

Figure 4.15 Closed-loop angular position control using d.c. motor and angle feedback from a servo-type potentiometer.

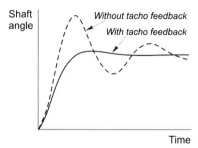

Figure 4.16 Typical step responses for a closed-loop position control system, showing the improved damping obtained by the addition of tacho feedback.

The dynamic performance of the simple scheme described above is very unsatisfactory as it stands. In order to achieve a fast response and to minimize position errors caused by static friction, the gain of the amplifier needs to be high, but this in turn leads to a highly oscillatory response which is usually unacceptable. For some fixed-load applications matters can be improved by adding a compensating network at the input to the amplifier, but the best solution is to use 'tacho' (speed) feedback (shown dashed in Figure 4.15) in addition to the main position feedback loop. Tacho feedback clearly has no effect on the static behavior (since the voltage from the tacho is proportional to the speed of the motor), but has the effect of increasing the damping of the transient response. The gain of the amplifier can therefore be made high in order to give a fast response, and the degree of tacho feedback can then be adjusted to provide the required damping (see Figure 4.16). Many servo motors have an integral tachogenerator for this purpose. (This is a particular example of the general principle by which the response can be improved by adding 'derivative of output' feedback: in this case the speed signal is the rate of change (or derivative) of the angular position.)

The example above dealt with an analogue scheme in the interests of simplicity, but digital position control schemes (with an encoder instead of a potentiometer, and no tacho) are now much more common, especially when brushless motors (see Chapter 9) are used. Complete 'controllers on a card' are available as off-the-shelf items, and these offer ease of interface to other systems as well as providing for improved flexibility in shaping the dynamic response.

6. DIGITALLY CONTROLLED DRIVES

As in all forms of industrial and precision control, digital implementations have replaced analogue circuitry in the vast majority of electric drive systems, but there are few instances where this has resulted in any real change to the basic structure of the drive with respect to the motor control, and in most cases understanding how the drive functions is still best approached in the first instance by studying the analogue version. Digital control electronics has brought with it a considerable advance in the auxiliary control and protection functions which are now routinely found on a drive system. Digital control electronics has also facilitated the commercial implementation of advanced a.c. motor control strategies, which will be discussed in Chapters 7 and 8. However, as far as understanding d.c. drives is concerned, users who have developed a sound understanding of how the analogue version operates will find little to trouble them when considering the digital equivalent. Accordingly this section is limited to the consideration of a few of the advantages offered by digital implementations, and readers seeking more are recommended to consult a book such as the *Control Techniques Drives and Controls Handbook*, 2nd edition, by W. Drury.

Many drives use digital speed feedback, in which a pulse train generated from a shaft-mounted encoder is compared (using a phase-locked loop) with a reference pulse train whose frequency corresponds to the desired speed, or where the reference is transmitted to the drive in the form of a synchronous serial word. Consequently, the feedback is more accurate and drift-free and noise in the encoder signal is easily rejected, so that very precise speed holding can be guaranteed. This is especially important when a number of independent motors must all be driven at identical speed. Phase-locked loops are also used in the firing-pulse synchronizing circuits, to overcome the problems caused by noise on the mains waveform.

Digital controllers offer freedom from drift, added flexibility (e.g. programmable ramp-up, ramp-down, maximum and minimum speeds, etc.), ease of interfacing and linking to other drives and host computers and controllers, and self-tuning. User-friendly diagnostics represents another benefit, providing the local or remote user with current and historical data on the state of all the key drive variables. Digital drives also offer many more functions, including user programmable functions as are found on programmable logic controllers as well as a host of communications interfaces to allow incorporation into industrial automation systems.

CHAPTER FIVE

Induction Motors – Rotating Field, Slip and Torque

1. INTRODUCTION

Judged in terms of fitness for purpose coupled with simplicity, the induction motor must rank alongside the screwthread as one of mankind's best inventions. It is not only supremely elegant as an electromechanical energy converter, but is also by far the most important, with something like half of all the electricity generated being converted back to mechanical energy in induction motors. Despite playing a key role in industrial society, it remains largely unnoticed because of its workaday role driving machinery, pumps, fans, compressors, conveyors, hoists, and a host of other routine but vital tasks. It will doubtless continue to dominate fixed-speed applications, but, thanks to the availability of reliable variable-frequency inverters, it is now also the leader in the controlled-speed arena.

Like the d.c. motor, the induction motor develops torque by the interaction of axial currents on the rotor and a radial magnetic field produced by the stator. But whereas in the d.c. motor the 'work' current has to be fed into the rotor by means of brushes and a commutator, the torque-producing currents in the rotor of the induction motor are induced by electromagnetic action, hence the name 'induction' motor. The stator winding therefore not only produces the magnetic field (the 'excitation'), but also supplies the energy which is converted to mechanical output. The absence of any sliding mechanical contacts and the consequent saving in terms of maintenance is a major advantage of the induction motor over the d.c. machine.

Other differences between the induction motor and the d.c. motor are first that the supply to the induction motor is a.c. (usually 3-phase, but in smaller sizes single-phase); secondly that the magnetic field in the induction motor rotates relative to the stator, while in the d.c. motor it is stationary; and thirdly that both stator and rotor in the induction motor are non-salient (i.e. effectively smooth) whereas the d.c. motor stator has projecting poles or saliencies which define the position of the field windings.

Given these differences we might expect to find major contrasts between the performance of the two types of motor, and it is true that their inherent characteristics exhibit distinctive features. But there are also many aspects of behavior which are similar, as we shall see. Perhaps most important from the user's point of view is that there is no dramatic difference in size or weight between an induction motor and a d.c. motor giving the same power at the same base speed, though the induction

© 2013 Austin Hughes and William Drury.
Published by Elsevier Ltd.
All rights reserved.

motor will usually be cheaper. The similarity in size is a reflection of the fact that both types employ similar amounts of copper and iron, while the difference in price stems from the simpler construction and production volume of the induction motor.

1.1 Outline of approach

Throughout this chapter we will be concerned with how the induction motor behaves in the steady state, i.e. the supply voltage and frequency are constant, the load is steady, and any transients have died away. We will aim to develop a sound qualitative understanding of the steady-state behavior, based on the ideas we have discussed so far (magnetic flux, m.m.f., reluctance, electromagnetic force, motional e.m.f.). But despite many similarities with the d.c. motor, most readers will probably find that the induction motor is more difficult to understand. This is because we are now dealing with alternating rather than steady quantities (so, for example, inductive reactance becomes very significant), and also because (as mentioned earlier) a single winding acts simultaneously as the producer of the flux and the supplier of the converted energy.

In the next chapter, we will extend our qualitative understanding to look at how motor performance depends on design parameters: we will again be following an approach that has served well since the early days of the induction motor, and was developed to reflect the fact that motors were operated at a fixed voltage and frequency. It turns out that under these 'utility supply' conditions, the transient performance is poor and fast control of torque is not possible, and so the induction motor was considered unable to compete with the d.c. motor in controlled-speed drives.

All this changed rapidly beginning in the 1970s. The full set of governing equations (describing not only the steady state but also the much more complex dynamic behavior) had become tractable with computer simulation, which in turn led the way to understanding how the stator currents would have to be manipulated to obtain fast control of torque. The hardware for implementing rapid current control became available with the development of pulse-width modulation (PWM) inverters, but it was not until digital signal processing finally became cheap and fast enough to deal with the complex control algorithms that so-called 'field-oriented' or 'vector' control emerged as a practicable commercial proposition. We will defer consideration of this spectacularly successful system until later, because experience has shown that a solid grounding based on the classical approach is invaluable before getting to grips with more demanding ideas, which are introduced in Chapter 7.

2. THE ROTATING MAGNETIC FIELD

To understand how an induction motor operates, we must first unravel the mysteries of the rotating magnetic field. We will see later that the rotor is effectively dragged along by the rotating field, but that it can never run quite as fast as the field.

Our look at the mechanism of the rotating field will focus on the stator windings because they act as the source of the flux. In this part of the discussion we will ignore the presence of the rotor conductors. This makes it much easier to understand what governs the speed of rotation and the magnitude of the field, which are the two factors that most influence motor behavior.

Having established how the rotating field is set up, and what its speed and strength depend on, we move on to examine the rotor, concentrating on how it behaves when exposed to the rotating field, and discovering how the induced rotor currents and torque vary with rotor speed. In this section we assume – again for the sake of simplicity – that the rotating flux set up by the stator is not influenced by the rotor.

Finally we turn attention to the interaction between the rotor and stator, verifying that our earlier assumptions are well justified. Having done this we are in a position to examine the 'external characteristics' of the motor, i.e. the variation of motor torque and stator current with speed. These are the most important characteristics from the point of view of the user.

Readers who are unfamiliar with routine a.c. circuit theory, including reactance, impedance, phasor diagrams (but not, at this stage, 'j' notation) and basic ideas about 3-phase systems will have to do some preparatory work[1] before tackling the later sections of this chapter.

Before we investigate how the rotating magnetic field is produced, we should be clear what it actually is. Because both the rotor and stator iron surfaces are smooth (apart from the regular slotting), and are separated by a small air-gap, the flux produced by the stator windings crosses the air-gap radially. The behavior of the motor is dictated by this radial flux, so we will concentrate first on establishing a mental picture of what is meant by the 'flux wave' in an induction motor.

The pattern of flux in an ideal 4-pole motor supplied from a balanced 3-phase source is shown in Figure 5.1(a). The top sketch corresponds to time $t = 0$; the middle one shows the flux pattern one-quarter of a cycle of the supply later (i.e. 5 ms if the frequency is 50 Hz); and the lower one corresponds to a further quarter-cycle later. We note that the pattern of flux lines is repeated in each case, except that the middle and lower ones are rotated by 45° and 90°, respectively, with respect to the top sketch.

The term '4-pole' reflects the fact that flux leaves the stator from two N poles, and returns at two S poles. Note, however, that there are no physical features of the stator iron that mark it out as being 4-pole, rather than say 2-pole or 6-pole. As we will see, it is the layout and interconnection of the stator coils which set the pole-number.

If we plot the variation of the radial air-gap flux density with respect to distance round the stator, at each of the three instants of time, we get the patterns shown in

---

[1] The revised book *Electrical and Electronic Technology*, 10th Edition by Edward Hughes (no relation) is a tried and tested favorite.

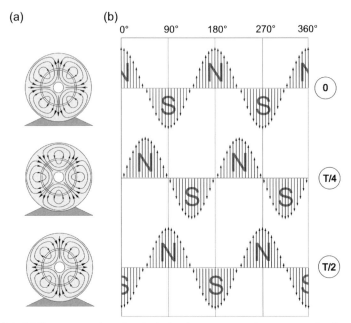

Figure 5.1 (a) Flux pattern in a 4-pole induction motor at three successive instants of time, each one-quarter of a cycle apart; (b) radial flux density distribution in the air-gap at the three instants shown in Figure 5.1(a).

Figure 5.1(b). The first feature to note is that the radial flux density varies sinusoidally in space. There are two N peaks and two S peaks, but the transition from N to S occurs in a smooth sinusoidal way, giving rise to the term 'flux wave'. The distance from the center of one N pole to the center of the adjacent S pole is called the pole-pitch, for obvious reasons.

Staying with Figure 5.1(b), we note that after one-quarter of a cycle of the mains frequency, the flux wave retains its original shape, but has moved round the stator by half a pole-pitch, while after half a cycle it has moved round by a full pole-pitch. If we had plotted the patterns at intermediate times, we would have discovered that the wave maintained a constant shape, and progressed smoothly, advancing at a uniform rate of two pole-pitches per cycle of the supply. The term 'traveling flux wave' is thus an appropriate one to describe the air-gap field.

For the 4-pole wave here, one complete revolution takes two cycles of the supply, so the speed is 25 rev/s (1500 rev/min) with a 50 Hz supply, or 30 rev/s (1800 rev/min) at 60 Hz. The general expression for the speed of the field (which is known as the synchronous speed) N_s, in rev/min is

$$N_s = \frac{120f}{p} \tag{5.1}$$

Table 5.1 Synchronous speeds, in rev/min

Pole-number	50 Hz	60 Hz
2	3000	3600
4	1500	1800
6	1000	1200
8	750	900
10	600	720
12	500	600

where p is the pole-number. The pole-number must be an even integer, since for every N pole there must be an S pole. Synchronous speeds for commonly used pole-numbers are given in Table 5.1.

We can see from the table that if we want the field to rotate at intermediate speeds, we will have to be able to vary the supply frequency, and this is what happens in inverter-fed motors, which are dealt with in Chapter 7.

2.1 Production of rotating magnetic field

Now that we have a picture of the field, we turn to how it is produced. If we inspect the stator winding of an induction motor we find that it consists of a uniform array of identical coils, located in slots. The coils are in fact connected to form three identical groups or phase-windings, distributed around the stator, and symmetrically displaced with respect to one another. The three phase-windings are connected either in star (wye) or delta (mesh), as shown in Figure 5.2.

The three phase-windings are connected to a balanced 3-phase a.c. supply, and so the currents (which produce the m.m.f. that sets up the flux) are of equal amplitude but differ in time-phase by one-third of a cycle (120°), forming a balanced 3-phase set.

2.2 Field produced by each phase-winding

The aim of the winding designer is to arrange the layout of the coils so that each phase-winding, acting alone, produces an m.m.f. wave (and hence an air-gap flux

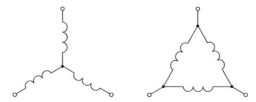

Figure 5.2 Star (wye) and delta (mesh) connection of the three phase-windings of a 3-phase induction motor.

wave) of the desired pole-number, and with a sinusoidal variation of amplitude with angle. Getting the desired pole-number is not difficult: we simply have to choose the right number and pitch of coils, as shown by the diagrams of an elementary 4-pole winding in Figure 5.3.

In Figure 5.3(a) we see that by positioning two coils (each of which spans one pole-pitch) 180° apart we obtain the correct number of poles (i.e. 4). However, the air gap field – shown by only two flux lines per pole for the sake of clarity – is uniform between each go and return coil-side, not sinusoidal.

A clearer picture of the air-gap flux wave is presented in the developed view in Figure 5.3(b), where more equally spaced flux lines have been added to emphasize the uniformity of the flux density between the 'go' and 'return' sides of the coils. Finally, the plot of the air-gap flux density underlines the fact that this very basic arrangement of coils produces a rectangular flux density wave, whereas what we are seeking is a sinusoidal wave.

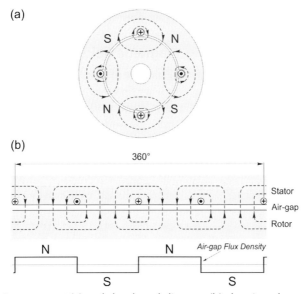

Figure 5.3 Arrangement (a) and developed diagram (b) showing elementary 4-pole, single-layer stator winding consisting of four conductors spaced by 90°. The 'go' side of each coil (shown by the plus symbol) carries current into the paper at the instant shown, while the 'return' side (shown by the dot) carries current out of the paper.

We can improve matters by adding more coils in the adjacent slots, as shown in Figure 5.4. All the coils have the same number of turns, and carry the same current. The addition of the extra slightly displaced coils gives rise to the stepped waveform of m.m.f. and air-gap flux density shown in Figure 5.4. It is still not sinusoidal, but is much better than the original rectangular shape.

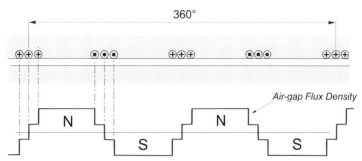

Figure 5.4 Developed diagram showing flux density produced by one phase of a single-layer winding having three slots per pole per phase.

It turns out that if we were to insist on having a perfect sinusoidal flux density waveform, we would have to distribute the coils of one phase in a smoothly varying sinusoidal pattern over the whole periphery of the stator. This is not a practicable proposition, first because we would also have to vary the number of turns per coil from point to point, and secondly because we want the coils to be in slots, so it is impossible to avoid some measure of discretization in the layout. For economy of manufacture we are also obliged to settle for all the coils being identical, and we must make sure that the three identical phase-windings fit together in such a way that all the slots are fully utilized. (See Plate 5.1)

Despite these constraints we can get remarkably close to the ideal sinusoidal pattern, especially when we use a 'two-layer' winding (in which case the stator slots may contain turns from more than one phase winding). A typical arrangement of one phase is shown in Figure 5.5. The upper expanded sketch shows how each coil sits with its 'go' side in the top of a slot while the 'return' side occupies the bottom of a slot rather less than one pole-pitch away. Coils which span less than a full pole-pitch are known as short-pitch or short-chorded: in this particular case the coil pitch is six slots and the pole-pitch is nine slots, so the coils are short-pitched by three slots.

This type of winding is almost universal in all but small induction motors, the coils in each phase being grouped together to form 'phase-bands' or 'phase-belts'. Since we are concentrating on the field produced by only one of the phase-windings (or 'phases'), only one-third of the coils in Figure 5.5 are shown carrying current. The remaining two-thirds of the coils form the other two phase-windings, as discussed below.

Returning to the flux density plot in Figure 5.5 we see that the effect of short-pitching is to increase the number of steps in the waveform, and that as a result the field produced by one phase is a fair approximation to a sinusoid.

The current in each phase pulsates at the supply frequency, so the field produced by say phase A, pulsates in sympathy with the current in phase A, the axis of each 'pole' remaining fixed in space, but its polarity changing from N to S and back once

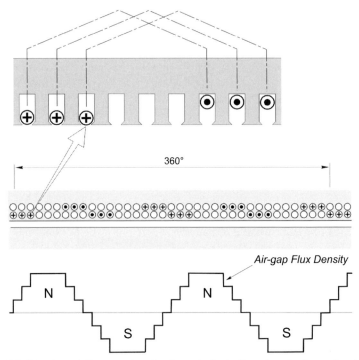

Figure 5.5 Developed diagram showing layout of windings in a 3-phase, 4-pole, two-layer induction motor winding, together with the flux density wave produced by one phase acting alone. The upper detail shows how the coil-sides form upper and lower layers in the slots.

per cycle. There is no hint of any rotation in the field of one phase, but when the fields produced by each of the three phases are combined, matters change dramatically.

2.3 Resultant 3-phase field

The layout of coils for the complete 4-pole winding is shown in Figure 5.6(a). The 'go' sides of each coil are represented by the capital letters (A, B, C) and the 'return' sides are identified by bars over the letters $(\overline{A}, \overline{B}, \overline{C})$. (For the sake of comparison, a 6-pole winding layout that uses the same stator slotting is shown in Figure 5.6(b): here the pole-pitch is six slots and the coils are short-pitched by one slot.)

Returning to the 4-pole winding, we can see that the windings of phases B and C are identical with that of phase A apart from the fact that they are displaced in space by plus and minus two-thirds of a pole-pitch, respectively. Phases B and C therefore also produce pulsating fields, along their own fixed axes in space. But the currents in phases B and C also differ in time-phase from the current in phase A, lagging by one-third and two-thirds of a cycle, respectively. To find the resultant

(a)
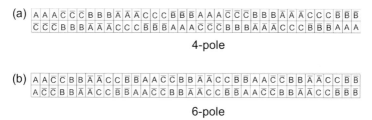

4-pole

(b)

6-pole

Figure 5.6 Developed diagram showing arrangement of 3-phase, two-layer windings in a 36-slot stator. A 4-pole winding with three slots/pole/phase is shown in (a), and a 6-pole winding with two slots/pole/phase is shown in (b).

field we must therefore superimpose the fields of the three phases, taking account not only of the spatial differences between windings, but also the time differences between the currents. This is a tedious process, so the intermediate steps have been omitted and instead we move straight to the plot of the resultant field for the complete 4-pole machine, for three discrete times during one complete cycle, as shown in Figure 5.7.

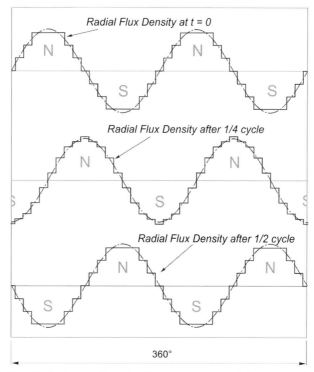

Figure 5.7 Resultant air–gap flux density wave produced by a complete 3-phase, 4-pole winding at three successive instants in time.

We see that the three pulsating fields combine beautifully and lead to a resultant 4-pole field which rotates at a uniform rate, advancing by two pole-pitches for every cycle of the supply. The resultant field is not exactly sinusoidal in shape (though it is actually more sinusoidal than the field produced by the individual phase-windings), and its shape varies a little from instant to instant; but these are minor worries. The resultant field is amazingly close to the ideal traveling wave and yet the winding layout is simple and easy to manufacture. This is an elegant engineering achievement, however one looks at it.

2.4 Direction of rotation

The direction of rotation depends on the order in which the currents reach their maxima, i.e. on the phase-sequence of the supply. Reversal of direction is therefore simply a matter of interchanging any two of the lines connecting the windings to the supply.

2.5 Main (air-gap) flux and leakage flux

Broadly speaking the motor designer shapes the stator and rotor teeth to encourage as much as possible of the flux produced by the stator windings to pass right down the rotor teeth, so that before completing its path back to the stator it is fully linked with the rotor conductors (see later) which are located in the rotor slots. We will see later that this tight magnetic coupling between stator and rotor windings is necessary for good running performance, and the field which provides the coupling is of course the main or air-gap field, which we are in the midst of discussing.

In practice the vast majority of the flux produced by the stator is indeed main or 'mutual' flux. But there is some flux which bypasses the rotor conductors, linking only with the stator winding, and known as stator leakage flux. Similarly not all the flux produced by the rotor currents links the stator, but some (the rotor leakage flux) links only the rotor conductors.

The use of the pejorative-sounding term 'leakage' suggests that these leakage fluxes are unwelcome imperfections, which we should go out of our way to minimize. However, while the majority of aspects of performance are certainly enhanced if the leakage is as small as possible, others (notably the large and unwelcome current drawn from the mains when the motor is started from rest directly on the utility supply) are made much worse if the coupling is too good. So we have the somewhat paradoxical situation in which the designer finds it comparatively easy to lay out the windings to produce a good main flux, but is then obliged to juggle the detailed design of the slots in order to obtain just the right amount of leakage flux to give acceptable all-round performance. (In contrast, as we will see later, an inverter-fed induction motor can avoid such issues as excessive starting current and, ideally, could be designed with much lower leakage than its utility-fed counterpart. It has to be said, however, that the majority of induction motors are still designed for general-purpose use, and in this respect they lose out in comparison with other forms of motor that are specifically designed for operation with a drive.)

The weight which attaches to the matter of leakage flux is reflected in the prominent part played by the associated leakage reactance in equivalent circuit models of the induction motor (see Appendix 2). However, such niceties are of limited importance to the user, so in this and the next chapters we will limit references to leakage reactance to well-defined contexts, and, in general, where the term 'flux' is used, it will refer to the main air-gap field.

2.6 Magnitude of rotating flux wave

We have already seen that the speed of the flux wave is set by the pole-number of the winding and the frequency of the supply. But what is it that determines the amplitude of the field?

To answer this question we can continue to neglect the fact that under normal conditions there will be induced currents in the rotor. We might even find it easier to imagine that the rotor conductors have been removed altogether: this may seem a drastic assumption, but will prove justified later. The stator windings are assumed to be connected to a balanced 3-phase a.c. supply so that a balanced set of currents flow in the windings. We denote the phase voltage by V, and the current in each phase by I_m, where the subscript m denotes 'magnetizing' or flux-producing current.

From the discussion in Chapter 1 we know that the magnitude of the flux wave (B_m) is proportional to the winding m.m.f., and is thus proportional to I_m. But what we really want to know is how the flux density depends on the supply voltage and frequency, since these are the only two parameters over which we have control.

To guide us to the answer, we must first ask what effect the traveling flux wave will have on the stator winding. Every stator conductor will of course be cut by the rotating flux wave, and will therefore have an e.m.f. induced in it. Since the flux wave varies sinusoidally in space, and cuts each conductor at a constant velocity, a sinusoidal e.m.f. is induced in each conductor. The magnitude of the e.m.f. is proportional to the magnitude of the flux wave (B_m), and to the speed of the wave (i.e. to the supply frequency f). The frequency of the induced e.m.f. depends on the time taken for one N pole and one S pole to cut the conductor. We have already seen that the higher the pole-number, the slower the field rotates, but we found that the field always advances by two pole-pitches for every cycle of the supply. The frequency of the e.m.f. induced in the stator conductors is therefore the same as the supply frequency, regardless of the pole-number. (This conclusion is what we would have reached intuitively, since we would expect any linear system to react at the same frequency at which we excited it.)

The e.m.f. in each complete phase winding (E) is the sum of the e.m.f.s in the phase coils, and thus will also be at supply frequency. (The alert reader will realize that while the e.m.f. in each coil has the same magnitude, it will differ in time phase, depending on the geometrical position of the coil. Most of the coils in each phase-band are close together, however, so their e.m.f.s – though slightly out of phase – will more or less add up directly.)

If we were to compare the e.m.f.s in the three complete phase windings, we would find that they were of equal amplitude, but out of phase by one-third of a cycle (120°), thereby forming a balanced 3-phase set. This result could have been anticipated from the overall symmetry. This is very helpful, as it means that we need only consider one of the phases in the rest of the discussion.

So we find that when an alternating voltage V is applied, an alternating e.m.f. E is induced. We can represent this state of affairs by the primitive a.c. equivalent circuit for one phase shown in Figure 5.8.

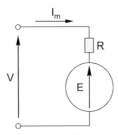

Figure 5.8 Simple equivalent circuit for the induction motor under no-load conditions.

The resistance shown in Figure 5.8 is the resistance of one complete phase-winding. Note that the e.m.f. E is shown as opposing the applied voltage V. This must be so, otherwise we would have a runaway situation in which the voltage V produced the magnetizing current I_m which in turn set up an e.m.f. E, which added to V, which further increased I_m and so on *ad infinitum*.

Applying Kirchhoff's voltage law to the a.c. circuit in Figure 5.8 yields

$$V = I_m R + E \tag{5.2}$$

We find in practice that the term $I_m R$ (which represents the volt-drop due to winding resistance) is usually very much less than the applied voltage V. In other words most of the applied voltage is accounted for by the opposing e.m.f. E. Hence we can make the approximation

$$V \approx E \tag{5.3}$$

But we have already seen that the e.m.f. is proportional to B_m and to f, i.e.

$$E \propto B_m f \tag{5.4}$$

So by combining equations (5.3) and (5.4) we obtain

$$B_m = k \frac{V}{f} \tag{5.5}$$

where the constant k depends on the number of turns per coil, the number of coils per phase and the distribution of the coils.

Equation (5.5) is of fundamental importance in induction motor operation. It shows that if the supply frequency is constant, the flux in the air-gap is directly proportional to the applied voltage, or in other words the voltage sets the flux. We can also see that if we raise or lower the frequency (in order to increase or reduce the speed of rotation of the field), we will have to raise or lower the voltage in proportion if, as is usually the case, we want the magnitude of the flux to remain constant. (We will see in Chapters 7 and 8 that the early inverter drives used this so-called 'V/f control' to keep the flux constant at all speeds.)

It may seem a paradox that having originally homed-in on the magnetizing current I_m as being the source of the m.m.f. which in turn produces the flux, we find that the actual value of the flux is governed only by the applied voltage and frequency, and I_m does not appear at all in equation (5.5). We can see why this is by looking again at Figure 5.8 and asking what would happen if, for some reason, the e.m.f. (E) were to reduce. We would find that I_m would increase, which in turn would lead to a higher m.m.f., more flux, and hence to an increase in E. There is clearly a negative feedback effect taking place, which continually tries to keep E equal to V. It is rather like the d.c. motor (Chapter 3) where the speed of the unloaded motor always adjusted itself so that the back e.m.f. almost equaled the applied voltage. Here, the magnetizing current always adjusts itself so that the induced e.m.f. is almost equal to the applied voltage.

Needless to say this does not mean that the magnetizing current is arbitrary, but to calculate it we would have to know the number of turns in the winding, the length of the air-gap (from which we could calculate the gap reluctance) and the reluctance of the iron paths. From a user point of view there is no need to delve further in this direction. We should, however, recognize that the reluctance will be dominated by the air-gap, and that the magnitude of the magnetizing current will therefore depend mainly on the size of the gap: the larger the gap, the bigger the magnetizing current. Since the magnetizing current contributes to stator copper loss, but not to useful output power, we would like it to be as small as possible, so we find that induction motors usually have the smallest air-gap which is consistent with providing the necessary mechanical clearances. Despite the small air-gap the magnetizing current can be appreciable: in a 4-pole motor, it may be typically 50% of the full-load current, and even higher in 6-pole and 8-pole designs.

2.7 Excitation power and volt-amps

The setting up of the traveling wave by the magnetizing current amounts to the provision of 'excitation' for the motor. Some energy is stored in the magnetic field, but since the amplitude remains constant once the field has been established, no net power input is needed to sustain the field. We therefore find that under the

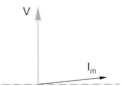

Figure 5.9 Phasor diagram for the induction motor under no-load conditions, showing magnetizing current I_m.

conditions discussed so far, i.e. in the absence of any rotor currents, the power input to the motor is very small. (We should perhaps note that the rotor currents in a real motor are very small when it is running light, so the hypothetical situation we are looking at is not as far removed from reality as we may have supposed.)

Ideally the only source of power losses would be the copper losses in the stator windings, but to this must be added the 'iron losses' which arise from eddy currents and hysteresis in the laminated steel cores of rotor and stator. However, we have seen that the magnetizing current can be quite large, its value being largely determined by the air-gap, so we can expect an unloaded induction motor to draw appreciable current from the supply, but very little real power. The volt-amps will therefore be substantial, but the power-factor will be very low, the magnetizing current lagging the supply voltage by almost 90°, as shown in the time phasor diagram (Figure 5.9).

Viewed from the supply the stator looks more or less like a pure inductance, a fact which we would expect intuitively given that – having ignored the rotor circuit – we are left with only an arrangement of flux-producing coils surrounded by a good magnetic circuit.

2.8 Summary

When the stator is connected to a 3-phase supply, a sinusoidally distributed, radially directed rotating magnetic flux density wave is set up in the air-gap. The speed of rotation of the field is directly proportional to the frequency of the supply, and inversely proportional to the pole-number of the winding. The magnitude of the flux wave is proportional to the applied voltage, and inversely proportional to the frequency.

When the rotor circuits are ignored (i.e. under no-load conditions), the real power drawn is small, but the magnetizing current itself can be quite large, giving rise to a significant reactive power demand from the utility supply.

3. TORQUE PRODUCTION

In this section we begin with a brief description of rotor types, and introduce the notion of 'slip', before moving on to explore how the torque is produced, and investigate the variation of torque with speed. We will find that the behavior of the rotor varies widely according to the slip, and we therefore look separately at low and high values of slip. Throughout this section we will assume that the rotating magnetic field is unaffected by anything which happens on the rotor side of the air-gap. Later, we will see that this assumption is pretty well justified.

3.1 Rotor construction

Two types of rotor are used in induction motors. In both, the rotor 'iron' consists of a stack of silicon steel laminations with evenly spaced slots punched around the circumference. As with the stator laminations, the surface is coated with an oxide layer which acts as an insulator, preventing unwanted axial eddy-currents from flowing in the iron.

The cage rotor is by far the most common: each rotor slot contains a solid conductor bar and all the conductors are physically and electrically joined together at each end of the rotor by conducting 'end-rings' (Figure 5.10, plate 5.2 and see also Figure 8.4). In the larger sizes the conductors will be of copper, in which case the end-rings are brazed on. In small and medium sizes, the rotor conductors and end rings may be of copper or die-cast in aluminum.

The term squirrel cage was widely used at one time and the origin should be clear from Figure 5.10. The rotor bars and end-rings are reminiscent of the rotating cages used in bygone days to exercise small rodents (or rather to amuse their human captors).

The absence of any means for making direct electrical connection to the rotor underlines the fact that in the induction motor the rotor currents are induced by the air-gap field. It is equally clear that because the rotor cage comprises permanently short-circuited conductor bars, no external control can be exercised over the resistance of the rotor circuit once the rotor has been made. This is a significant drawback which can be avoided in the second type of rotor, which is known as the 'wound-rotor' or 'slipring' type.

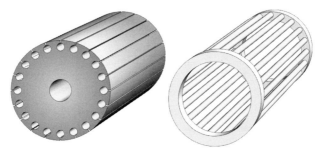

Figure 5.10 Cage rotor construction. The stack of pre-punched laminations is shown on the left, with the copper or aluminum rotor bars and end-rings on the right.

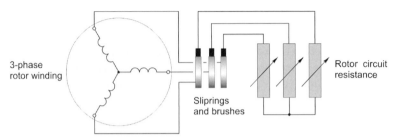

Figure 5.11 Schematic diagram of wound rotor for induction motor, showing sliprings and brushes to provide connection to the external (stationary) 3-phase resistance.

In the wound rotor, the slots accommodate a set of three phase-windings very much like those on the stator. The windings are connected in star, with the three ends brought out to three sliprings (Figure 5.11). The rotor circuit is thus open, and connection can be made via brushes bearing on the sliprings. In particular, the resistance of each phase of the rotor circuit can be increased by adding external resistances, as indicated in Figure 5.11. Adding resistance can be beneficial in some circumstances, as we will see.

Cage-rotors are usually cheaper to manufacture, and are very robust and reliable. Until the advent of variable-frequency inverter supplies, however, the superior control which was possible from the slipring type meant that the extra expense of the wound rotor and its associated control gear was frequently justified, especially for high-power machines. Nowadays comparatively few are made, and then only in large sizes. But many old motors remain in service, so they are included in Chapter 6.

3.2 Slip

A little thought will show that the behavior of the rotor depends very much on its relative velocity with respect to the rotating field. If the rotor is stationary, for example, the rotating field will cut the rotor conductors at synchronous speed, thereby inducing a high e.m.f. in them. On the other hand, if the rotor was running at the synchronous speed, its relative velocity with respect to the field would be zero, and no e.m.f.s would be induced in the rotor conductors.

The relative velocity between the rotor and the field is known as the slip speed. If the speed of the rotor is N, the slip speed is $N_s - N$, where N_s is the synchronous speed of the field, usually expressed in rev/min. The slip (as distinct from slip speed) is the normalized quantity defined by

$$s = \frac{N_s - N}{N_s} \tag{5.6}$$

and is usually expressed either as a ratio as in equation (5.6), or as a percentage. A slip of 0 therefore indicates that the rotor speed is equal to the synchronous speed, while a slip of 1 corresponds to zero speed. (When tests are performed on induction motors with their rotor deliberately held stationary so that the slip is 1, the test is said to be under 'locked-rotor' conditions. The same expression is often used loosely to mean zero speed, even when the rotor is free to move, e.g. when it is started from rest.)

3.3 Rotor-induced e.m.f. and current

The rate at which the rotor conductors are cut by the flux – and hence their induced e.m.f. – is directly proportional to the slip, with no induced e.m.f. at synchronous speed ($s = 0$) and maximum induced e.m.f. when the rotor is stationary ($s = 1$).

The frequency of the rotor e.m.f. (the slip frequency) is also directly proportional to slip, since the rotor effectively slides with respect to the flux wave, and the higher the relative speed, the more times in a second each rotor conductor is cut by an N and an S pole. At synchronous speed (slip $= 0$) the slip frequency is zero, while at standstill (slip $= 1$), the slip frequency is equal to the supply frequency. These relationships are shown in Figure 5.12.

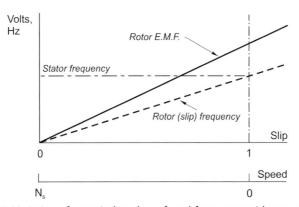

Figure 5.12 Variation of rotor-induced e.m.f. and frequency with speed and slip.

Although the e.m.f. induced in every rotor bar will have the same magnitude and frequency, they will not be in phase. At any particular instant, bars under the peak of the N poles of the field will have maximum positive voltage in them, those under the peak of the S poles will have maximum negative voltage (i.e. 180° phase shift), and those in between will have varying degrees of phase shift. The pattern of instantaneous voltages in the rotor is thus a replica of the flux density wave, and the rotor-induced 'voltage wave' therefore moves relative to the rotor at slip speed, as shown in Figure 5.13.

All the rotor bars are short-circuited by the end-rings, so the induced voltages will drive currents along the rotor bars, the currents forming closed paths through the end-rings, as shown in the developed diagram (Figure 5.14).

In Figure 5.14 the variation of instantaneous e.m.f. in the rotor bars is shown in the upper sketch, while the corresponding instantaneous currents flowing in the

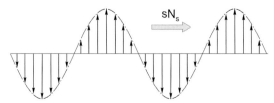

Figure 5.13 Pattern of induced e.m.f.s in rotor conductors. The rotor 'voltage wave' moves at a speed of sN_s with respect to the rotor surface.

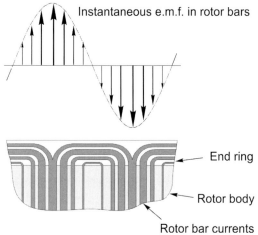

Figure 5.14 Instantaneous sinusoidal pattern of rotor currents in rotor bars and end-rings. Only one pole-pitch is shown, but the pattern is repeated.

rotor bars and end-rings are shown in the lower sketch. The lines representing the currents in the rotor bars have been drawn so that their width is proportional to the instantaneous currents in the bars.

3.4 Torque

The axial currents in the rotor bars will interact with the radial flux wave to produce the driving torque of the motor, which will act in the same direction as the rotating field, the rotor being dragged along by the field. We note that slip is essential to this mechanism, so that it is never possible for the rotor to catch up with the field, as there would then be no rotor e.m.f., no current, and no torque. The fact that motor action is only possible if the speed is less than the synchronous speed explains why the induction machine is described as 'asynchronous'. Finally, we can see that the cage rotor will automatically adapt to whatever pole–number is impressed by the stator winding, so that the same rotor can be used for a range of different stator pole–numbers.

3.5 Rotor currents and torque – small slip

When the slip is small (say between 0 and 10%), the frequency of induced e.m.f. is also very low (between 0 and 5 Hz if the supply frequency is 50 Hz). At these low frequencies the impedance of the rotor circuits is predominantly resistive, the inductive reactance being small because the rotor frequency is low.

The current in each rotor conductor is therefore in time-phase with the e.m.f. in that conductor, and the rotor current-wave is therefore in space-phase with the rotor e.m.f. wave, which in turn is in space-phase with the flux wave. This situation was assumed in the previous discussion, and is represented by the space waveforms shown in Figure 5.15.

To calculate the torque we first need to evaluate the 'BI_rl_r' product (see equation (1.2)) in order to obtain the tangential force on each rotor conductor. The torque is then given by the total force multiplied by the rotor radius. We can see from Figure 5.15 that where the flux density has a positive peak, so does the rotor current, so that particular bar will contribute a high tangential force to the total torque. Similarly, where the flux has its maximum negative peak, the induced current is maximum and negative, so the tangential force is again positive. We don't need to work out the torque in detail, but it should be clear that the resultant will be given by an equation of the form

$$T = kBI_r \tag{5.7}$$

where B and I_r denote the amplitudes of the flux density wave and the rotor current wave, respectively. Provided that there are a large number of rotor bars (which is a safe bet in practice), the waves shown in Figure 5.15 will remain the same at all instants of time, so the torque remains constant as the rotor rotates.

If the supply voltage and frequency are constant, the flux will be constant (see equation (5.5)). The rotor e.m.f. (and hence I_r) is then proportional to slip, so we can see from equation (5.7) that the torque is directly proportional to slip. We must remember that this discussion relates to low values of slip only, but since this is the normal running condition, it is extremely important.

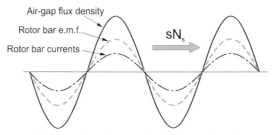

Figure 5.15 Pattern of air-gap flux density, induced e.m.f. and current in cage rotor bars at low values of slip.

The torque–speed (and torque–slip) relationship for small slips is thus approximately a straight line, as shown by the section of line AB in Figure 5.16.

If the motor is unloaded, it will need very little torque to keep running – only enough to overcome friction in fact – so an unloaded motor will run with a very small slip at just below the synchronous speed, as shown at A in Figure 5.16.

When the load is increased, the rotor slows down, and the slip increases, thereby inducing more rotor e.m.f. and current, and thus more torque. The speed will settle when the slip has increased to the point where the developed torque equals the load torque – e.g. point B in Figure 5.16.

Induction motors are usually designed so that their full-load torque is developed for small values of slip. Small ones typically have a full-load slip of 8%, large ones around 1%. At the full-load slip, the rotor conductors will be carrying their safe maximum continuous current, and if the slip is any higher, the rotor will begin to overheat. This overload region is shown by the dotted line in Figure 5.16.

The torque–slip (or torque–speed) characteristic shown in Figure 5.16 is a good one for most applications, because the speed only falls a little when the load is raised from zero to its full value. We note that, in this normal operating region, the torque–speed curve is very similar to that of a d.c. motor (see Figure 3.9).

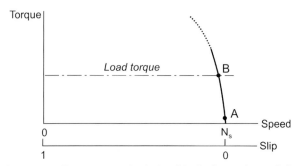

Figure 5.16 Torque–speed relationship for low values of slip.

3.6 Rotor currents and torque – large slip

As the slip increases, the rotor e.m.f. and rotor frequency both increase in direct proportion to the slip. At the same time the rotor inductive reactance, which was negligible at low slip (low rotor frequency), begins to be appreciable in comparison with the rotor resistance. Hence although the induced current continues to increase with slip, it does so more slowly than at low values of slip, as shown in Figure 5.17.

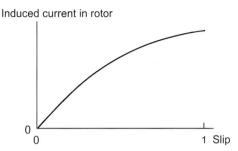

Figure 5.17 Magnitude of current induced in rotor over the full (motoring) range of slip.

At high values of slip, the rotor current also lags behind the rotor e.m.f. because of the inductive reactance. The alternating current in each bar reaches its peak well after the induced voltage, and this in turn means that the rotor current wave has a space-lag with respect to the rotor e.m.f. wave (which is in space-phase with the flux wave). This space-lag is shown by the angle ϕ_r in Figure 5.18.

The space-lag means that the peak radial flux density and peak rotor currents no longer coincide, which is bad news from the point of view of torque production, because although we have high values of both flux density and current, they do not occur simultaneously at any point around the periphery. What is worse is that at some points we even have flux density and currents of opposite sign, so over those regions of the rotor surface the torque contributed will actually be negative. The overall torque will still be positive, but is much less than it would be if the flux and current waves were in phase. We can allow for the unwelcome space-lag by modifying equation (5.7), to obtain a more general expression for torque as

$$T = kBI_r \cos \phi_r \qquad (5.8)$$

Equation (5.7) is merely a special case of equation (5.8), which only applies under low-slip conditions where $\cos\phi_r \approx 1$.

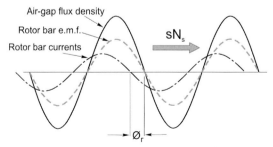

Figure 5.18 Pattern of air-gap flux density, induced e.m.f. and current in cage rotor bars at high values of slip. (These waveforms should be compared with the corresponding ones when the slip is small, see Figure 5.15.)

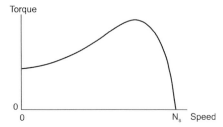

Figure 5.19 Typical complete torque–speed characteristic for motoring region of cage induction motor.

For most cage rotors, it turns out that as the slip increases the term cos ϕ_r reduces more quickly than the current (I_r) increases, so that at some slip between 0 and 1 the developed torque reaches a maximum value. This is illustrated in the typical torque–speed characteristic shown in Figure 5.19. The peak torque actually occurs at a slip at which the rotor inductive reactance is equal to the rotor resistance, so the motor designer can position the peak torque at any slip by varying the reactance to resistance ratio.

3.7 Generating – negative slip

When we explored the steady-state characteristics of the d.c. machine (see section 4 in Chapter 3) we saw that at speeds less than that at which it runs when unloaded the machine acts as a motor, converting electrical energy into mechanical energy. But if the speed is above the no-load speed (for example, when driven by a prime-mover), the machine generates and converts mechanical energy into electrical form.

The inherently bi-directional energy converting property of the d.c. machine seems to be widely recognized. But in the experience of the authors the fact that the induction machine behaves in the same way is far less well accepted, and indeed it is not uncommon to find users expressing profound scepticism at the thought that their 'motor' could possibly generate.

In fact, the induction machine behaves in essentially the same way as the d.c. machine, and if the rotor is driven by an external torque such that its speed is above the synchronous speed (i.e. the slip becomes negative), the electromagnetic torque reverses direction, and the power becomes negative, with energy fed back to the utility supply. It is important to note that, just as with the d.c. machine, this transition from motoring to generating takes place naturally, without intervention on our part.

When the speed is greater than synchronous, we can see from equation (5.6) that the slip is negative, and in this negative slip region the torque is also negative, the torque–speed curve broadly mirroring that in the motoring region, as shown in Figure 5.20. We will discuss this further in Chapter 6, but it is worth noting that for both, motoring and generating continuous operation will be confined to low values of slip, as indicated by the heavy line in Figure 5.20.

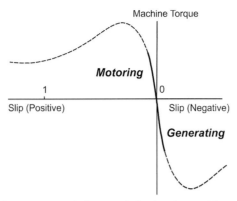

Figure 5.20 Typical torque–speed characteristic showing stable motoring and generating regions.

4. INFLUENCE OF ROTOR CURRENT ON FLUX

Up to now all our discussion has been based on the assumption that the rotating magnetic field remains constant, regardless of what happens on the rotor. We have seen how torque is developed, and that mechanical output power is produced. We have focused attention on the rotor, but the output power must be provided from the stator winding, so we must turn attention to the behavior of the whole motor, rather than just the rotor. Several questions spring to mind.

First, what happens to the rotating magnetic field when the motor is working? Won't the m.m.f. of the rotor currents cause it to change? Secondly, how does the stator know when to start supplying real power across the air-gap to allow the rotor to do useful mechanical work? And finally, how will the currents drawn by the stator vary as the slip is changed?

These are demanding questions, for which full treatment is beyond our scope. But we can deal with the essence of the matter without too much difficulty. Further illumination can be obtained from the equivalent circuit, which is discussed in Appendix 2.

4.1 Reduction of flux by rotor current

We should begin by recalling that we have already noted that when the rotor currents are negligible ($s = 0$), the e.m.f. which the rotating field induces in the stator winding is very nearly equal to the applied voltage. Under these conditions a reactive current (which we termed the magnetizing current) flows into the windings, to set up the rotating flux. Any slight tendency for the flux to fall is immediately detected by a corresponding slight reduction in e.m.f. which is reflected in a disproportionately large increase in magnetizing current, which thus opposes the tendency for the flux to fall.

Exactly the same feedback mechanism comes into play when the slip increases from zero, and rotor currents are induced. The rotor currents are at slip frequency, and they give rise to a rotor m.m.f. wave, which therefore rotates at slip speed (sN_s) relative to the rotor. But the rotor is rotating at a speed of $(1 - s)N_s$, so that when viewed from the stator, the rotor m.m.f. wave always rotates at synchronous speed, regardless of the speed of the rotor.

The rotor m.m.f. wave would, if unchecked, cause its own 'rotor flux wave', rotating at synchronous speed in the air-gap, in much the same way that the stator magnetizing current originally set up the flux wave. The rotor flux wave would oppose the original flux wave, causing the resultant flux wave to reduce.

However, as soon as the resultant flux begins to fall, the stator e.m.f. reduces, thereby admitting more current to the stator winding, and increasing its m.m.f. A very small drop in the e.m.f. induced in the stator is sufficient to cause a large increase in the current drawn from the supply because the e.m.f. E (see Figure 5.8) and the supply voltage V are both very large in comparison with the stator resistance volt-drop IR. The 'extra' stator m.m.f. produced by the large increase in stator current effectively 'cancels' the m.m.f. produced by the rotor currents, leaving the resultant m.m.f. (and hence the rotating flux wave) virtually unchanged.

There must be a small drop in the resultant m.m.f. (and flux) of course, to alert the stator to the presence of rotor currents. But because of the delicate balance between the applied voltage and the induced e.m.f. in the stator the change in flux with load is very small, at least over the normal operating speed-range, where the slip is small. In large motors, the drop in flux over the normal operating region is typically less than 1%, rising to perhaps 10% in a small motor.

The discussion above should have answered the question as to how the stator knows when to supply mechanical power across the air-gap. When a mechanical load is applied to the shaft, the rotor slows down, the slip increases, rotor currents are induced and their m.m.f. results in a modest (but vitally important) reduction in the air-gap flux wave. This in turn causes a reduction in the e.m.f. induced in the stator windings and therefore an increase in the stator current drawn from the supply. We can anticipate that this is a stable process (at least over the normal operating range) and that the speed will settle when the slip has increased sufficiently that the motor torque equals the load torque.

As far as our conclusions regarding torque are concerned, we see that our original assumption that the flux was constant is near enough correct when the slip is small. We will find it helpful and convenient to continue to treat the flux as constant (for given stator voltage and frequency) when we turn later to methods of controlling the normal running speed.

It has to be admitted, however, that at high values of slip (i.e. low rotor speeds), we cannot expect the main flux to remain constant, and in fact we would find in practice that when the motor was first switched on to the utility supply (50 or 60 Hz), with the rotor stationary, the main flux might typically be only half what it was when the motor was at full speed. This is because at high slips, the leakage fluxes assume a much greater importance than under normal low-slip conditions. The simple arguments we have advanced to predict torque would therefore need to be modified to take account of the reduction of main flux if we wanted to use them quantitatively at high slips. There is no need for us to do this explicitly, but it will be reflected in any subsequent curves portraying typical torque–speed curves for real motors. Such curves are of course used when selecting a motor to run directly from the utility supply, since they provide the easiest means of checking whether the starting and run-up torque is adequate for the job in hand. Fortunately, we will see in Chapter 7 that when the motor is fed from an inverter, we can avoid the undesirable effects of high-slip operation, and guarantee that the flux is at its optimum value at all times.

5. STATOR CURRENT–SPEED CHARACTERISTICS

To conclude this chapter we will look at how the stator current behaves, remembering that we are assuming that the machine is directly connected to a utility supply of fixed voltage and frequency. Under these conditions the maximum current likely to be demanded and the power factor at various loads are important matters that influence the running cost.

In the previous section, we argued that as the slip increased, and the rotor did more mechanical work, the stator current increased. Since the extra current is associated with the supply of real (i.e. mechanical output) power (as distinct from the original magnetizing current which was seen to be reactive), this additional 'work' component of current is more or less in phase with the supply voltage, as shown in the phasor diagrams (Figure 5.21).

The resultant stator current is the sum of the magnetizing current, which is present all the time, and the load component, which increases with the slip. We can see that as the load increases, the resultant stator current also increases, and moves more nearly into phase with the voltage. But because the magnetizing current is appreciable, the difference in magnitude between no-load and full-load currents may not be all that great. (This is in sharp contrast to the d.c. motor, where the no-load current in the armature is very small in comparison with the full-load current. Note, however, that in the d.c. motor, the excitation (flux) is provided by a separate field circuit, whereas in the induction motor the stator winding furnishes both the excitation and the work currents. If we consider the behavior of the work components of current only, both types of machine look very similar.)

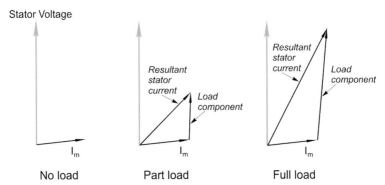

Figure 5.21 Phasor diagrams showing stator current at no-load, part-load and full-load. The resultant current in each case is the sum of the no-load (magnetizing) current and the load component.

The simple ideas behind Figure 5.21 are based on an approximation, so we cannot push them too far: they are fairly close to the truth for the normal operating region, but break down at higher slips, where the rotor and stator leakage reactances become significant. A typical current locus over the whole range of slips for a cage motor is shown in Figure 5.22. We note that the power factor is poor when the motor is lightly loaded, and becomes worse again at high slips, and also that the current at standstill (i.e. the 'starting' current) is perhaps five times the full-load value.

Very high currents when started direct–on–line are one of the worst features of the cage induction motor. They not only cause unwelcome volt-drops in the supply system, but also call for heavier switchgear than would be needed to cope with full-load conditions. Unfortunately, for reasons discussed earlier, the high starting

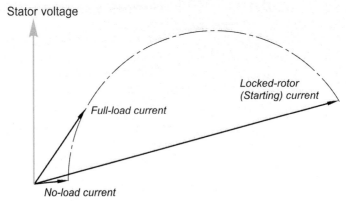

Figure 5.22 Phasor diagram showing the locus of stator current over the full range of speeds from no-load (full speed) down to the locked-rotor (starting) condition.

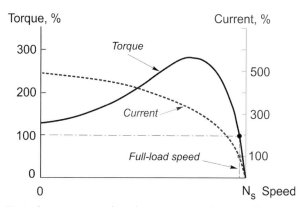

Figure 5.23 Typical torque–speed and current–speed curves for a cage induction motor. The torque and current axes are scaled so that 100% represents the continuously rated (full-load) value.

currents are not accompanied by high starting torques, as we can see from Figure 5.23, which shows current and torque as functions of slip for a general-purpose cage motor.

We note that the torque per ampere of current drawn from the mains is typically very low at start-up, and only reaches a respectable value in the normal operating region, i.e. when the slip is small. This matter is explored further in Chapter 6, and also in Appendix 2.

CHAPTER SIX

Induction Motors – Operation from 50/60 Hz Supply

1. INTRODUCTION

This chapter is concerned with how the induction motor behaves when connected to a supply of constant voltage and frequency. Despite the onward march of the inverter-fed motor, this remains the most widely used and important mode of operation, the motor running directly connected to a utility supply.

The key operating characteristics are considered, and we look at how these can be modified to meet the needs of some applications through detailed design. The limits of operation are investigated for the induction machine operating as both a motor and a generator. Methods of speed control which are not dependent on changing the frequency of the stator supply are also explored. Finally, while the majority of industrial applications utilize the 3-phase induction motor, the role played by single-phase motors is acknowledged with a review of the types and characteristics of this variant.

2. METHODS OF STARTING CAGE MOTORS

2.1 Direct starting – problems

Our everyday domestic experience is likely to lead us to believe that there is nothing more to starting a motor than closing a switch, and indeed for most low-power machines (up to a few kW) – of whatever type – that is indeed the case. By simply connecting the motor to the supply we set in train a sequence of events which sees the motor draw power from the supply while it accelerates to its target speed. When it has absorbed and converted sufficient energy from electrical to kinetic form, the speed stabilizes and the power drawn falls to a low level until the motor is required to do useful mechanical work. In these low-power applications acceleration to full speed may take less than a second, and we are seldom aware of the fact that the current drawn during the acceleration phase is often higher than the continuous rated current.

For motors over a few kW, however, it is necessary to assess the effect on the supply system before deciding whether or not the motor can be started simply by switching directly on to the supply. If supply systems were ideal (i.e. the supply voltage remained unaffected regardless of how much current was drawn) there would be no problem starting any induction motor, no matter how large. The problem is that the heavy current drawn while the motor is running up to speed may cause a large drop in the system supply voltage, annoying other customers on the same supply and perhaps taking it outside statutory limits.

It is worthwhile reminding ourselves about the influence of supply impedance at this point, as this is at the root of the matter, so we begin by noting that any supply system, no matter how complicated, can be modeled by means of the delightfully simple Thévenin equivalent circuit shown in Figure 6.1. (We are assuming balanced 3-phase operation, so a 1-phase equivalent circuit will suffice.)

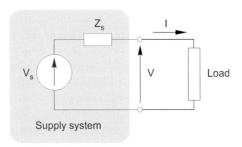

Figure 6.1 Equivalent circuit of supply system.

The supply is represented by an ideal voltage source (V_s) in series with the supply impedance Z_s. When no load is connected to the supply, and the current is zero, the terminal voltage is V_s; but as soon as a load is connected the load current (I) flowing through the source impedance results in a volt-drop, and the output voltage falls from V_s to V, where

$$V = V_s - IZ_s \qquad (6.1)$$

For most industrial supplies the source impedance is predominantly inductive, so that Z_s is simply an inductive reactance, X_s. Typical phasor diagrams relating to a supply with a purely inductive reactance are shown in Figure 6.2: in (a) the load is also taken to be purely reactive, while the load current in (b) has the same magnitude as in (a) but the load is resistive. The output (terminal) voltage in each case is represented by the phasor labeled V.

For the inductive load (a) the current lags the terminal voltage by 90°, while for the resistive load (b) the current is in phase with the terminal voltage. In both cases the volt-drop across the supply reactance (IX_s) leads the current by 90°.

The first point to note is that, for a given magnitude of load current, the volt-drop is in phase with V_s when the load is inductive, whereas with a resistive load the volt-drop is almost at 90° to V_s. This results in a much greater fall in the magnitude of the output voltage when the load is inductive than when it is resistive. The second – obvious – point is that the larger the current, the more the drop in voltage.

Unfortunately, when we try to start a large cage induction motor we face a double-whammy because not only is the starting current typically five or six times

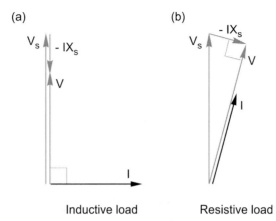

Inductive load   Resistive load

Figure 6.2 Phasor diagrams showing the effect of supply-system impedance on the output voltage with (a) inductive load and (b) resistive load.

rated current, but it is also at a low power-factor, i.e. the motor looks predominantly inductive when the slip is high. (In contrast, when the machine is up to speed and fully loaded, its current is perhaps only one-fifth of its starting current and it presents a predominantly resistive appearance as seen by the supply. Under these conditions the supply voltage is hardly any different from no-load.)

Since the drop in voltage is attributable to the supply impedance, if we want to be able to draw a large starting current without upsetting other consumers it would clearly be best for the supply impedance to be as low as possible, and preferably zero. But from the supply authority viewpoint a very low supply impedance brings the problem of how to cope in the event of an accidental short-circuit across the terminals. The short-circuit current is inversely proportional to the supply impedance, and tends to infinity as Z_s approaches zero. The cost of providing the switchgear to clear such a large fault current would be prohibitive, so a compromise always has to be reached, with values of supply impedances being set by the supply authority to suit the anticipated demands.

Systems with low internal impedance are known as 'stiff' supplies, because the voltage is almost constant regardless of the current drawn. (An alternative way of specifying the nature of the supply is to consider the fault current that would flow if the terminals were short-circuited: a system with a low impedance would have a high fault current or 'fault level'.) Starting on a stiff supply requires no special arrangements and the three motor leads are simply switched directly onto the mains. This is known as 'direct-on-line' (DOL) or 'direct-to-line' (DTL) starting. The switching will usually be done by means of a contactor, incorporating fuses and other overload/thermal protection devices, and operated manually by local or remote pushbuttons, or interfaced to permit operation from a programmable controller or computer.

In contrast, if the supply impedance is high (i.e. a low fault level) an appreciable volt-drop will occur every time the motor is started, causing lights to dim and interfering with other apparatus on the same supply. With this 'weak' supply, some form of starter is called for to limit the current at starting and during the run-up phase, thereby reducing the magnitude of the volt-drop imposed on the supply system. As the motor picks up speed, the current falls, so the starter is removed as the motor approaches full speed. Naturally enough the price to be paid for the reduction in current is a lower starting torque, and a longer run-up time.

Whether or not a starter is required depends on the size of the motor in relation to the capacity or fault level of the supply, the prevailing regulations imposed by the supply authority, and the nature of the load.

The references above to 'low' and 'high' supply impedances must therefore be interpreted in relation to the impedance of the motor when it is stationary. A large (and therefore low impedance) motor could well be started quite happily direct-on-line in a major industrial plant, where the supply is 'stiff', i.e. the supply impedance is very much less than the motor impedance. But the same motor would need a starter when used in a rural setting remote from the main power system, and fed by a relatively high impedance or 'weak' supply. Needless to say, the stricter the rules governing permissible volt-drop, the more likely it is that a starter will be needed.

Motors which start without significant load torque or inertia can accelerate very quickly, so the high starting current is only drawn for a short period. A 10 kW motor would be up to speed in a second or so, and the volt-drop may therefore be judged as acceptable. Clutches are sometimes fitted to permit 'off-load' starting, the load being applied after the motor has reached full speed. Conversely, if the load torque and/or inertia are/is high, the run-up may take many seconds, in which case a starter may prove essential. No strict rules can be laid down, but obviously the bigger the motor, the more likely it is to require a starter.

2.2 Star/delta (wye/mesh) starter

This is the simplest and most widely used method of starting. It provides for the windings of the motor to be connected in star (wye) to begin with, thereby reducing the voltage applied to each phase to 58% (i.e. $1/\sqrt{3}$) of its direct-on-line value. Then, when the motor speed approaches its running value, the windings are switched to delta (mesh) connection. The main advantage of the method is its simplicity, while its main drawbacks are that the starting torque is reduced (see below), and the sudden transition from star to delta gives rise to a second shock – albeit of lesser severity – to the supply system and to the load. For star/delta switching to be possible both ends of each phase of the motor windings must be brought out to the terminal box. This requirement is met in the majority of motors, except small ones which are usually permanently connected in delta.

With a star/delta starter the current drawn from the supply is approximately one-third of that drawn in a direct-on-line start, which is very welcome, but at the same time the starting torque is also reduced to one-third of its direct-on-line value. Naturally we need to ensure that the reduced torque will be sufficient to accelerate the load, and bring it up to a speed at which it can be switched to delta without an excessive jump in the current.

Various methods are used to detect when to switch from star to delta. Historically, in manual starters, the changeover is determined by the operator watching the ammeter until the current has dropped to a low level, or listening to the sound of the motor until the speed becomes steady. Automatic versions are similar in that they detect either falling current or speed rising to a threshold level, but for all but large motors they operate after a preset time.

2.3 Autotransformer starter

A 3-phase autotransformer is usually used where star/delta starting provides insufficient starting torque. Each phase of an autotransformer consists of a single winding on a laminated core. The incoming supply is connected across the ends of the coils, and one or more tapping points (or a sliding contact) provide a reduced voltage output, as shown in Figure 6.3.

The motor is first connected to the reduced voltage output, and when the current has fallen to the running value, the motor leads are switched over to the full voltage.

If the reduced voltage is chosen so that a fraction α of the line voltage is used to start the motor, the starting torque is reduced to approximately α^2 times its direct-on-line value, and the current drawn from the mains is also reduced to α^2 times its direct value. As with the star/delta starter, the torque per ampere of supply current is the same as for a direct start.

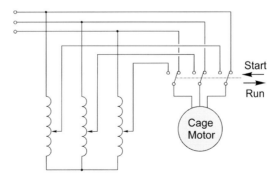

Figure 6.3 Autotransformer starter for cage induction motor.

2.4 Resistance or reactance starter

By inserting three resistors or inductors of appropriate value in series with the motor, the starting current can be reduced by any desired extent, but only at the expense of a disproportionate reduction in starting torque.

For example, if the current is reduced to half its direct-on-line value, the motor voltage will be halved, so the torque (which is proportional to the square of the voltage – see later) will be reduced to only 25% of its direct-on-line value. This approach is thus less attractive in terms of torque per ampere of supply current than the star/delta method. One attractive feature, however, is that as the motor speed increases and its effective impedance rises, the volt-drop across the extra impedance reduces, so the motor voltage rises progressively with the speed, thereby giving more torque. When the motor is up to speed, the added impedance is shorted out by means of a contactor.

2.5 Solid-state soft starting

This method is now the most widely used. It provides a smooth build-up of current and torque, the maximum current and acceleration time are easily adjusted, and it is particularly valuable where the load must not be subjected to sudden jerks. The only real drawback over conventional starters is that the mains currents during run-up are not sinusoidal, which can lead to interference with other equipment on the same supply.

The most widely used arrangement comprises three pairs of back-to-back thyristors (or triacs) connected in series with the three supply lines, as shown in Figure 6.4(a).

Each thyristor is fired once per half-cycle, the firing being synchronized with the utility supply and the firing angle being variable so that each pair conducts for a varying proportion of a cycle. Typical current waveforms are shown in Figure 6.4(b): they are clearly not sinusoidal but the motor will tolerate them quite happily.

A wide variety of control philosophies can be found, with the degree of complexity and sophistication being reflected in the price. The cheapest open-loop systems simply alter the firing angle linearly with time, so that the voltage applied to the motor increases as it accelerates. The 'ramp-time' can be set by trial and error to give an acceptable start, i.e. one in which the maximum allowable current from the supply is not exceeded at any stage. This approach is reasonably satisfactory when the load remains the same, but requires resetting each time the load changes. Loads with high static friction are a problem because nothing happens for the first part of the ramp, during which time the motor torque is insufficient to move the load. When the load finally moves, its acceleration is often too rapid. The more advanced open-loop versions allow the level of current at the start of the ramp to be chosen, and this is helpful with 'sticky' loads.

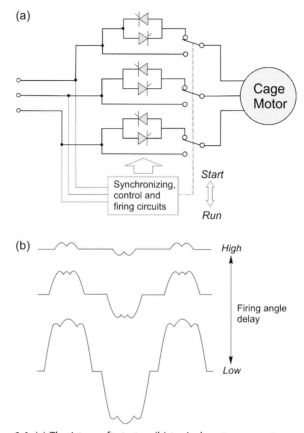

Figure 6.4 (a) Thyristor soft-starter, (b) typical motor current waveforms.

More sophisticated systems – usually with on-board digital controllers – provide for tighter control over the acceleration profile by incorporating an inner current-control loop. After an initial ramping up to the start level (over the first few cycles), the current is held constant at the desired level throughout the accelerating period, the firing angle of the thyristors being continually adjusted to compensate for the changing effective impedance of the motor. By keeping the current at the maximum value which the supply can tolerate, the run-up time is minimized. As with the open-loop systems the velocity–time profile is not necessarily ideal, since with constant current the motor torque exhibits a very sharp rise as the pull-out slip is reached, resulting in a sudden surge in speed. Some systems also include a motor model which estimates speed and allows the controller to follow a ramp or other speed–time profile.

Prospective users need to be wary of some of the promotional literature: claims are sometimes made that massive reductions in starting current can be achieved without corresponding reductions in starting torque. This is nonsense: the current can certainly be limited, but as far as torque per line amp is concerned soft-start systems are no better than series reactor systems, and not as good as the auto-transformer and star/delta methods. Caution should also be exercised in relation to systems that use only one or two triacs: these are fairly common in smaller sizes (<50 kW). Although they do limit the current in one or two phases as compared with direct-on-line starting, the unbalanced currents distort the air-gap flux and this gives rise to uneven (pulsating) torque.

2.6 Starting using a variable-frequency inverter

Operation of induction motors from variable-frequency inverters is discussed in Chapters 7 and 8, but it is appropriate to mention here that one of the advantages of inverter-fed operation is that starting is not a problem because it is usually possible to obtain rated torque from standstill up to rated speed without drawing an excessive current from the supply. None of the other starting methods we have looked at has this ability, so in some applications it may be that the comparatively high cost of the inverter is justified solely on the grounds of its starting and run-up potential.

3. RUN-UP AND STABLE OPERATING REGIONS

In addition to having sufficient torque to start the load it is obviously necessary for the motor to bring the load up to full speed. To predict how the speed will rise after switching on we need the torque–speed curves of the motor and the load, and the total inertia.

By way of example, we can look at the case of a motor with two different loads (Figure 6.5). The solid line is the torque–speed curve of the motor, while the broken lines represent two different load characteristics. Load (A) is typical of a simple hoist, which applies constant torque to the motor at all speeds, while load

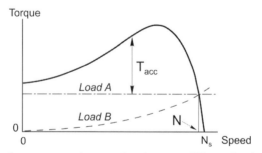

Figure 6.5 Typical torque–speed curve showing two different loads which have the same steady running speed (N).

(B) might represent a fan. For the sake of simplicity, we will assume that the load inertias (as seen at the motor shaft) are the same.

The speed–time curves for run-up are shown in Figure 6.6. Note that the gradient of the speed–time curve (i.e. the acceleration) is obtained by dividing the accelerating torque T_{acc} (which is the difference between the torque developed by the motor and the torque required to run the load at that speed) by the total inertia.

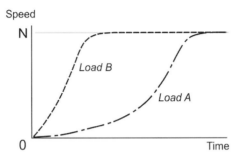

Figure 6.6 Speed–time curves during run-up, for motor and loads shown in Figure 6.5.

In this example, both loads ultimately reach the same steady speed, N (i.e. the speed at which motor torque equals load torque), but B reaches full speed much more quickly because the accelerating torque is higher during most of the run-up. Load A picks up speed slowly at first, but then accelerates hard (often with a characteristic 'whoosh' produced by the ventilating fan) as it passes through the peak torque speed and approaches equilibrium conditions.

It should be clear that the higher the total inertia, the slower the acceleration, and vice versa. The total inertia means the inertia as seen at the motor shaft, so if gearboxes or belts are employed the inertia must be 'referred', as discussed in Chapter 11.

An important qualification ought to be mentioned in the context of the motor torque–speed curves shown by the solid line in Figure 6.5. This is that curves like this represent the torque developed by the motor when it has settled down at the speed in question, i.e. they are the true steady-state curves. In reality, a motor will generally only be in a steady-state condition when it settles at its normal running speed, so for most of the speed range the motor will be accelerating.

In particular, when the motor is first switched on, there will be a transient period as the three currents gradually move towards a balanced 3-phase pattern. During this period the torque can fluctuate wildly at the supply frequency, and, with brief negative excursions, typically as shown in Figure 6.7 which relates to a small unloaded motor. During the transient period the average torque may be very low (as in Figure 6.7) in which case acceleration only begins in earnest after the first few cycles. In this particular example the transient persists long enough to cause an overshoot of the steady-state speed and an oscillation before settling.

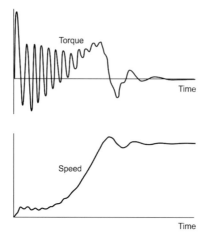

Figure 6.7 Run-up of unloaded motor, showing torque transients persisting for the first few cycles of the supply.

Fortunately, the average torque during run-up can be fairly reliably obtained from the steady-state curves (usually available from the manufacturer), particularly if the inertia is high and the motor takes many cycles to reach full speed, in which case we would consider the torque–speed curve as being 'quasi-steady state'.

3.1 Harmonic effects – skewing

A further cautionary note in connection with the torque–speed curves shown in this and most other books relates to the effects of harmonic air-gap fields. In Chapter 5 it was explained that despite the limitations imposed by slotting, the stator winding m.m.f. is remarkably close to the ideal of a pure sinusoid. Unfortunately, because it is not a perfect sinusoid, Fourier analysis reveals that in addition to the predominant fundamental component, there are always additional unwanted 'space harmonic' fields. These harmonic fields have synchronous speeds that are inversely proportional to their order. For example, a 4-pole, 50 Hz motor will have a main field rotating at 1500 rev/min, but in addition there may be a 5th harmonic (20-pole) field rotating in the reverse direction at 300 rev/min, a 7th harmonic (28-pole) field rotating forwards at 214 rev/min, etc. These space harmonics are minimized by stator winding design, but not all can be eliminated.

If the rotor has a very large number of bars it will react to the harmonic field in much the same way as to the fundamental, producing additional induction-motor torques centered on the synchronous speed of the harmonic, and leading to unwanted dips in the torque speed, typically as shown in Figure 6.8.

Users should not be too alarmed as in most cases the motor will ride through the harmonic during acceleration, but in extreme cases a motor might, for example,

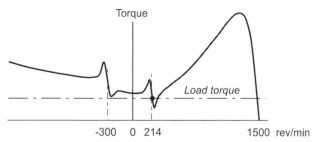

Figure 6.8 Torque–speed curve showing the effect of space harmonics, and illustrating the possibility of a motor 'crawling' on the 7th harmonic.

stabilize on the 7th harmonic, and 'crawl' at about 214 rev/min, rather than running up to 4-pole speed (1500 rev/min at 50 Hz), as shown by the dot in Figure 6.8.

To minimize the undesirable effects of space harmonics the rotor bars in the majority of induction motors are not parallel to the axis of rotation, but instead they are skewed along the rotor length, (typically by around one or two slot-pitches, see later Figure 8.4). This has very little effect as far as the fundamental field is concerned, but can greatly reduce the response of the rotor to harmonic fields.

Because the overall influence of the harmonics on the steady-state curve is barely noticeable, and their presence might worry users, they are rarely shown, the accepted custom being that 'the' torque–speed curve represents the behavior due to the fundamental component only.

3.2 High inertia loads – overheating

Apart from accelerating slowly, high inertia loads pose a particular problem of rotor heating which can easily be overlooked by the unwary user. Every time an induction motor is started from rest and brought up to speed, the total energy dissipated as heat in the motor windings is equal to the stored kinetic energy of the motor plus load. Hence with high inertia loads, very large amounts of energy are released as heat in the windings during run-up, even if the load torque is negligible when the motor is up to speed. With totally enclosed motors the heat ultimately has to find its way to the finned outer casing of the motor, which is cooled by air from the shaft-mounted external fan. Cooling of the rotor is therefore usually much worse than that of the stator, and the rotor is thus most likely to overheat during high inertia run-ups.

No hard and fast rules can be laid down, but manufacturers usually work to standards which specify how many starts per hour can be tolerated. Actually, this information is useless unless coupled with reference to the total inertia, since doubling the inertia makes the problem twice as bad. However, it is usually assumed that the total inertia is not likely to be more than twice the motor inertia, and this is certainly the case for most loads. If in doubt, the user should consult the

manufacturer, who may recommend a larger motor than might seem necessary if it
was simply a matter of meeting the full-load power requirements.

3.3 Steady-state rotor losses and efficiency

The discussion above is a special case which highlights one of the less attractive
features of induction machines. This is that it is never possible for all the power
crossing the air-gap from the stator to be converted to mechanical output, because
some is always lost as heat in the rotor circuit resistance. In fact, it turns out that at
slip s the total power (P_r) crossing the air-gap always divides, so a fraction sP_r is lost as
heat, while the remainder $(1 - s)P_r$ is converted to useful mechanical output (see
also Appendix 2).

Hence when the motor is operating in the steady-state the energy-conversion
efficiency of the rotor is given by

$$\eta_r = \frac{\text{Mechanical output power}}{\text{Power into rotor}} = (1 - s) \tag{6.2}$$

This result is very important, and shows us immediately why operating at small
values of slip is desirable. With a slip of 5% (or 0.05), for example, 95% of the air-gap
power is put to good use. But if the motor was run at half the synchronous speed
($s = 0.5$), 50% of the air-gap power would be wasted as heat in the rotor.

We can also see that the overall efficiency of the motor must always be
significantly less than $(1 - s)$, because in addition to the rotor copper losses there are
stator copper losses, iron losses and windage and friction losses. This fact is sometimes forgotten, leading to conflicting claims such as 'full-load slip = 5%, overall
efficiency = 96%', which is clearly impossible.

3.4 Steady-state stability – pull-out torque and stalling

We can check stability by asking what happens if the load torque suddenly changes
for some reason. The load shown by the lower broken line in Figure 6.9 is stable at
speed X: for example, if the load torque increased from T_a to T_b, the load torque

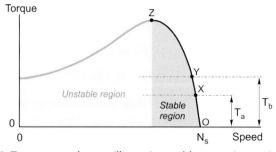

Figure 6.9 Torque–speed curve illustrating stable operating region (OXYZ).

would be greater than the motor torque, so the motor would decelerate. As the speed dropped, the motor torque would rise, until a new equilibrium was reached, at the slightly lower speed (Y). The converse would happen if the load torque reduced, leading to a higher stable running speed.

But what happens if the load torque is increased more and more? We can see that as the load torque increases, beginning at point X, we eventually reach point Z, at which the motor develops its maximum torque. Quite apart from the fact that the motor is now well into its overload region, and will be in danger of overheating, it has also reached the limit of stable operation. If the load torque is further increased, the speed falls (because the load torque is more than the motor torque), and as it does so the shortfall between motor torque and load torque becomes greater and greater. The speed therefore falls faster and faster, and the motor is said to be 'stalling'. With loads such as machine tools (a drilling machine, for example), as soon as the maximum or 'pull-out' torque is exceeded, the motor rapidly comes to a halt, making an angry humming sound. With a hoist, however, the excess load would cause the rotor to be accelerated in the reverse direction, unless it was prevented from doing so by a mechanical brake.

4. TORQUE–SPEED CURVES – INFLUENCE OF ROTOR PARAMETERS

We saw earlier that the rotor resistance and reactance influenced the shape of the torque–speed curve. Both of these parameters can be varied by the designer, and we will explore the pros and cons of the various alternatives. To limit the mathematics the discussion will be mainly qualitative, but it is worth mentioning that the whole matter can be dealt with rigorously using the equivalent circuit approach, as discussed in Appendix 2.

We will deal with the cage rotor first because it is the most important, but the wound rotor allows a wider variation of resistance to be obtained, so it is discussed later.

4.1 Cage rotor

For small values of slip, i.e. in the normal running region, the lower we make the rotor resistance the steeper the slope of the torque–speed curve becomes, as shown in Figure 6.10. We can see that at the rated torque (shown by the horizontal dotted line in Figure 6.10) the full-load slip of the low-resistance cage is much lower than that of the high-resistance cage. But we saw earlier that the rotor efficiency is equal to $(1 - s)$, where s is the slip. So we conclude that the low resistance rotor not only gives better speed holding, but is also much more efficient. There is of course a limit to how low we can make the resistance: copper allows us to achieve a lower resistance than aluminum, but we can't do any better than fill the slots with solid copper bars.

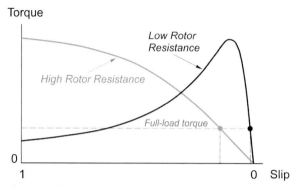

Figure 6.10 Influence of rotor resistance on the torque–speed curve of a cage motor. The full-load running speeds are indicated by the vertical dotted lines.

As we might expect there are drawbacks with a low resistance rotor. The direct-on-line starting torque is reduced (see Figure 6.10), and worse still the starting current is increased. The lower starting torque may prove insufficient to accelerate the load, while increased starting current may lead to unacceptable volt-drops in the supply.

Altering the rotor resistance has little or no effect on the value of the peak (pull-out) torque, but the slip at which the peak torque occurs is directly proportional to the rotor resistance. By opting for a high enough resistance (by making the cage from bronze, brass or other relatively high-resistivity material) we can if we wish arrange for the peak torque to occur at or close to starting, as shown in Figure 6.10. The snag in doing this is that the full-load efficiency is inevitably low because the full-load slip will be high (see Figure 6.10).

There are some applications for which high-resistance motors were traditionally well suited, an example being for metal punching presses, where the motor accelerates a flywheel which is used to store energy. In order to release a significant amount of energy, the flywheel slows down appreciably during impact, and the motor then has to accelerate it back up to full speed. The motor needs a high torque over a comparatively wide speed range, and does most of its work during acceleration. Once up to speed the motor is effectively running light, so its low efficiency is of little consequence. (We should note that this type of application is now often met by drives, but because the induction motor is so robust and long-lived, significant numbers of 'heritage' installations will persist, particularly in the developing world.)

High-resistance motors are sometimes used for speed control of fan-type loads, and this is taken up again later when we explore speed control.

To sum up, a high rotor resistance is desirable when starting and at low speeds, while a low resistance is preferred under normal running conditions. To get the best of both worlds, we need to be able to alter the resistance from a high value at starting to a lower value at full speed. Obviously we can't change the actual resistance of the cage once it has been manufactured, but it is possible to achieve the desired effect with either a 'double cage' or a 'deep bar' rotor.

4.2 Double cage and deep bar rotors

Double cage rotors have an outer cage made of relatively high-resistivity material such as bronze, and an inner cage of low resistivity, usually copper, as shown on the left in Figure 6.11.

The inner cage of low-resistance copper is sunk deep into the rotor, so that it is almost completely surrounded by iron. This causes the inner bars to have a much higher leakage inductance than if they were near the rotor surface, so that under starting conditions (when the induced rotor frequency is high) their inductive reactance is very high and little current flows in them. In contrast, the bars of the outer cage (of higher resistance bronze) are placed so that their leakage fluxes face a much higher reluctance path, leading to a low leakage inductance. Hence under starting conditions, rotor current is concentrated in the outer cage, which, because of its high resistance, produces a high starting torque.

At the normal running speed the roles are reversed. The rotor frequency is low, so both cages have low reactance and most of the current therefore flows in the low-resistance inner cage. The torque–speed curve is therefore steep, and the efficiency is high.

Considerable variation in detailed design is possible in order to shape the torque–speed curve to particular requirements. In comparison with a single-cage rotor, the double cage gives much higher starting torque, substantially less starting current, and marginally worse running performance.

The deep bar rotor has a single cage, usually of copper, formed in slots which are deeper and narrower than in a conventional single-cage design. Construction is simpler and therefore cheaper than in a double-cage rotor, as shown on the right in Figure 6.11.

The deep bar approach ingeniously exploits the fact that the effective resistance of a conductor is higher under a.c. conditions than under d.c. conditions. With a typical copper bar of the size used in an induction motor rotor, the difference in effective resistance between d.c. and say 50 or 60 Hz (the so-called 'skin effect') would be negligible if the conductor were entirely surrounded by air. But when it is almost completely surrounded by iron, as in the rotor slots, its effective resistance at mains frequency may be two or three times its d.c. value.

Figure 6.11 Double cage (left) and deep bar (right) rotors.

Under starting conditions, when the rotor frequency is equal to the supply frequency, the skin effect is very pronounced, and the rotor current is concentrated towards the top of the slots. The effective resistance is therefore increased, resulting in a high starting torque from a low starting current. When the speed rises and the rotor frequency falls, the effective resistance reduces towards its d.c. value, and the current distributes itself more uniformly across the cross-section of the bars. The normal running performance thus approaches that of a low-resistance single-cage rotor, giving a high efficiency and stiff torque–speed curve. The pull-out torque is, however, somewhat lower than for an equivalent single-cage motor because of the rather higher leakage reactance.

Most small and medium motors are designed to exploit the deep bar effect to some extent, reflecting the view that for most applications the slightly inferior running performance is more than outweighed by the much better starting behavior. A typical torque–speed curve for a general-purpose medium-size (55 kW) motor is shown in Figure 6.12. Such motors are unlikely to be described by the maker specifically as 'deep bar' but they nevertheless incorporate a measure of the skin effect and consequently achieve the 'good' torque–speed characteristic shown by the solid line in Figure 6.12.

The current–speed relationship is shown by the dashed line in Figure 6.12, both torque and current scales being expressed in per-unit (p.u.). This notation is widely used as a shorthand, with 1 p.u. (or 100%) representing rated value. For example, a torque of 1.5 p.u. simply means one and a half times rated value, while a current of 400% means a current of four times rated value.

4.3 Starting and run-up of slipring motors

By adding external resistance in series with the rotor windings the starting current can be kept low but at the same time the starting torque is high. This was the major advantage of the wound-rotor or slipring motor, which made it well suited for loads

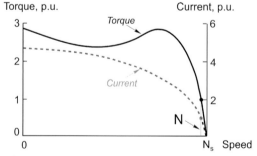

Figure 6.12 Typical torque–speed and current–speed curves for a general-purpose industrial cage motor.

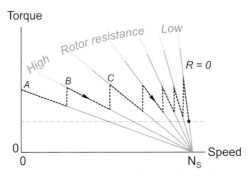

Figure 6.13 Torque–speed curves for a wound-rotor (slipring) motor showing how the external rotor-circuit resistance (R) can be varied in steps to provide an approximately constant torque during acceleration.

with heavy starting duties such as stone-crushers, cranes and conveyor drives, for most of which the inverter-fed cage motor is now preferred.

The influence of rotor resistance is shown by the set of torque–speed curves in Figure 6.13.

Typically the resistance at starting would be selected to give full-load torque together with rated current from the utility supply. The starting torque is then as indicated by point A in Figure 6.13.

As the speed rises, the torque would fall more or less linearly if the resistance remained constant, so in order to keep close to full torque the resistance is gradually reduced, either in steps, in which case the trajectory ABC, etc. is followed (Figure 6.13), or continuously so that maximum torque is obtained throughout. Ultimately the external resistance is made zero by shorting out the sliprings, and thereafter the motor behaves like a low-resistance cage motor, with a high running efficiency.

5. INFLUENCE OF SUPPLY VOLTAGE ON TORQUE–SPEED CURVE

We established earlier that at any given slip, the air-gap flux density is proportional to the applied voltage, and the induced current in the rotor is proportional to the flux density. The torque – which depends on the product of the flux and the rotor current – therefore depends on the square of the applied voltage. This means that a comparatively modest fall in the voltage will result in a much larger reduction in torque capability, with adverse effects which may not be apparent to the unwary until too late.

To illustrate the problem, consider the torque–speed curves for a cage motor shown in Figure 6.14. The curves (which have been expanded to focus attention on

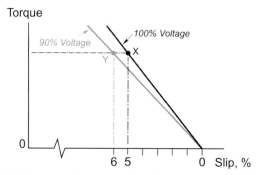

Figure 6.14 Influence of stator supply voltage on torque–speed curves.

the low-slip region) are drawn for full voltage (100%), and for a modestly reduced voltage of 90%. With full voltage and full-load torque the motor will run at point X, with a slip of say 5%. Since this is the normal full-load condition, the rotor and stator currents will be at their rated values.

Now suppose that the voltage falls to 90%. The load torque is assumed to be constant so the new operating point will be at Y. Since the air-gap flux density is now only 0.9 of its rated value, the rotor current will have to be about 1.1 times rated value to develop the same torque, so the rotor e.m.f. is required to increase by 10%. But the flux density has fallen by 10%, so an increase in slip of 20% is called for. The new slip is therefore 6%.

The drop in speed from 95% of synchronous to 94% may well not be noticed, and the motor will apparently continue to operate quite happily. But the rotor current is now 10% above its rated value, so the rotor heating will be 21% more than is allowable for continuous running. The stator current will also be above rated value, so if the motor is allowed to run continuously, it will overheat. This is one reason why all large motors are fitted with protection which is triggered by over-temperature. Many small and medium motors do not have such protection, so it is important to guard against the possibility of undervoltage operation.

Another potential danger arises if the supply voltage is unbalanced, i.e. the three line-to-line voltages are unequal. This is most likely where the supply impedance is high and the line currents are unequal due to unbalanced loads elsewhere on the system.

Electrical engineers use the technique of 'symmetrical components' to analyze unbalanced 3-phase voltages. It is a method whereby the effect of unbalanced voltages on, say, a motor is quantified by finding how the motor behaves when subjected, independently, to three sets of balanced voltages that together simulate the actual unbalanced voltages.

The first balanced set is the positive sequence component: its phase-sequence is the normal one, e.g. UVW or ABC; the second is the negative sequence, having the phase sequence WVU; and the third is the zero sequence, in which the 3-phase voltage components are co-phasal. There are simple analytic formulae for finding the components from the original three unbalanced voltages.

If the supply is balanced, the negative sequence and zero sequence components are both zero. Any unbalance gives rise to a negative sequence component, which will set up a rotating magnetic field traveling in the opposite direction to that of the main (positive sequence) component, and thus cause a braking torque and increase the losses, especially in the rotor. The zero sequence components produce a stationary field with three times the pole-number of the main field, but this can only happen when the motor has a star point connection.

Perhaps the most illuminating example of the application of symmetrical components as far as we are concerned is in the case of single-phase machines, where a single winding produces a pulsating field. This is an extreme case of unbalance because two of the phases are non-existent! It turns out that the positive sequence and negative sequence components are equal, which leads naturally to the idea that the pulsating field can be resolved into two counter-rotating fields. We will see later in this chapter that we can extend this picture to understand how the single-phase induction motor works.

Returning to 3-phase motors, a quite modest unbalance can be serious in terms of overheating. For example, to meet international standards motors are expected to tolerate only 1% negative sequence voltage continuously. Large motors often have negative sequence protection fitted, while small ones will rely on their thermal protection device to prevent overheating. Alternatively, if continuous operation under unbalanced conditions is required the motor must be de-rated significantly, e.g. to perhaps 80% of full-load with a voltage unbalance of 4%.

---

6. GENERATING

Having explored the torque–speed curve for the normal motoring region, where the speed lies between zero and just below synchronous, we must ask what happens if the speed is above the synchronous speed, i.e. the slip is negative. The unambiguous answer to this question is that the machine will switch from motoring to generating. Strangely, as mentioned previously, the authors have often encountered users who express deep scepticism, or even outright disbelief at the prospect of induction machines generating, so it is important that we attempt to counter what is clearly a widely held misconception.

A typical torque–speed curve for a cage motor covering the full range of speeds which are likely in practice is shown in Figure 6.15.

We can see from Figure 6.15 that the decisive factor as far as the direction of the torque is concerned is the slip, rather than the speed. When the slip is positive the torque is positive, and vice versa. The torque therefore always acts so as to urge the rotor to run at zero slip, i.e. at the synchronous speed. If the rotor is tempted to

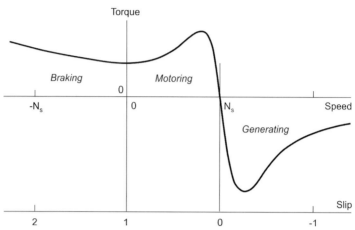

Figure 6.15 Torque–speed curve over motoring region (slip between 0 and 1), braking region (slip greater than 1) and generating region (negative slip).

run faster than the field it will be slowed down, while if it is running below synchronous speed it will be urged to accelerate forwards. In particular, we note that for slips greater than 1, i.e. when the rotor is running backwards (i.e. in the opposite direction to the field), the torque will remain positive, so that if the rotor is unrestrained it will first slow down and then change direction and accelerate in the direction of the field.

6.1 Generating region

For negative slips, i.e. when the rotor is turning in the same direction, but at a higher speed than the traveling field, the 'motor' torque is in fact negative. In other words, the machine develops a torque that opposes the rotation, which can therefore only be maintained by applying a driving torque to the shaft. In this region the machine acts as an induction generator, converting mechanical power from the shaft into electrical power into the supply system. Cage induction machines are used in this way in wind-power generation schemes, as described in a later section.

It is worth stressing that, just as with the d.c. machine, we do not have to make any changes to an induction motor to turn it into an induction generator. In both cases, all that is needed is a source of mechanical power to turn the rotor faster than it would run if there was zero load or friction torque. For the d.c. motor, its ideal no-load speed is that at which its back e.m.f. equals the supply voltage, whereas for the induction motor, it is the synchronous speed.

On the other hand, we should be clear that, unlike the d.c. machine, the induction machine can only generate when it is connected to the supply. If we disconnect an induction motor from the utility supply and try to make it generate simply by turning the rotor we will not get any output because there is nothing to set up the working flux: the flux (excitation) is not present until the motor is supplied with magnetizing current from the supply. It seems likely that this apparent inability to generate in isolation is what gave rise to the myth that induction machines cannot generate at all: a widely held view, but wholly incorrect!

There are comparatively few applications in which utility-supplied motors find themselves in the generating region, though as we will see later it is quite common in inverter-fed drives. We will, however, look at one example of a utility supply-fed motor in the so-called regenerative mode to underline the value of the motor's inherent ability to switch from motoring to generating automatically, without the need for any external intervention.

Consider a cage motor driving a simple hoist through a reduction gearbox, and suppose that the hook (unloaded) is to be lowered. Because of the static friction in the system, the hook will not descend on its own, even after the brake is lifted, so on pressing the 'down' button the brake is lifted and power is applied to the motor so that it rotates in the lowering direction. The motor quickly reaches full speed and the hook descends. As more and more rope winds off the drum, a point is reached where the lowering torque exerted by the hook and rope is greater than the running friction, and a restraining torque is then needed to prevent a runaway. The necessary stabilizing torque is automatically provided by the motor acting as a generator as soon as the synchronous speed is exceeded, as shown in Figure 6.15. The speed will therefore be held at just above the synchronous speed, provided of course that the peak generating torque (see Figure 6.15) is not exceeded.

6.2 Self-excited induction generator

In previous sections we have stressed that the rotating magnetic field or excitation is provided by the magnetizing current drawn from the supply, so it would seem obvious that the motor could not generate unless a supply was provided to furnish the magnetizing current. However, it is possible to make the machine 'self-excite' if the conditions are right, and given the robustness of the cage motor this can make it an attractive proposition, especially for small-scale isolated installations.

We saw in Chapter 5 that when the induction motor is running at its normal speed, the rotating magnetic field that produces the currents and torque on the rotor also induced balanced 3-phase induced e.m.f.s in the stator windings, the magnitude of the e.m.f.s being not a great deal less than the voltage of the utility supply. So to act as an independent generator what we want to do is to set up the rotating magnetic field without having to connect to an active voltage source.

We discussed a similar matter in Chapter 3, in connection with self-excitation of the shunt d.c. machine. We saw that if enough residual magnetic flux remained in the field poles after the machine had been switched off, the e.m.f. produced when the shaft was rotated could begin to supply current to the field winding, thereby increasing the flux, further raising the e.m.f. and initiating a positive feedback (or bootstrap) process which was ultimately stabilized by the saturation characteristic of the iron in the magnetic circuit.

Happily, much the same can be achieved with an isolated induction motor. We aim to capitalize on the residual magnetism in the rotor iron, and, by turning the rotor, generate an initial voltage in the stator to kick-start the process. The e.m.f. induced must then drive current to reinforce the residual field and promote the positive feedback to build up the traveling flux field. Unlike the d.c. machine, however, the induction motor has only one winding that provides both excitation and energy converting functions, so given that we want to get the terminal voltage to its rated level before we connect whatever electrical load we plan to supply, it is clearly necessary to provide a closed path for the would-be excitation current. This path should encourage the build-up of magnetizing current – and hence terminal voltage.

'Encouraging' the current means providing a very low impedance path, so that a small voltage drives a large current, and, since we are dealing with a.c. quantities, we naturally seek to exploit the phenomenon of resonance, by placing a set of capacitors in parallel with the (inductive) windings of the machine, as shown in Figure 6.16.

The reactance of a parallel circuit consisting of pure inductance (L) and capacitance (C) at angular frequency ω is given by $X = \omega L - 1/\omega C$, so at low and high frequencies the reactance is very large, but at the so-called resonant frequency ($\omega_0 = 1/\sqrt{LC}$), the reactance becomes zero. Here the inductance is the magnetizing inductance of each phase of the induction machine, and C is the added capacitance, the value being chosen to give resonance at the desired frequency of generation. Of course the circuit is not ideal because there is resistance in the windings, but nevertheless the inductive reactance can be 'tuned out' by choice of capacitance, leaving a circulating path of very low resistance. Hence by turning the rotor at the speed at which the desired frequency is produced by the residual

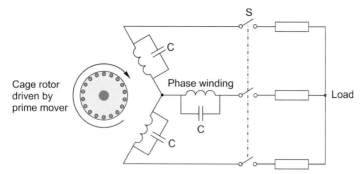

Figure 6.16 Self-excited induction generator. The load is connected only after the stator voltage has built up.

magnetism (e.g. 1800 rev/min for a 4-pole motor to generate 60 Hz), the initial modest e.m.f. produces a disproportionately high current and the flux builds up until limited by the non-linear saturation characteristic of the iron magnetic circuit. We then get balanced 3-phase voltages at the terminals, and the load can be applied by closing switch S (Figure 6.16).

The description above gives only a basic outline of the self-excitation mechanism. Such a scheme would only be satisfactory for a very limited range of driven speeds and loads, and in practice further control features are required to vary the effective capacitance (typically using triac control) in order to keep the voltage constant when the load and/or speed vary widely.

6.3 Doubly-fed induction machine for wind-power generation

The term doubly-fed refers to an induction machine in which both stator and rotor windings are connected to an a.c. power source: we are therefore talking about wound rotor (or slipring) motors, where the rotor windings are accessed through insulated rotating sliprings.

Traditionally, large slipring machines were used in Kramer[1] drives to recover the slip energy in the rotor and return it to the supply, so that efficient operation was possible at much higher slips than would otherwise have been possible. Some Kramer drives remain, but in the twenty-first century by far the major application for the doubly-fed induction machine is in wind-power generation where the wind turbine feeds directly into the utility grid.

Before we see why the doubly-fed motor is favored, we should first acknowledge that in principle we could take a cage motor, connect it directly to the utility grid, and drive its rotor from the wind turbine (via a gearbox with an output speed a little above synchronous speed) so that it supplied electrical power to the grid. However, the range of speeds over which stable generation is possible would be only a few per cent above synchronous, and this is a poor match as far as the turbine characteristics are concerned. Ideally, in order to abstract the maximum power from the wind, the speed of rotation of the blades must vary according to the conditions, so being forced to remain at a more-or-less constant speed by being connected to a cage motor is not good news. In addition, when there are rapid fluctuations in wind speed that produce bursts of power, the fact that the speed is constant means that there are rapid changes in torque which produce unwelcome fatigue loading in the gearbox. What is really wanted is a generator in which the generated frequency can be maintained constant over a wider speed range, and this is where the doubly-fed system scores.

---

[1] See, for example, *The Control Techniques Drives and Controls Handbook*, 2nd edition, by W. Drury.

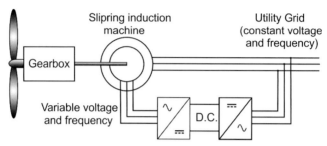

Figure 6.17 Doubly-fed wound-rotor induction machine for wind-power generation.

The stator is connected directly to the utility grid, while the rotor windings are also linked to the grid, but via a pair of a.c./d.c. converters, as shown in Figure 6.17. The converters – connected via a d.c. link – allow power to flow to or from the fixed-frequency grid into or out of the rotor circuit, the frequency of which varies according to the shaft speed (see below). The rating of these converters will be substantially less than the rated power output of the induction machine, depending on the speed range that is to be accepted. For example, if the speed range is to be $\pm 1.3 N_s$, where N_s is the synchronous speed of the induction motor connected to the grid, and full torque is to be available over the full speed range, the rating of the converters will be only 30% of the rating of the machine. (This is a major advantage over the alternative of having a conventional synchronous generator and a frequency converter of 100% rating.)

Understanding all the details of how the doubly-fed induction generator operates is not easy, but we can get to the essence by picturing the relationship between the rotating magnetic field and the stator and rotor.

Given that the stator is permanently connected to the utility grid, we know that the magnitude and speed of the rotating magnetic field cannot vary, so the synchronous speed of a 4-pole machine connected to the 60 Hz supply will be 1800 rev/min. If we want to generate power to the grid at this speed, we feed d.c. (i.e. zero frequency) into the rotor, and we then have what amounts to a conventional synchronous machine, which can motor or generate, and its power-factor can be controlled via the rotor current (see section 2 of Chapter 9). The 4-pole field produced by the rotor is stationary with respect to the rotor, but because the rotor is turning at 1800 rev/min, it rotates at the same speed as the traveling field produced by the stator windings, the two effectively being locked together so that torque can be transmitted and power can be converted. Energy conversion is only possible at the speed of exactly 1800 rev/min.

Now suppose that the wind turbine speed falls so that the speed at the machine shaft is only 1500 rev/min. For the rotor to be able to lock on to the 1800 rev/min stator field, the field that it produces must rotate at 300 rev/min (in a positive sense) relative to the rotor, so that its speed relative to the stator is $1500 + 300 = 1800$. This is achieved by supplying the rotor with 3-phase current at a frequency of

10 Hz. Conversely, if the turbine drives the machine shaft speed at 2100 rev/min, the rotor field must rotate at 300 rev/min in a negative sense relative to the rotor; i.e. the rotor currents must again be at 10 Hz, but with reversed phase sequence.

It turns out that if the input speed is below synchronous (1500 rev/min in the example above), electrical power has to be fed into the rotor circuit. This power is taken from the utility grid, but (neglecting losses, which are small) it then emerges from the stator, together with the mechanical power supplied by the turbine. Thus we can think of the electrical input power to the rotor as merely 'borrowed' from the grid to allow the energy conversion to take place; of course, the net power supplied to the grid all comes from the turbine. In this example, if the turbine torque is at rated value, the overall power output will be 1500/1800P, i.e. 5P/6, where P is the rated power at normal (synchronous) speed. This will be made up of an output power of P from the stator, from which the rotor converters take 300/1800P, i.e. P/6 into the rotor.

When the driven speed is above synchronous (e.g. 2100 rev/min), both stator and rotor circuits export power to the grid. In this case, with rated turbine torque, the power to the grid will be 2100/1800P, i.e. 7P/6, comprising P from the stator and P/6 from the rotor converter.

We should be clear that there is no magic in being able to exceed the original rated power of our machine. At full torque, the electric and magnetic loadings will be at their rated values, and the increase in power is therefore entirely due to the higher speed. We discussed this towards the end of Chapter 1.

Just as with a conventional (single-speed) synchronous machine, we can control the extent to which the stator and rotor sides contribute to the setting up of the resultant flux, and this means that we can control the grid power-factor via the rotor circuit; this can be very advantageous where there is a system requirement to export or absorb reactive volt-amperes.

7. BRAKING

7.1 Plug reversal and plug braking

Because the rotor always tries to catch up with the rotating field, it can be reversed rapidly simply by interchanging any two of the supply leads. The changeover is usually obtained by having two separate 3-pole contactors, one for forward and one for reverse. This procedure is known as plug reversal or plugging, and is illustrated in Figure 6.18.

The motor is initially assumed to be running light (and therefore with a very small positive slip) as indicated by point A on the dashed torque–speed curve in Figure 6.18(a). Two of the supply leads are then reversed, thereby reversing the direction of the field, and bringing the mirror-image torque–speed curve shown by the solid line into play. The slip of the motor immediately after reversal is

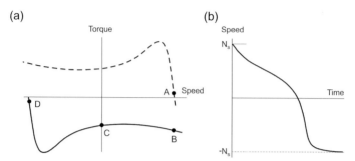

Figure 6.18 Torque–speed and speed–time curves for plug reversal of cage motor.

approximately 2, as shown by point B on the solid curve. The torque is thus negative, and the motor decelerates, the speed passing through zero at point C and then rising in the reverse direction before settling at point D, just below the synchronous speed.

The speed–time curve is shown in Figure 6.18(b). We can see that the deceleration (i.e. the gradient of the speed–time graph) reaches a maximum as the motor passes through the peak torque (pull-out) point, but thereafter the final speed is approached gradually, as the torque tapers down to point D.

Very rapid reversal is possible using plugging; for example, a 1 kW motor will typically reverse from full speed in under one second. But large cage motors can only be plugged if the supply can withstand the very high currents involved, which are even larger than when starting from rest. Frequent plugging will also cause serious overheating, because each reversal involves the 'dumping' of four times the stored kinetic energy as heat in the windings.

Plugging can also be used to stop the rotor quickly, but obviously it is then necessary to disconnect the supply when the rotor comes to rest, otherwise it will run up to speed in reverse. A shaft-mounted reverse-rotation detector is therefore used to trip out the reverse contactor when the speed reaches zero.

We should note that whereas in the regenerative mode (discussed in the previous section) the slip was negative, allowing mechanical energy from the load to be converted to electrical energy and fed back to the mains, plugging is a wholly dissipative process in which all the kinetic energy ends up as heat in the motor.

7.2 Injection braking

This is the most widely used method of electrical braking. When the 'stop' signal occurs the 3-phase supply is interrupted, and a d.c. current is fed into the stator via two of its terminals. The d.c. supply is usually obtained from a rectifier fed via a low-voltage high-current transformer.

We saw earlier that the speed of rotation of the air-gap field is directly proportional to the supply frequency, so it should be clear that since d.c. is

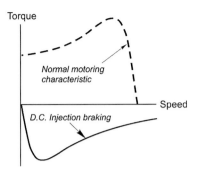

Figure 6.19 Torque–speed curve for d.c. injection braking of cage motor.

effectively zero frequency, the air-gap field will be stationary. We also saw that the rotor always tries to run at the same speed as the field. So if the field is stationary, and the rotor is not, a braking torque will be exerted. A typical torque–speed curve for braking a cage motor is shown in Figure 6.19, from which we see that the braking (negative) torque falls to zero as the rotor comes to rest.

This is in line with what we would expect, since there will only be induced currents in the rotor (and hence torque) when the rotor is 'cutting' the flux. As with plugging, injection (or dynamic) braking is a dissipative process, all the kinetic energy being turned into heat inside the motor.

8. SPEED CONTROL

We have seen that to operate efficiently an induction motor must run with a small slip. It follows that any efficient method of speed control must be based on varying the synchronous speed of the field, rather than the slip. The two factors which determine the speed of the field are the supply frequency and the pole-number (see equation (5.1)).

The pole-number has to be an even integer, so where continuously adjustable speed control over a wide range is called for, the best approach by far is to provide a variable-frequency supply. This method is very important, and is dealt with separately in Chapters 7 and 8. In this chapter we are concerned with constant-frequency (utility-connected) operation, so we are limited to either pole-changing, which can provide discrete speeds only, or slip-control which can provide continuous speed control, but is inherently inefficient.

8.1 Pole-changing motors

For some applications continuous speed control may be an unnecessary luxury, and it may be sufficient to be able to run at two discrete speeds. Among many instances where this can be acceptable and economic are pumps, lifts and hoists, fans and some machine tool drives.

We established in Chapter 5 that the pole-number of the field was determined by the layout and interconnection of the stator coils, and that once the winding has been designed, and the frequency specified, the synchronous speed of the field is fixed. If we wanted to make a motor which could run at either of two different speeds, we could construct it with two separate stator windings (say 4-pole and 6-pole), and energize the appropriate one. There is no need to change the cage rotor since the pattern of induced currents can readily adapt to suit the stator pole-number. Early two-speed motors did have two distinct stator windings, but were bulky and inefficient.

It was soon realized that if half of the phase-belts within each phase-winding could be reversed in polarity, the effective pole-number could be halved. For example, a 4-pole m.m.f. pattern (N–S–N–S) would become (N–N–S–S), i.e. effectively a 2-pole pattern with one large N and one large S pole. By bringing out six leads instead of three, and providing switching contactors to effect the reversal, two discrete speeds in the ratio 2:1 are therefore possible from a single winding. The performance at the high (e.g. 2-pole) speed is relatively poor, which is not surprising in view of the fact that the winding was originally optimized for 4-pole operation.

It was not until the advent of the more sophisticated pole amplitude modulation (PAM) method in the 1960s that two-speed single-winding high-performance motors with more or less any ratio of speeds became available from manufacturers. This subtle technique allows close ratios such as 4/6, 6/8, 8/10 or wide ratios such as 2/24 to be achieved. The beauty of the PAM method is that it is not expensive. The stator winding has more leads brought out, and the coils are connected to form non-uniform phase-belts, but otherwise construction is the same as for a single-speed motor. Typically six leads will be needed, three of which are supplied for one speed, and three for the other, the switching being done by contactors. The method of connection (star or delta) and the number of parallel paths within the winding are arranged so that the air-gap flux at each speed matches the load requirement. For example, if constant torque is needed at both speeds, the flux needs to be made the same, whereas if reduced torque is acceptable at the higher speed the flux can obviously be lower.

8.2 Voltage control of high-resistance cage motors

Where efficiency is not of paramount importance, the torque (and hence the running speed) of a cage motor can be controlled simply by altering the supply voltage. The torque at any slip is approximately proportional to the square of the voltage, so we can reduce the speed of the load by reducing the voltage. The method is not suitable for standard low-resistance cage motors, because their stable

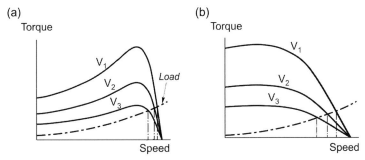

Figure 6.20 Speed control of cage motor by stator voltage variation; (a) low-resistance rotor, (b) high-resistance rotor.

operating speed range is very restricted, as shown in Figure 6.20(a). But if special high-rotor resistance motors are used, the gradient of the torque–speed curve in the stable region is much less, and a somewhat wider range of steady-state operating speeds is available, as shown in Figure 6.20(b).

The most unattractive feature of this method is the low efficiency which is inherent in any form of slip-control. We recall that the rotor efficiency at slip s is $(1 - s)$, so if we run at say 70% of synchronous speed (i.e. $s = 0.3$), 30% of the power crossing the air-gap is wasted as heat in the rotor conductors. The approach is therefore only practicable where the load torque is low at low speeds, so a fan-type characteristic is suitable, as shown in Figure 6.20(b). Voltage control became feasible only when relatively cheap thyristor a.c. voltage regulators arrived on the scene during the 1970s, and although it enjoyed some success it is now seldom seen. The hardware required is essentially the same as discussed earlier for soft starting, and a single piece of kit can therefore serve for both starting and speed control.

8.3 Speed control of wound-rotor motors

The fact that the rotor resistance can be varied easily allows us to control the slip from the rotor side, with the stator supply voltage and frequency constant. Although the method is inherently inefficient it is still sometimes used because of its simplicity and comparatively low cost.

A set of torque–speed characteristics is shown in Figure 6.21, from which it should be clear that by appropriate selection of the rotor circuit resistance, any torque up to typically 1.5 times full-load torque can be achieved at any speed.

8.4 Slip energy recovery

Instead of wasting rotor-circuit power in an external resistance, it can be converted and returned to the mains supply. Frequency conversion is necessary because the rotor circuit operates at slip frequency, so it cannot be connected directly to the utility supply. These systems (which are known as the static Kramer drive) have

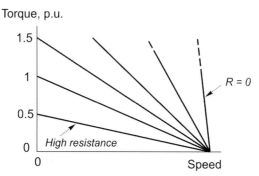

Figure 6.21 Influence of external rotor resistance (R) on torque–speed curve of wound-rotor motor.

largely been superseded by inverter-fed cage motors, but some survive so a brief mention is in order.

In a slip energy recovery system, the slip-frequency a.c. from the rotor is first rectified in a 3-phase diode bridge and smoothed before being returned to the utility supply via a 3-phase thyristor bridge converter operating in the inverting mode (see Chapter 4). A transformer is usually required to match the output from the controlled bridge to the mains voltage. Since the cost of both converters depends on the slip power they have to handle, this system is most often used where only a modest range of speeds (say from 70% of synchronous and above) is required, such as in large pump and compressor drives.

9. POWER-FACTOR CONTROL AND ENERGY OPTIMIZATION

In addition to their use for soft-start and speed control, thyristor voltage regulators provide a means for limited control of power-factor for cage motors, and this can allow a measure of energy economy. However, the fact is that there are comparatively few situations where considerations of power-factor and/or energy economy alone are sufficient to justify the expense of a voltage controller. Only when the motor operates for very long periods running light or at low load can sufficient savings be made to cover the outlay. There is certainly no point in providing energy economy when the motor spends most of its time working at or near full-load.

Both power-factor control and energy optimization rely on the fact that the air-gap flux is proportional to the supply voltage, so that by varying the voltage, the flux can be set at the best level to cope with the prevailing load. We can see straightaway that nothing can be achieved at full-load, since the motor needs full flux (and hence full voltage) to operate as intended. Some modest savings in losses can be achieved at reduced load, as we will see.

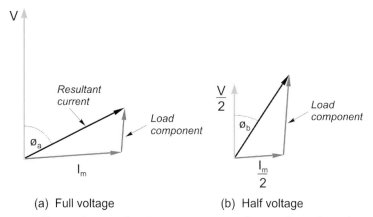

(a) Full voltage   (b) Half voltage

Figure 6.22 Phasor diagram showing improvement of power-factor by reduction of stator voltage.

If we imagine the motor to be running with a low load torque and full voltage, the flux will be at its full value, and the magnetizing component of the stator current will be larger than the work component, so the input power-factor ($\cos \phi_a$) will be very low, as shown in Figure 6.22(a).

Now suppose that the voltage is reduced to say half (by phasing back the thyristors), thereby halving the air–gap flux and reducing the magnetizing current by at least a factor of two. With only half the flux, the rotor current must double to produce the same torque, so the work current reflected in the stator will also double. The input power-factor ($\cos \phi_b$) will therefore improve considerably (see Figure 6.22(b)). Of course the slip with 'half-flux' operation will be higher (by a factor of four), but with a low–resistance cage it will still be small, and the drop in speed will therefore be slight.

The success (or otherwise) of the energy economy obtained depends on the balance between the iron losses and the copper losses in the motor. Reducing the voltage reduces the flux, and hence reduces the eddy current and hysteresis losses in the iron core. But as we have seen above, the rotor current has to increase to produce the same torque, so the rotor copper loss rises. The stator copper loss will reduce if (as in Figure 6.22) the magnitude of the stator current falls. In practice, with average general-purpose motors, a net saving in losses only occurs for light loads, say at or below 25% of full-load, though the power-factor will always increase.

10. SINGLE-PHASE INDUCTION MOTORS

Single-phase induction motors are simple, robust and reliable, and continue to be used in large numbers especially in domestic and commercial applications where 3-phase supplies are not available. Although outputs of up to a few kW are possible, the majority are below 0.5 kW, and are used in straightforward applications such as refrigeration compressors, and dryers, pumps and fans, small machine tools, etc. However, these traditional applications can now benefit from the superior control and low cost provided by the simple inverter and 3-phase induction or permanent magnet motor, so the single-phase motor looks set for a niche role in future.

10.1 Principle of operation

If one of the leads of a 3-phase motor is disconnected while it is running light, it will continue to run with a barely perceptible drop in speed, and a somewhat louder hum. With only two leads remaining there can only be one current, so the motor must be operating as a single-phase machine. If load is applied the slip increases more quickly than under 3-phase operation, and the stall torque is much less, perhaps one-third. When the motor stalls and comes to rest it will not restart if the load is removed, but remains at rest drawing a heavy current and emitting an angry hum. It will burn out if not disconnected rapidly.

It is not surprising that a truly single-phase cage induction motor will not start from rest, because as we saw in Chapter 5 the single winding, fed with a.c., simply produces a pulsating flux in the air-gap, without any suggestion of rotation. It is, however, surprising to find that if the motor is given a push in either direction it will pick up speed, slowly at first but then with more vigor, until it settles with a small slip, ready to take up load. Once turning, a rotating field is evidently brought into play to continue propelling the rotor.

We can understand how this comes about by first picturing the pulsating m.m.f. set up by the current in the stator winding as being the resultant of two identical traveling waves of m.m.f., one in the forward direction and the other in reverse. (This equivalence is not self-evident, but can be demonstrated by applying the method of symmetrical components, discussed earlier in this chapter.)

When the rotor is stationary, it reacts equally to both traveling waves, and no torque is developed. When the rotor is turning, however, the induced rotor currents are such that their m.m.f. opposes the reverse stator m.m.f. to a greater extent than they oppose the forward stator m.m.f. The result is that the forward flux wave (which is what develops the forward torque) is bigger than the reverse flux wave (which exerts a drag). The difference widens as the speed increases, the forward flux wave becoming progressively bigger as the speed rises while the reverse flux wave simultaneously reduces. This 'positive feedback' effect explains why the speed builds slowly at first, but later zooms up to just below synchronous speed. At the normal running speed (i.e. small slip), the forward flux is many times larger than the backward flux, and the drag torque is only a small percentage of the forward torque.

As far as normal running is concerned, a single winding is therefore sufficient. But all motors must be able to self-start, so some mechanism has to be provided to produce a rotating field even when the rotor is at rest. Several methods are employed, all of them using an additional winding.

The second winding usually has less copper than the main winding, and is located in the slots which are not occupied by the main winding, so that its m.m.f. is displaced in space relative to that of the main winding. The current in the second winding is supplied from the same single-phase source as the main winding current, but is caused to have a phase-lag, by a variety of means which are discussed later. The combination of a space displacement between the two windings together with a time displacement between the currents produces a 2-phase machine. If the two windings were identical, displaced by 90°, and fed with currents with 90° phase-shift, an ideal rotating field would be produced. In practice we can never achieve a 90° phase-shift between the currents, and it turns out to be more economic not to make the windings identical. Nevertheless, a decent rotating field is set up, and entirely satisfactory starting torque can be obtained. Reversal is simply a matter of reversing the polarity of one of the windings, and performance is identical in both directions.

The most widely used methods are described below. At one time it was common practice for the second or auxiliary winding to be energized only during start and run-up, and for it to be disconnected by means of a centrifugal switch mounted on the rotor, or sometimes by a time-switch. This practice gave rise to the term 'starting winding'. Nowadays it is more common to find both windings in use all the time.

10.2 Capacitor run motors

A capacitor is used in series with the auxiliary winding (Figure 6.23) to provide a phase-shift between the main and auxiliary winding currents. The capacitor (usually of a few μF, and with a voltage rating which may well be higher than the mains voltage) may be mounted piggyback fashion on the motor, or located elsewhere. Its value represents a compromise between the conflicting requirements of high starting torque and good running performance.

A typical torque–speed curve is also shown in Figure 6.21; the modest starting torque indicates that capacitor run motors are generally best suited to fan-type loads. Where higher starting torque is needed, two capacitors can be used, one being switched out when the motor is up to speed.

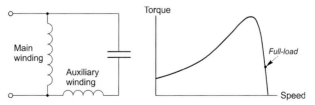

Figure 6.23 Single-phase capacitor-run induction motor.

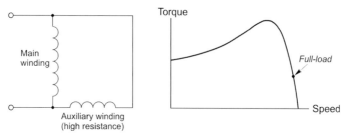

Figure 6.24 Single-phase split-phase induction motor.

As mentioned above, the practice of switching out the starting winding altogether is no longer favored, but many old ones remain, and where a capacitor is used they are known as 'capacitor start' motors.

10.3 Split-phase motors

The main winding is of thick wire, with a low resistance and high reactance, while the auxiliary winding is made of fewer turns of thinner wire with a higher resistance and lower reactance (Figure 6.24). The inherent difference in impedance is sufficient to give the required phase-shift between the two currents without needing any external elements in series. Starting torque is good at typically 1.5 times full-load torque, as also shown in Figure 6.24. As with the capacitor type, reversal is accomplished by changing the connections to one of the windings.

10.4 Shaded pole motors

There are several variants of this extremely simple, robust and reliable cage motor, which is used for low-power applications such as hair-dryers, oven fans, office equipment, and display drives. A 2-pole version from the cheap end of the market is shown in Figure 6.25.

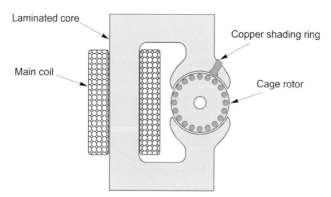

Figure 6.25 Shaded pole induction motor.

The rotor, typically between 1 and 4 cm diameter, has a die-cast aluminum cage, while the stator winding is a simple concentrated coil wound round the laminated core. The stator pole is slotted to receive the 'shading ring', which is a single short-circuited turn of thick copper or aluminum.

Most of the pulsating flux produced by the stator winding by-passes the shading ring and crosses the air-gap to the rotor. But some of the flux passes through the shading ring, and because it is alternating it induces an e.m.f. and current in the ring. The opposing m.m.f. of the ring current diminishes and retards the phase of the flux through the ring, so that the flux through the ring reaches a peak after the main flux, thereby giving what amounts to a rotation of the flux across the face of the pole. This far from perfect traveling wave of flux produces the motor torque by inter-action with the rotor cage. Efficiencies are low because of the rather poor magnetic circuit and the losses caused by the induced currents in the shading ring, but this is generally acceptable when the aim is to minimize first cost. Series resistance can be used to obtain a crude speed control, but this is only suitable for fan-type loads. The direction of rotation depends on whether the shading ring is located on the right or left side of the pole, so shaded pole motors are only suitable for uni-directional loads.

11. POWER RANGE

Having praised the simplicity and elegance of the induction motor, it is perhaps not surprising that they are successful over an extraordinarily wide power range from multi-megawatts down to a few tens of watts. (Indeed we cannot think of any other non-electric energy converter that can exceed a span of six orders of magnitude!)

At the upper end the limit is largely a question of low demand, there being few applications that need a shaft that delivers many tens of megawatts. But at the lower end we may wonder why there are no very small ones. Industrial (3-phase) induction motors are rarely found below about 200 W, and single-phase versions rarely extend below about 50 W, yet applications in this range are legion.

We will see that when we scale down a successful design, the excitation or flux-producing function of the windings becomes more and more demanding until eventually the heat produced in the windings by the excitation current causes the permissible temperature to be reached. There is then no spare capacity for the vital function of supplying the mechanical output power, so the machine is of no use.

11.1 Scaling down – the excitation problem

We can get to the essence of the matter by imagining that we take a successful design and scale all the linear dimensions by half. We know that in order to fully utilize the iron of the magnetic circuit we would want the air-gap flux density to be the same as in the original design, so because the air-gap length has been halved the stator m.m.f. needs to be half of what it was. The number of coils and the turns in each coil remain as before, so if the original magnetizing current was I_m, the magnetizing current of the half-scale motor will be $I_m/2$.

Turning now to what happens to the resistance of the winding, we will assume that the resistance of the original winding was R. In the half-scale motor, the total length of wire is half of what it was, but the cross-sectional area of the wire is only a quarter of the original. As a result the new resistance is twice as great, i.e. $2R$.

The power dissipated in providing the air-gap flux in the original motor is given by $I_m^2 R$, while the corresponding excitation power in the half-scale motor is given by

$$\left(\frac{I_m}{2}\right)^2 \times 2R = \frac{1}{2}I_m^2 R$$

When we consider what determines the steady temperature rise of a body in which heat is dissipated, we find that the equilibrium condition is reached when the rate of loss of heat to the surroundings is equal to the rate of production of heat inside the body. And, not surprisingly, the rate of loss of heat to the surroundings depends on the temperature difference between the body and its surroundings, and on the surface area through which the heat escapes. In the case of the copper windings in a motor, the permissible temperature rise depends on the quality of insulation, so we will make the reasonable assumption that the same insulation is used for the scaled motor as for the original.

We have worked out that the power dissipation in the new motor is half of that in the original. However, the surface area of the new winding is only one-quarter, so clearly the temperature rise will be higher, and if all other things were equal, it will double. We might aim to ease matters by providing bigger slots so that the current density in the copper could be reduced, but as explained in Chapter 1 this means that there is less iron in the teeth to carry the working flux. A further problem arises because it is simply not practicable to go on making the air-gap smaller because the need to maintain clearances between the moving parts would require unacceptably tight manufacturing tolerances.

Obviously, there are other factors that need to be considered, not least that a motor is designed to reach its working temperature when the full current (not just the magnetizing current) is flowing. But the fact is that the magnetization problem we have highlighted is the main obstacle in small sizes, not only in induction motors but also in any motor that derives its excitation from the stator windings. Permanent magnets therefore become attractive for small motors, because they provide the working flux without producing unwelcome heat. Permanent magnet motors are discussed in Chapter 9.

CHAPTER SEVEN

Variable Frequency Operation of Induction Motors

1. INTRODUCTION

We saw in Chapter 6 the many attributes of the induction motor which have made it the preferred workhorse of industry. These include simple low-cost construction which lends itself to totally enclosed designs suitable for dirty or even hazardous environments; limited routine maintenance with no brushes; only three power connections; and good full-load efficiency. We have also seen that when operated from the utility supply there are a number of undesirable characteristics, the most notable being that there is only one speed of operation (or more precisely a narrow load-dependent speed range). In addition, starting equipment is often required to avoid excessive currents of up to six times rated current when starting direct-on-line; reversal requires two of the power cables to be interchanged; and the instantaneous torque cannot be controlled, so the transient performance is poor.

We will see in this chapter that all the good features of the mains operated induction motor are retained and all the bad characteristics detailed above can be avoided when the induction motor is supplied from a variable-frequency source; i.e. its supply comes from an inverter.

The chapter divides broadly into two parts, both dealing with the capabilities of the induction motor when supplied from an inverter. The first part (sections 2 and 3) deals with the steady-state behavior when the operating frequency is solely determined by the inverter, and is independent of what is happening at the motor. We will refer to this set-up as 'inverter-fed', and in the early days of converter-driven induction motors this was the norm, the frequency being set by an oscillator that controlled the sequential periodic switching of the devices in the inverter. We will see that by appropriate control of the frequency and voltage we are able to operate over a very wide range of the torque–speed plane, but we will also identify the factors that place limits on what can be achieved. On a steady-state basis, this arrangement proved able to compete with the d.c. drive, but even when incorporated into a closed-loop control scheme the transient performance remained inferior. It is well worth absorbing the main messages from this study because although most contemporary drives now operate on a different basis, the steady-state running conditions at the motor remain the same. Readers who were able to follow the material in Chapter 6 should find this part straightforward.

In the second major part of this chapter (from section 4), we explore both 'field-oriented' and 'direct torque' methods for control of the inverter/induction motor combination. We prefer not to use the term 'inverter-fed' in these circumstances because although the motor derives its supply from an inverter, the switching of the inverter devices is determined by the state of the flux and currents in the motor, rather than being imposed from a separate oscillator. Both methods allow us to achieve hitherto unheard of levels of transient performance, but it is important to note that they only became possible because of the development of fast, cheap, digital processors that can implement the high-speed calculations necessary to model and control the motor in real time.

Understanding field-oriented control is usually regarded as challenging, even for experienced drives personnel, not least because the subject tends to be highly mathematical. However, although we will adopt a largely graphical approach to get to the heart of the matter, it is likely that most readers will find it advisable to absorb the tutorial material in sections 4 and 5 first.

For most of this chapter we will assume that the motor is supplied from an ideal balanced sinusoidal voltage source. Our justification for doing this is that although the pulse-width-modulated voltage waveform supplied by the inverter will not be sinusoidal (see Figure 7.1), the motor performance depends principally on the fundamental (sinusoidal) component of the applied voltage. This is a somewhat surprising but extremely welcome simplification, because it allows us to make use of our knowledge of how the induction motor behaves with a sinusoidal supply to anticipate how it performs when fed from an inverter.

There will be more discussion of the various control arrangements and practicalities of inverter-fed operation in Chapter 8.

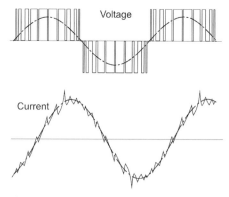

Figure 7.1 Typical voltage and current waveforms for PWM inverter-fed induction motor. (The fundamental-frequency component is shown by the dotted line.)

2. INVERTER-FED INDUCTION MOTOR DRIVES

It was explained in Chapter 6 that the induction motor can only run efficiently at low slips, i.e. close to the synchronous speed of the rotating field. The best method of speed control must therefore provide for continuous smooth variation of the synchronous speed, which in turn calls for variation of the supply frequency. This is readily achieved using a power electronic inverter (as discussed in Chapter 2) to supply the motor. A complete speed control scheme, which is illustrated with speed feedback, is shown in simplified block diagram form in Figure 7.2.

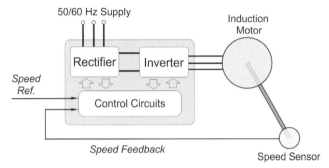

Figure 7.2 General arrangement of inverter-fed variable-frequency induction motor controlled-speed drive.

The arrangement shown in Figure 7.2 shows the motor with a speed sensor attached to the motor shaft. For all but the most demanding dynamic applications, or where full torque at standstill is a requirement, a speed sensor would not normally be required. This is good news as fitting a speed sensor to a standard induction motor involves significant additional cost and additional cabling.

We should recall that the function of the converter (i.e. rectifier and variable-frequency inverter) is to draw power from the fixed-frequency constant-voltage mains, rectify it and then convert it to variable frequency, variable voltage for driving the induction motor. Both the rectifier and the inverter employ switching strategies (see Chapter 2), so the power conversions are accomplished efficiently and the converter can be compact.

2.1 Steady-state operation – importance of achieving full flux

Three simple relationships need to be borne in mind in order to simplify understanding of how the inverter-fed induction motor behaves. First, we established in Chapter 5 that, for a given induction motor, the torque developed depends on the magnitude of the rotating flux density wave, and on the slip speed of the rotor, i.e. on the relative velocity of the rotor with respect to the flux wave. Secondly, the strength or amplitude of the flux wave depends directly on the supply voltage to the stator windings, and inversely on the supply frequency. Thirdly, the absolute speed of the flux wave depends directly on the supply frequency.

Recalling that the motor can only operate efficiently when the slip is small, we see that the basic method of speed control rests on the control of the speed of rotation of the flux wave (i.e. the synchronous speed), by control of the supply frequency. If the motor is a 4-pole one, for example, the synchronous speed will be 1500 rev/min when supplied at 50 Hz, 1800 rev/min at 60 Hz, 750 rev/min at 25 Hz, and so on. The no-load speed will therefore be almost exactly proportional to the supply frequency, because the torque at no load is small and the corresponding slip is also very small.

Turning now to what happens on load, we know that when a load is applied the rotor slows down, the slip increases, more current is induced in the rotor, and more torque is produced. When the speed has reduced to the point where the motor torque equals the load torque, the speed becomes steady. We normally want the drop in speed with load to be as small as possible, not only to minimize the drop in speed with load, but also to maximize efficiency: in short, we want to minimize the slip for a given load.

We saw in Chapter 5 that the slip for a given torque depends on the amplitude of the rotating flux wave: the higher the flux, the smaller the slip needed for a given torque. It follows that having set the desired speed of rotation of the flux wave by controlling the output frequency of the inverter we must also ensure that the magnitude of the flux is adjusted so that it is at its full (rated) value,[1] regardless of the speed of rotation. This is achieved, in principle, by making the output voltage from the inverter vary in the appropriate way in relation to the frequency.

Given that the amplitude of the flux wave is proportional to the supply voltage and inversely proportional to the frequency, it follows that if we arrange that the voltage supplied by the inverter varies in direct proportion to the frequency, the flux wave will have a constant amplitude. This simple mode of operation – where the V/f ratio is constant – was for many years the basis of the control strategy applied to most inverter-fed induction motors, and it can still be found in some commercial products.

Many inverters are designed for direct connection to the utility supply, without a transformer, and as a result the maximum inverter output voltage is limited to a value similar to that of the supply system. Since the inverter will normally be used to supply a standard induction motor designed, for example, for 400 V, 50 Hz operation, it is obvious that when the inverter is set to deliver 50 Hz, the voltage

---

[1] In general, operating at rated flux gives the best performance and on most loads the highest efficiency. Some commercial drives offer a mode of control in which the flux is reduced typically with the square of the speed: this can provide some benefit at low speeds for fan and pump-type loads where the magnetizing current accounts for a significant proportion of the motor losses.

should be 400 V, which is within the inverter's voltage range. But when the frequency is raised to say 100 Hz, the voltage should – ideally – be increased to 800 V in order to obtain full flux. The inverter cannot supply voltages above 400 V, and it follows that in this case full flux can only be maintained up to the base speed. Established practice is for the inverter to be capable of maintaining the 'V/f ratio', or rather the flux, constant up to the base speed (frequently 50 or 60 Hz), but to accept that at higher frequencies the voltage will be constant at its maximum value. This means that the flux is maintained constant at speeds up to base speed, but beyond that the flux reduces inversely with frequency. Needless to say the performance above base speed is adversely affected, as we will see.

Users are sometimes alarmed to discover that both voltage and frequency change when a new speed is demanded. Particular concern is expressed when the voltage is seen to reduce when a lower speed is called for. Surely, it is argued, it can't be right to operate say a 400 V induction motor at anything less than 400 V. The fallacy in this view should now be apparent: the figure of 400 V is simply the correct voltage for the motor when run directly from the utility supply, at say 50 Hz. If this full voltage were to be applied when the frequency was reduced to say 25 Hz, the implication would be that the flux would rise to twice its rated value. This would greatly overload the magnetic circuit of the machine, giving rise to excessive saturation of the iron, an enormous magnetizing current, and wholly unacceptable iron and copper losses. To prevent this from happening, and keep the flux at its rated value, it is essential to reduce the voltage in proportion to frequency. In the case above, for example, the correct voltage at 25 Hz would be 200 V.

It is worth stressing here that when considering a motor to be fed from an inverter there is no longer any special significance about the utility network frequency, and the motor can be wound for almost any base frequency. For example, a motor wound for 400 V, 100 Hz could in the above example operate with constant flux right up to 100 Hz.

3. TORQUE–SPEED CHARACTERISTICS

When the voltage at each frequency is adjusted so that the ratio of voltage to frequency (V/f) is kept constant up to base speed, a family of torque–speed curves as shown in Figure 7.3 is obtained. These curves are typical for a standard induction motor of several kW output.

As expected, the no-load speeds are directly proportional to the frequency, and if the frequency is held constant, e.g. at 25 Hz in Figure 7.3, the speed drops only modestly from no-load (point a) to full-load (point b). These are therefore good, useful open-loop characteristics, because the speed is held fairly well from no-load to full-load. If the application calls for the speed to be held precisely, this can clearly be achieved by raising the frequency so that the full-load operating point moves to point (c).

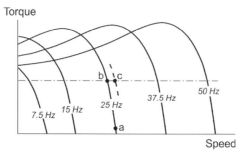

Figure 7.3 Torque–speed curves for inverter-fed induction motor with constant *V/f* ratio.

We note also that the pull-out torque and the torque stiffness (i.e. the slope of the torque–speed curve in the normal operating region) is more or less the same at all points below base speed, except at low frequencies where the voltage drop over the stator resistance becomes very significant as the applied voltage is reduced. A simple *V/f* control system would therefore suffer from significantly reduced flux and hence less torque at low speeds, as indicated in Figure 7.3.

The low-frequency performance can be improved by increasing the *V/f* ratio at low frequencies in order to restore full flux, a technique which is referred to as 'voltage boost'. In modern drive control schemes which calculate flux from a motor model (see section 8), the voltage is automatically boosted from the linear *V/f* characteristic that the approximate theory leads us to expect. A typical set of torque–speed curves for a drive with the improved low-speed torque characteristics obtained with voltage boost is shown in Figure 7.4.

The characteristics in Figure 7.4 have an obvious appeal because they indicate that the motor is capable of producing practically the same maximum torque at all speeds from zero up to the base (50 Hz) speed. This region of the characteristics is

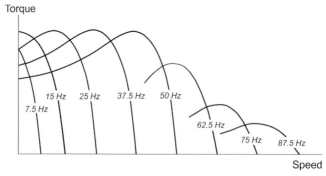

Figure 7.4 Typical torque–speed curves for inverter-fed induction motor with constant flux up to base speed (50 Hz) and constant voltage at higher frequencies.

known as the 'constant torque' region, which means that for frequencies up to base speed, the maximum possible torque which the motor can deliver is independent of the set speed. Continuous operation at peak torque will not be allowable because the motor will overheat, so an upper limit will be imposed by the controller, as discussed shortly. With this imposed limit, operation below base speed corresponds to the armature-voltage control region of a d.c. drive, as exemplified in Figure 3.9.

We should note that the availability of high torque at low speeds (especially at zero speed) means that we can avoid all the 'starting' problems associated with fixed-frequency operation (see Chapter 6). By starting off with a low frequency which is then gradually raised the slip speed of the rotor is always small, i.e. the rotor operates in the optimum condition for torque production all the time, thereby avoiding all the disadvantages of high slip (low torque and high current) that are associated with utility-frequency/direct-on-line (DOL) starting. This means that not only can the inverter-fed motor provide rated torque at low speeds, but − perhaps more importantly − it does so without drawing any more current from the utility supply than under full-load conditions, which means that we can safely operate from a weak supply without causing excessive voltage dips. For some essentially fixed-speed applications, the superior starting ability of the inverter-fed system alone may justify its cost.

Beyond the base frequency, the flux ('V/f ratio') reduces because V remains constant. The amplitude of the flux wave therefore reduces inversely with the frequency. Under constant flux operation, the pull-out torque always occurs at the same absolute value of slip, but in the constant-voltage region the peak torque reduces inversely with the square of the frequency and the torque–speed curve becomes less steep, as shown in Figure 7.4.

Although the curves in Figure 7.4 show what torque the motor can produce for each frequency and speed, they give no indication of whether continuous operation is possible at each point, yet this matter is of course extremely important from the user's viewpoint, and is discussed next.

3.1 Limitations imposed by the inverter − constant power and constant torque regions

A primary concern in the inverter is to limit the currents to a safe value as far as the main switching devices and the motor are concerned. The current limit will be typically set to the rated current of the motor, and the inverter control circuits will be arranged so that no matter what the user does the output current cannot exceed this safe (thermal) value, other than for clearly defined overload (e.g. 120% for 60 seconds) for which the motor and inverter will have been specified and rated. (For some applications involving a large number of starts and stops, the motor and drive may be specially designed for the specific duty.)

In modern control schemes (sections 8 and 9) it is possible to have independent control of the flux- and torque-producing components of the current, and in this way the current limit imposes an upper limit on the permissible torque. In the region below base speed, this will normally correspond to the rated torque of the motor, which is typically about half the pull-out torque, as indicated by the shaded region in Figure 7.5. Note that this is usually a thermal limit imposed by the motor design.

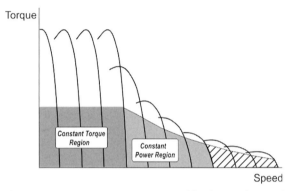

Figure 7.5 Constant torque, constant power and high-speed motoring regions.

Above base speed, it is not possible to increase the voltage and so the flux reduces inversely with the frequency. Since the stator (and therefore rotor) currents are again thermally limited (as we saw in the constant torque region), the maximum permissible torque also reduces inversely with the speed, as shown in Figure 7.5. This region is consequently known as the 'constant power' region. There is of course a close parallel with the d.c. drive here, both systems operating with reduced or weak field in the constant power region. In the constant power region, the flux is reduced and so the motor has to operate with higher slips than below base speed to develop the full (rated) rotor current and correspondingly reduced torque.

The voltage drop over the stator leakage inductance (see Appendix 2) increases with frequency. At typically twice base speed the extent of this voltage drop reduces the available voltage to such an extent that it is no longer possible for the motor to provide constant power operation, as indicated by the cross-hatched area in Figure 7.5.

3.2 Limitations imposed by the motor

The traditional practice in d.c. drives is to use a motor specifically designed for operation from a thyristor converter. The motor will have a laminated frame, will probably come complete with a tachogenerator, and – most important of all – will have been designed for through ventilation and equipped with an auxiliary air blower. Adequate ventilation is guaranteed at all speeds, and continuous operation with full torque (i.e. full current) at even the lowest speed is therefore in order.

By contrast, it is still common for inverter-fed systems to use a standard industrial induction motor. These motors are usually totally enclosed, with an external shaft-mounted fan which blows air over the finned outer case (and an internal stirring fan to circulate air inside the motor to avoid spot heating). They are designed first and foremost for continuous operation from the fixed frequency utility supply, and running at base speed.

As we have mentioned earlier, when such a motor is operated at a low frequency (e.g. 7.5 Hz), the speed is much lower than base speed and the efficiency of the cooling fan is greatly reduced. At the lower speed the motor will be able to produce as much torque as at base speed (see Figure 7.4) but in doing so the losses in both stator and rotor will also be more or less the same as at base speed, so it will overheat if operated for any length of time.

However, induction motors bearing the name 'inverter grade' or similar are readily available. As well as having reinforced insulation systems (see Chapter 8), they have been designed to offer a constant torque operating range below rated speed, typically down to 30% of base speed, without the need for an external cooling fan. In addition they may be offered with an external cooling fan to allow operation at constant (rated) torque down to standstill.

3.3 Four-quadrant capability

So far in this chapter it is natural that we have concentrated on motoring in quadrant 1 of the torque–speed plane (see Figure 3.12), because this is where the machine will spend most of its time running, but it is important to remind ourselves that the induction motor is equally at home as a generator, a role that it will frequently perform, even with an ordinary load, when a reduction in speed is called for. We should also recall that in this part of the chapter we are studying variable-frequency operation at the fundamental level, so we should bear in mind that in practice details of the control strategy will vary from drive to drive.

We can see how intermittent generation occurs with the aid of the torque–speed curves shown in Figure 7.6. These have been extended into quadrant 2, i.e. the negative-slip region, where the rotor speed is higher than synchronous, and a braking torque is exerted.

The family of curves indicates that for each set speed (i.e. each frequency) the speed remains reasonably constant because of the relatively steep torque–slip characteristic of the cage motor. If the load is increased beyond rated torque, an internal current limit comes into play to prevent the motor from reaching the unstable region beyond pull-out. Instead, the frequency and speed are reduced, and so the system behaves in a similar way to a d.c. drive.

Sudden changes in the speed reference are buffered so that the frequency is gradually increased or decreased. If the load inertia is low and/or the ramp time sufficiently long, the acceleration will be accomplished without the motor entering

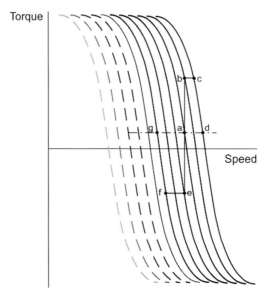

Figure 7.6 Acceleration and deceleration trajectories in the torque–speed plane.

the current-limit region. On the other hand, if the inertia is large and/or the ramp time was very short, the acceleration will take place as discussed below.

Suppose the motor is operating in the steady state with a constant load torque at point (a), when a new higher speed corresponding to point (d) is demanded. The frequency is increased, causing the motor torque to rise to point (b), where the current has reached the allowable limit. The rate of increase of frequency is then automatically reduced so that the motor accelerates under constant-current conditions to point (c), where the current falls below the limit: the frequency then remains constant and the trajectory follows the curve from (c) to settle finally at point (d).

A typical deceleration trajectory is shown by the path *aefg* in Figure 7.6. The torque is negative for much of the time, the motor operating in quadrant 2 and regenerating kinetic energy. Because we have assumed that the motor is supplied from an ideal voltage source, this excess energy will return to the supply automatically. In practice, however, we should note that many drives do not have the capability to return power to the a.c. supply, and the excess energy therefore has to be dissipated in a resistor inside the converter. (The resistor is usually connected across the d.c. link, and controlled by a chopper. When the level of the d.c. link voltage rises, because of the regenerated energy, the chopper switches the resistor on to absorb the energy. High inertia loads which are subjected to frequent deceleration can therefore pose problems of excessive power dissipation in this 'dump' resistor.)

To operate as a motor in quadrant 3 all that is required is for the phase sequence of the supply to be reversed, say from ABC to ACB. Unlike the utility-fed motor,

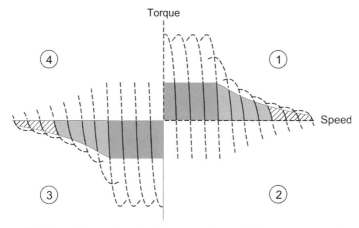

Figure 7.7 Operating regions in all four quadrants of the torque–speed plane.

there is no need to swap two of the power leads because the phase sequencing can be changed easily at the low-power logic level in the inverter. With reverse phase sequence, a mirror image set of 'motoring' characteristics is available, as shown in Figure 7.7. The shaded regions are as described for Figure 7.5, and the dashed lines indicate either short-term overload operation (quadrants 1 and 3) or regeneration during deceleration (quadrants 2 and 4).

Note that unlike the d.c. motor control strategies we examined in Chapter 4, neither the motor current, nor indeed any representation of torque, plays a role in the motor control strategy discussed so far (except when the current hits a limit, as discussed above).

4. INTRODUCTION TO FIELD-ORIENTED CONTROL

We now begin the second part of this chapter, which explores the contemporary approach to control of the inverter/induction motor combination. Field-oriented (or vector) control allows the induction motor/inverter combination to outperform conventional industrial d.c. drives, and its progressive refinement since the 1980s represents a major landmark in the history of electrical drives. It is therefore appropriate that its importance is properly reflected in this book, because one of our aims is to equip readers with sufficient understanding to allow them to converse intelligently with manufacturers and suppliers.

4.1 Outline of remainder of this chapter

Up to now, we have been able to cover topics without recourse to any very demanding mathematics, relying instead on physical explanations and diagrams, and our aim is to continue this approach. However, anyone who has consulted an article or textbook on the subject of field-oriented control will quickly become aware that most treatments involve extensive use of matrices and transform theory, and that many of the terms used will not be familiar to anyone not already schooled in the analysis of electrical machines. Fortunately, from our point of view, it is nevertheless possible to understand the underlying basis of field-oriented control via a relatively simple graphical approach, but even for this we have to make use of some ideas (such as flux linkage, and space phasors) that we have not discussed previously, so these are presented in the remainder of this section.

In section 5 we take a fresh look at the production of torque in induction motors, this time with the motor being supplied with controlled *currents* from a voltage-source inverter. We begin by taking a physical viewpoint that yields simple pictures that turn out to be accompanied by surprisingly simple formulae for torque. More importantly, the subsequent discussion in sections 6 and 7 makes clear what has to be done to achieve 'ideal' dynamic control of torque, something that was considered impossible until power electronics arrived.

The practical implementation of field-oriented control of torque is explored in section 8 through a detailed examination of the modus operandi of a typical sensorless control scheme. And finally, we look briefly at direct torque control, an alternative control strategy, in section 9.

In the remainder of this section we provide an introduction to some of the graphical and circuit-based techniques that we will need in order to understand torque control. The aim is to familiarize ourselves with the methodology and techniques that are used, after which we can sidestep the actual analysis and instead highlight the lessons that emerge from the torque-modeling exercise.

Readers who are comfortable with the distinction between transient and steady-state conditions and familiar with space phasors, transformation between reference frames, and the circuit modeling of electrical machines, may wish to skip this (inevitably rather long) treatment.

4.2 Transient and steady states in electric circuits

Field-oriented control allows us to obtain (almost) instantaneous (step) changes in torque on demand, and it does this by jumping directly from one steady-state condition to another. This simple statement is seldom given the prominence it deserves, but it is a simple truth, to be recalled whenever there is a danger of being bamboozled by a surfeit of technospeak.

Given the very poor inherent response of the induction motor to sudden changes in load or utility supply (see, for example, Figure 6.7), it will come as no surprise when we learn later that this sudden transition between steady states can only be achieved by precise control of the magnitude, frequency and instantaneous phase of the stator currents. As will emerge, the key requirement of a successful sudden transition is that it must not involve a step change in the stored energy of the system.

As an introduction to the underlying principle of changing from one steady state to another without any transient, we can look at the behavior of a series resistor and inductor circuit fed by an ideal voltage source (Figure 7.8). This is much simpler than the induction motor (it only has one energy storage element – the inductor) but it demonstrates the key requirement to be satisfied for transient-free switching.

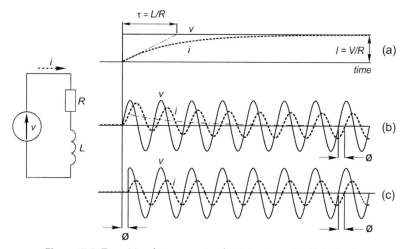

Figure 7.8 Transition between steady states in series R–L circuit.

First, we look at the current when the voltage is a step at $t = 0$ (Figure 7.8(a)). The steady-state current is simply V/R, but the current cannot rise instantaneously because that would require the energy stored in the inductor ($\frac{1}{2}Li^2$) to be supplied in zero time, which corresponds to an impulse of infinite power. So in addition to the steady-state term $i_{ss} = V/R$, there is a transient term given by $i_{tr} = -(V/R)e^{(-t/\tau)}$, where the time-constant, $\tau = L/R$. The total current is the sum of the steady-state and transient components, as shown by the lower dotted line in Figure 7.8(a).

Now consider a more relevant situation, where we wish the current to jump suddenly from a steady state at one frequency (in this case zero amplitude at zero frequency (i.e. d.c.) for $t < 0$) to a sinusoidal steady state for $t > 0$.

Figure 7.8(b) shows what happens if we make the sudden transition in the applied voltage (from zero d.c.) at a point where the new voltage waveform is zero but rising, i.e. at $t = 0$. We note that the current does not immediately assume its steady state, but displays the characteristic decaying transient, lasting for several cycles before the steady state is reached, with the current finally lagging the voltage by an angle ϕ. Examination of the steady-state current waveform shows that the current is negative as the voltage rises through zero, so this particular attempt to jump straight into the steady state is clearly doomed from the outset because it would have required the circuit to anticipate the arrival of the voltage by having a negative current already in existence!

The fundamental reason for the transient adjustment in Figure 7.8(b) is that we are seeking an instantaneous increase in the energy stored in the inductor from its initial value of zero, which is clearly impossible. It turns out that if we want to avoid the transient, we must make the jump without requiring a change in the stored energy, which in this example means at the point when the current passes through zero, as shown in Figure 7.8(c). The voltage that has to be applied therefore starts abruptly at a value $V \sin \phi$, as shown, and the current immediately enters its steady state, with no transient term.

We will see later that the principle of not disturbing the stored energy is the key requirement for obtaining step changes in torque from an induction motor.

4.3 Space phasor representation of m.m.f. waves

The space phasor (or space vector) provides a shorthand graphical way of representing sinusoidally distributed spatial quantities such as the m.m.f. and flux waves that we explored in Chapter 5. It avoids us having to consider individual currents by focusing on their combined effects, and thus makes things easier to understand.

We begin by taking a fresh look at the rotating stator m.m.f., making the reasonable assumption that each of the 3-phase windings produces a sinusoidally distributed m.m.f. with respect to distance around the air–gap, which in turn implies that the winding itself is sinusoidally distributed (rather than sitting in clearly defined groups of coils as in the real machine discussed previously). For convenience, we will consider a 2-pole winding.

We can represent the relative position of the windings *in space* as shown in Figure 7.9. In Figure 7.9(a) phases B and C are on open-circuit so that we can focus on how the m.m.f. of phase A is represented. When the current in phase A is positive (i.e. current flows into the dotted end), we have chosen to represent its sinusoidal m.m.f. pattern by a vector along the axis of the winding and pointing away from it (Figure 7.9(a)), and so when the current is negative the vector points towards the coil (Figure 7.9(b)). The length of the vector is directly proportional to the instantaneous value of the current, as indicated by the relative sizes of the two vectors.

In Figure 7.9(c), phase A has its maximum positive current, while phases B and C both have negative currents of half the maximum value. Because each m.m.f. is distributed sinusoidally in space, we can find their resultant (R) using the approach that is probably more familiar in the context of a.c. circuits, i.e. by adding the three components vectorially. In this particular example, the resultant m.m.f. (Figure 7.9(d)) is co-phasal with the m.m.f. of phase A, but one and half times larger.

We can now use the approach outlined above to find the resultant m.m.f. when the windings are supplied with balanced 3-phase currents of equal amplitude but

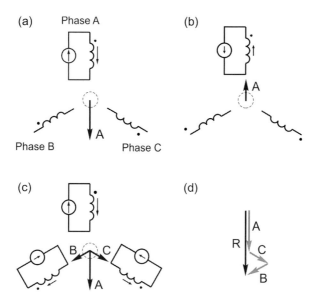

Figure 7.9 Space phasor representation of m.m.f. waves.

displaced in time by one-third of a cycle (i.e. 120°). The axes of the phases are displaced in *space* as shown in Figure 7.9, and the three currents are shown as functions of *time* in the upper part of Figure 7.10. Four consecutive times are identified, separated by one-twelfth of a complete cycle, or 30° in angle terms.

The lower part of the diagram represents the m.m.f.s in a space phasor diagram. At each instant the three individual phase m.m.f.s are shown in magnitude and position, together with the resultant m.m.f. At time t_0, for example, the situation is the same as in Figure 7.9, with phase A at maximum positive current and phases B and C having equal negative currents of half the magnitude of that in phase A; at t_1 phase B is zero while phases A and C have equal but opposite currents; and so on.

The four sketches suggest that the resultant m.m.f. describes an arc of constant radius, and it can easily be shown analytically that this is true. So although each phase produces a pulsating m.m.f. along its axis, the overall m.m.f. is constant in amplitude (with a value equal to 1.5 times the phase peak), and it rotates at a uniform rate, completing one revolution per cycle if the field is 2-pole (as here), half a revolution if 4-pole, etc. This is in line with our findings in Chapter 5.

We should note that although we have developed the idea of space phasors by focusing on steady-state sinusoidal operation, the approach is equally valid for any set of instantaneous currents, and is therefore applicable under transient conditions, for example during acceleration when the instantaneous frequency of the currents may change continuously.

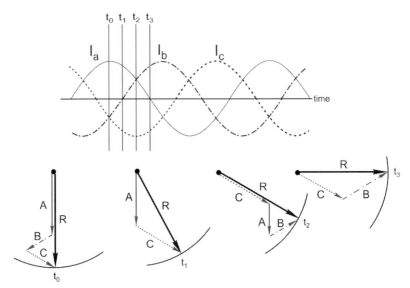

Figure 7.10 Resultant m.m.f. space phasor for balanced 3-phase operation at four discrete times, each separated by one-twelfth of a cycle (i.e. 30°).

An alternative way of representing the *resultant m.m.f. pattern* produced by a set of balanced 3-phase windings follows naturally from the discussion above. We imagine a hypothetical *single m.m.f. vector* that has a constant magnitude but *rotates relative to the stator* at the synchronous speed. This turns out to be an exceptionally useful mental picture when we come to study the behavior of the inverter-fed induction motor, because the currents are then under our control and we are able to specify precisely the magnitude, speed and angular position of the stator m.m.f. vector in order to achieve precise control of torque.

4.4 Transformation of reference frames

In the previous section we saw that the resultant m.m.f. was of constant amplitude and rotated at a constant angular velocity with respect to a reference frame fixed to the stator. As far as an observer in the stationary reference frame is concerned, the same m.m.f. could equally well be produced by a set of sinusoidally distributed windings fed with constant (d.c.) current and mounted on a structure that rotated at the same angular velocity as the actual m.m.f. wave. On the other hand, if we were attached to a reference frame rotating with the m.m.f., the space phasor would clearly appear to us to be constant.

Transformations between reference frames have long been used to simplify the analysis of electrical machines, especially under dynamic conditions, but until fast signal-processing became available that approach was seldom used for live control

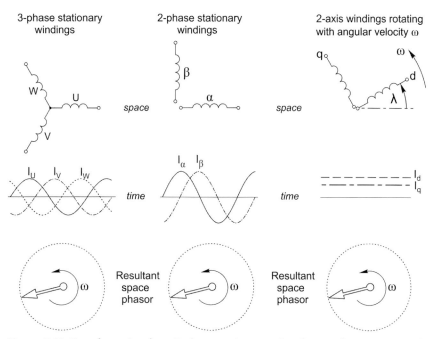

Figure 7.11 Transformation from 3-phase stationary axis reference frame to two-axis (d–q) rotating reference frame.

purposes. We will see later in this chapter that the method is used in field–oriented control schemes to transform the stator currents into a rotating reference frame locked to the rotating rotor flux space phasor, thereby making them amenable for control purposes.

Transformation is usually accomplished in two stages, as shown in Figure 7.11.

The first stage involves replacing the three windings by two in quadrature, with balanced sinusoidal currents of the same frequency but having a 90° phase shift. In this case the '$\alpha\beta$' stationary reference frame has phase α aligned with phase U. To produce the same m.m.f., the two windings need either more turns or more current, or a combination of both. This is known as the Clarke transformation.

The second stage (the Park transformation) is more radical as the new variables I_d and I_q are in a rotating reference frame, and they remain constant under steady-state conditions, as shown in Figure 7.11. Again we need to specify the turns ratio and/or the current scaling. (Strictly speaking there is no need for the intermediate (2-phase) transformation, because we can transform directly from 3-phase to two-axis, but we have included it because it is often mentioned in the literature.)

It should be clear that the magnitude of the currents I_d and I_q will depend on the angle λ, which is the angle between the two reference frames at a specified instant, typically at $t = 0$. As far as we are concerned, it is sufficient to note that there are well-established formulae relating the input and output variables, both for the forward transformation (U, V, W to d, q) and for the inverse transformation, so it is straightforward to construct algorithms to perform the transformations. We should also note that while we have considered the transformation of sinusoidal currents, the technique is equally valid for instantaneous values.

4.5 Circuit modeling of the induction motor

Up to now in this book we have developed our understanding by starting with a physical picture of the interactions between the magnetic field and current-carrying conductors, but we quickly realized that in the case of the d.c. machine (and the utility-fed induction motor) there was a lot to be gained by making use of an 'equivalent circuit' model, particularly in terms of performance prediction. In so doing we were representing all the distributed interactions of the motor by way of their ultimate effect as manifested at the electrical terminals and the mechanical 'terminal', i.e. the output shaft.

As long ago as the early nineteenth century it was known that the a.c. transformer could be analyzed as a pair of magnetically linked coils, and it did not take long to show that all of the important types of a.c. electrical machine can also be analyzed by regarding them as a set of circuits, the electrical parameters (resistance, inductance) of which were either measured or calculated. The vital difference compared with the static transformer is that in the machine, the coils on the rotor move with respect to those on the stator, thereby causing a variation in the extent of the magnetic interaction between the rotor and stator. This variation turns out to be the essential requirement for the machine to produce torque and to be capable of energy conversion.

4.6 Coupled circuits, induced e.m.f. and flux linkage

By 'coupled circuits' we mean two or more circuits, often in the form of multi-turn coils sharing a magnetic circuit, where the magnetic flux produced by the current in one coil not only links with its own winding, but also with those of the other coils. The coupling medium is the magnetic field, and as we will see the electrical effect of the coupling is manifested when the flux changes.

We know from Faraday's law that when the magnetic flux (ϕ) linking a coil changes, an e.m.f. (e) is induced in the coil, given by

$$e = -N \frac{d\phi}{dt},$$

i.e. the e.m.f. is proportional to the rate of change of the flux. (The minus sign indicates that if the induced e.m.f. is allowed to drive a current, the m.m.f. produced by the current will be in opposition to that which produced the original changing flux.) This equation only applies if all the flux links all N turns of the coil, the situation most commonly approached in transformer windings that share a common magnetic circuit, and are thus very tightly coupled.

We have seen that windings for induction motors are distributed, and the flux wave produced by the current in the winding is also distributed around the air–gap. As a result not all of the flux produced by one winding links with all of its turns, and we have to perform a summation (integration) of all the 'turns times flux that links them' contributions to find the 'total effective self flux linkage', which we denote by the symbol psi (ψ). The e.m.f. induced when the self-produced flux linkage changes in, say, a stator winding (subscript S) is then given by

$$e_s = \frac{d\psi_S}{dt}$$

In an induction motor there are three distributed windings on the stator, and either a cage or three more distributed windings on the rotor, and some of the flux produced by current in any one of the windings will link all of the others. We term this 'mutual flux linkage', and often denote it by a double suffix: for example, the symbol ψ_{SR} is the mutual flux linkage between a stator winding and a rotor winding.

In the same way that an e.m.f. is induced in a winding when its self-produced flux changes, so also are e.m.f.s induced in all other windings that are mutually coupled to it. For example, if the flux produced by the stator winding changes, the e.m.f. in the rotor (subscript R) is given by

$$e_R = \frac{d\psi_{SR}}{dt}$$

4.7 Self and mutual inductance

The self and mutual flux linkages produced by a winding are proportional to the current in the winding. The ratio of flux linkage to the current that produces it is therefore a constant, and is defined as the inductance of the winding. The self inductance (L) is given by

$$L = \frac{\text{Self flux linkage}}{\text{Current}} = \frac{\psi_S}{i_S},$$

while the mutual inductance (M) is defined as

$$M_{SR} = \frac{\text{Mutual flux linkage}}{\text{Current}} = \frac{\psi_{SR}}{i_S}$$

The self and mutual inductances therefore depend on the design of the magnetic circuit and the layout of the windings.

We can now recast the expressions for e.m.f. derived above so that they involve the rates of change of the currents and the inductances, rather than the fluxes. This is a very important simplification because it allows us to represent the distributed effects of the magnetic coupling in single lumped–parameter electric circuit terms. The self-induced and mutually induced e.m.f.s are now given by

$$e_S = L\frac{di_S}{dt},$$

and

$$e_R = M_{SR}\frac{di_S}{dt}$$

4.8 Obtaining torque from a circuit model

We represent the two sets of 3-phase distributed windings of the induction motor by means of the six fictitious 'equivalent' coils shown in Figure 7.12. (We are using the well-proven fact that a cage rotor behaves in essentially the same way as one with a wound rotor, as explained in Chapter 5.) The three stator coils remain stationary, while those on the rotor obviously move when the angle θ changes.

Because the air-gap is smooth, and the rotor is assumed to be magnetically homogeneous, all the self inductances are independent of the rotor position, as are the mutual inductances between pairs of stator coils and between pairs of rotor coils. Symmetry also means that the mutuals between any two stator or rotor phases are the same.

However, it is obvious that the mutual inductance between a stator and a rotor winding will vary with the position of the rotor: when stator and rotor windings are aligned, the flux linkage will be maximum, and when they are positioned at right angles, the flux linkage will be zero. With windings that are distributed so as to produce sinusoidal flux waves, the mutual inductances vary sinusoidally with the angle θ.

To a circuit theory practitioner, it is this variation of mutual inductance with position that immediately signals that torque production is possible. In fact it is straightforward (if somewhat intellectually demanding) to show that the torque

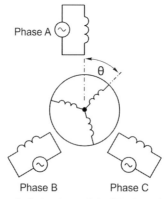

Figure 7.12 Coupled-circuit model of 3-phase induction motor.

produced when the sets of windings in Figure 7.12 carry currents is given by the rather fearsome-looking expression

$$T = \sum i_S i_R \frac{\mathrm{d}M_{SR}}{\mathrm{d}\theta}$$

What this means is that to find the total torque we have to find the summation of nine terms, each of which represents a contribution to the total torque from one of the nine stator–rotor pairs. So we need the instantaneous value of each of the six currents, and the rate of change of inductance with rotor position for each stator–rotor pair. For example, the term representing the contribution to torque made by stator coil A interacting with rotor coil B is given by

$$T_{SARB} = i_{SA} i_{RB} \frac{\mathrm{d}M_{SARB}}{\mathrm{d}\theta}$$

In practice we can use various expedients to simplify the torque expression, for example we know that mutual inductance is a reciprocal property, i.e. $M_{AB} = M_{BA}$, and we can exploit symmetry, but the important thing to note here is that it is a straightforward business to find the torque from the circuit model, provided that we know the currents and the inductances.

4.9 Finding the rotor currents

The rotor currents are induced, and to find them we have to solve the set of six equations relating them to the applied stator voltages, using Kirchhoff's voltage law.

So, for example, the voltage equation below relating to rotor phase A includes a term representing the resistive volt-drop, another representing the self-induced e.m.f. and five others representing the mutual coupling with the other windings. There are two more rotor equations and three similar ones for the stator windings.

$$v_{RA} = i_{RA} R_R + L_{RA} \frac{\mathrm{d}i_{RA}}{\mathrm{d}t} + M_{RARB} \frac{\mathrm{d}i_{RB}}{\mathrm{d}t} + M_{RARC} \frac{\mathrm{d}i_{RC}}{\mathrm{d}t} + M_{RASA} \frac{\mathrm{d}i_{SA}}{\mathrm{d}t}$$
$$+ M_{RASB} \frac{\mathrm{d}i_{SB}}{\mathrm{d}t} + M_{RASC} \frac{\mathrm{d}i_{SC}}{\mathrm{d}t}$$

In the induction motor the rotor windings are usually short-circuited, so there is no applied voltage and the left-hand side of each rotor equation is zero.

If we have to solve these six simultaneous differential equations when the stator terminal *voltages* are specified (typical of utility-fed constant-frequency conditions), we have a very challenging task that demands computer assistance, even under steady-state conditions. However, when the stator *currents* are specified (as we will see is the norm in an inverter-fed motor under vector control), the equations can be solved much more readily. Indeed under steady-state locked rotor conditions we can employ an armory of techniques such as j notation and phasor diagrams to solve the equations by hand.

We have now seen in principle how to predict the torque, and how to solve for the rotor currents, when the stator currents are specified. So we are now in a position to see what can be learned from a study of the known outcomes under two specific conditions.

In the next section, we look at how the torque varies when the stator windings are fed with a balanced set of 3-phase a.c. currents of constant amplitude but variable frequency, and the rotor is stationary. Although this is not of practical importance, it is very illuminating, and it points the way to the second and much more significant mode of operation, in which the net rotor flux is kept constant at all frequencies; this is dealt with in section 6.

5. STEADY-STATE TORQUE UNDER CURRENT-FED CONDITIONS

Historically there was little interest in analysis under current-fed conditions because we had no means of direct control over the stator currents, but the inverter-fed drive allows the stator currents to be forced very rapidly to whatever value we want, regardless of the induced e.m.f.s in the windings. Fortunately, we will see that knowing the currents from the outset makes quantifying the torque very much easier, and it also allows us to derive simple quantitative expressions that indicate how the machine should be controlled to achieve precise torque control, even under dynamic conditions.

To simplify our mental picture we will begin with the rotor at rest, and we will assume that we have a wound rotor with balanced 3-phase windings that for the moment are open-circuited, i.e. that no current can flow in them. With balanced 3-phase stator currents of amplitude I_s we know from the discussion above that the traveling stator m.m.f. wave can be represented by a single space phasor that rotates at the synchronous speed, and that in the absence of any currents in the rotor (and neglecting saturation of the iron) the flux wave will be in phase with the m.m.f. and proportional to it. This is shown Figure 7.13(a): in this sketch the rotor and stator are stationary, but all the patterns rotate at synchronous speed.

On the left of Figure 7.13(a) is a graphical representation of the sinusoidal distribution of resultant current around the stator at a given instant, and the corresponding flux pattern (dashed lines). Note that there is no rotor current. On the right of Figure 7.13(a) is a phasor that can represent both the stator m.m.f. and what we will call the resultant mutual flux linkage, both of which are proportional to the stator current. The expression 'mutual flux linkage' in Figure 7.13(a) thus represents the total effective flux linkages with the rotor due to the stator traveling flux wave. In circuit terms, this flux linkage is proportional to the mutual inductance between the stator and rotor windings (M), and to the stator current (I_S), i.e. MI_S.

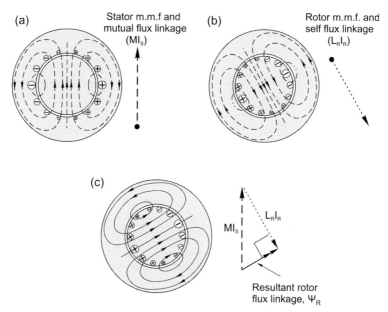

Figure 7.13 Space phasors of m.m.f. and flux linkage under locked-rotor conditions.

Now we short-circuit the rotor windings, and solve the set of equations for the rotor currents in the steady state. In view of the symmetry it comes as no surprise to find that they also form a balanced 3-phase set, at the same frequency as those of the stator, but displaced in time-phase. The resultant pattern of currents in the rotor is shown on the left of Figure 7.13(b), together with the flux pattern (dashed lines) that they would set up if they acted alone. Note that the stator currents that are responsible for inducing the rotor currents have been deliberately suppressed in Figure 7.13(b), because we want to highlight the rotor's reaction separately.

The m.m.f. due to the rotor currents is represented by the phasor shown on the right of Figure 7.13(b), and again this can also serve to represent the rotor self flux linkages $(L_R I_R)$ attributable to the induced currents. It is clear that the time phase shift between stator and rotor currents causes a space phase shift between stator and rotor m.m.f.s, with the rotor m.m.f. broadly tending to oppose the stator m.m.f. If the rotor had zero resistance, the rotor m.m.f. would directly oppose that of the stator. The finite rotor resistance displaces the angle, as shown in Figure 7.13(b). We will see shortly that this phase angle varies widely and is determined by the frequency.

To find the resultant m.m.f. acting on the rotor we simply add the stator and rotor m.m.f. vectors, as shown in Figure 7.13(c). It is this m.m.f. that produces the resultant flux at the rotor, and we can therefore also use it to represent the net rotor flux linkage (ψ_R). The flux pattern at the rotor is shown by the solid lines in Figure 7.13(c, left). (But we should note that the number of flux lines shown in Figure 7.13(a–c) is not intended to reflect the relative magnitudes of flux densities, which, if saturation was not present, would be higher in the two upper sketches.) We should also note that, as expected, the behavior is independent of the rotor position angle θ, because the rotor symmetry means that when viewed from the stator, the rotor always looks the same overall.

Close examination of Figure 7.13(c) reveals an extremely important fact. The resultant rotor flux vector (ψ_R) is perpendicular to the rotor current vector. This means that the rotor current wave (shown on the left of Figure 7.13(c)) is oriented in the ideal position in space to maximize the torque production, because the largest current is coincident with the maximum flux density. If we look back to Figures 3.1 and 3.2, we will see that this is exactly how the flux and current are disposed in the d.c. machine, the N pole facing the positive currents and the S pole opposite the negative currents.

When we evaluate the torque under these conditions, a very simple analytical result emerges: the torque turns out to be given by the product of the rotor flux linkage and the rotor current, i.e.

$$T = \psi_R I_R$$

The similarity of this expression to the torque expression for a d.c. machine is self-evident, and further underlines the fundamental unity of machines exploiting the 'BIl' mechanism discussed in Chapter 1. We note also that in Figure 7.13(c, right), one side of the right-angle triangle is ψ_R while the other is proportional to the rotor current I_R. Hence the area of the triangle is proportional to the torque, which makes it easy to visualize how torque varies with frequency, which we look at shortly.

(The keen reader may recall that the mental pictures we employed in Chapter 5 were based on the calculation of torque from the product of the air-gap flux wave and the rotor current wave, and that these were not in phase, except at very low slip frequency. The much simpler picture which has now been revealed – in which the flux and current waves are always ideally disposed as far as torque production is concerned – arises because we have chosen to focus on the resultant rotor flux linkage, not the air-gap flux: we are discussing the same mechanism as in Chapter 5, but the new viewpoint has thrown up a much simpler picture of torque production.)

We will see later that the rotor flux linkage is the central player in the field-oriented or vector control methods that now dominate in inverter-fed drives. To make full use of the flux-carrying capacity of the rotor iron, and to achieve step changes in torque, we will keep the amplitude of ψ_R constant, and we will explore this shortly. But first we will look at how the torque depends on slip when the amplitude of the stator current is kept constant.

5.1 Torque vs slip frequency – constant stator current

An alert reader might question why the title of this section includes reference to slip frequency, when we have specified so far that the rotor is stationary, in which case the effective slip is 1 and the frequency induced in the rotor will always be the same as the stator frequency. The reason for referring to slip frequency is that, as far as the reaction of the rotor is concerned, the only thing that matters is the relative speed of the traveling stator field with respect to the rotor.

So if we study the static model with an induced rotor frequency of 2 Hz, the torque that we predict can represent locked rotor conditions with 2 Hz on the stator; or the rotor running with a slip of 0.1 with 20 Hz on the stator; or a slip of 0.04 with 50 Hz on the stator; and so on. In short, under current-fed conditions, our model correctly predicts the rotor behavior, including the torque, when we supply the stator windings at the slip frequency. (Note that all other aspects of behavior on the stator side are not represented in this model.)

The variation in the flux linkage triangle with slip frequency, assuming that the amplitude of the stator current is constant, is shown in Figure 7.14. The locus of the resultant rotor flux linkage as the slip is varied is shown by the semi-circles. The right-hand side relates to low values of slip frequency, where the rotor self flux linkage is much less than the stator mutual flux linkage, so the resultant rotor flux is not much less than when the slip is zero. In other words, at low slips the presence of the rotor currents has little effect on the magnitude of the resultant flux, as we saw in Chapter 5. Low-slip operation is the norm in controlled drives. The left-hand drawing relates to high values of slip, where the large induced currents in the rotor lead to a rotor m.m.f. that is almost able to wipe out the stator m.m.f., leaving a very small resultant flux in the rotor. We will not be concerned with this end of the diagram in an inverter drive.

There is a simple formula for the angle ϕ, which is given by

$$\tan \phi = \omega_s \tau \qquad (7.1)$$

where $\tau = L_R/R_R$, the rotor time-constant.

We noted earlier that the torque is proportional to the area of the triangle, so it should be clear that the peak torque is reached when the slip increases from the

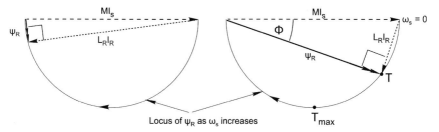

Figure 7.14 Locus of rotor flux linkage space phasor as slip frequency varies.

point T and moves to $T_{\max}$. At this point, $\phi = \pi/4$ and the slip frequency is given by $\omega_S = 1/\tau = R_R/L_R$. Under these constant-current conditions, the slip at which maximum torque occurs is much less than under constant-voltage conditions, because the rotor self inductance is much larger than the rotor leakage inductance.

6. TORQUE VS SLIP FREQUENCY – CONSTANT ROTOR FLUX LINKAGE

As already mentioned, it is clear that to make full use of the flux-carrying capacity of the rotor iron, we will want to keep the amplitude of the rotor flux ψ_R constant. Given that the majority of the rotor flux links the stator (see Figure 7.13(c)), keeping the rotor flux constant also means that for most operating conditions, the stator flux is also more or less constant, as we assumed in Chapter 5.

In this section we explore how steady-state torque varies with slip when the rotor flux is maintained constant: this is illuminating, but much more importantly it prepares us for the final section, which deals with how we are able to achieve precise control of torque even under dynamic conditions.

We can see from Figure 7.14 that to keep the rotor flux constant we will have to increase the stator current with slip. This is illustrated graphically in Figure 7.15, in which the rotor flux linkage ψ_R is shown vertically to make it easier to see that it remains constant. In the left-hand diagram, the slip is very small, so the induced rotor current and the torque (which is proportional to the area of the triangle) are both small. The rotor flux is more or less in phase with the applied stator flux linkage because the 'opposing' influence of the rotor m.m.f. is small.

In the middle and right-hand diagrams the slip is progressively higher, so the induced rotor current is larger and the stator current has to increase in order to keep the rotor flux constant.

There is a simple analytical relationship that gives the required value of stator current as a function of slip, but of more interest is the relationship between the

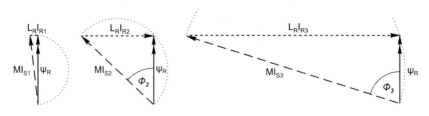

Lower slip frequency   Higher slip frequency

Figure 7.15 Constant rotor flux linkage space phasors at low, medium and high values of slip, showing variation of stator current required to keep rotor flux constant.

induced rotor current and the slip. From Figure 7.15, we can see that the tangent of the angle ϕ is given by

$$\tan \phi = \frac{L_R I_R}{\psi_R}$$

Combining this with equation (7.1) we find that the rotor current is given by

$$I_R = \left(\frac{\psi_R}{R_R}\right) \omega_{slip} \tag{7.2}$$

The bracketed term is constant; therefore the rotor current is directly proportional to the slip. Hence the horizontal sides of the triangles in Figure 7.15 are proportional to slip, and since the vertical side is constant, the area of each triangle (and torque) is also proportional to slip. To emphasize this simple relationship, the right-hand diagram in Figure 7.15 has been drawn to correspond to a slip three times higher than that of the middle one, so the horizontal side of the right-hand sketch is three times as long, and the area of the triangle (and torque) is trebled.

We note that when the rotor flux is maintained constant, the torque–speed curve becomes identical to that of the d.c. motor. In this respect the behavior differs markedly from that under both constant-voltage and constant-current conditions, where a peak or pull-out torque is reached at some value of slip. With constant rotor flux there is no theoretical limit to the torque, but in practice the maximum will be governed by thermal limits on the rotor and stator currents.

For those who prefer the physical viewpoint it is worth noting that the results discussed in this section could have been deduced directly from Figure 7.13(c), which indicates that the resultant rotor flux and rotor current waves are always aligned (i.e. the peak flux density coincides with the peak current density) so that if the flux is held constant, the torque is proportional to the rotor current. The rotor current is proportional to the motionally induced e.m.f., which in turn depends on the velocity of the flux wave relative to the rotor, i.e. the slip speed.

6.1 Flux and torque components of stator current

If we resolve the stator flux-linkage phasor MI_s into its components parallel and perpendicular to the rotor flux, the significance of the terms 'flux component' and 'torque component' of the stator current becomes obvious (Figure 7.16).

We can view the 'flux' component as being responsible for setting up the rotor flux, and this is the component that we must keep constant in order to maintain the working flux of the machine at a constant value for all slips. It is clearly analogous to the field current that sets up the flux in a d.c. motor.

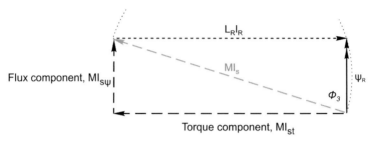

Figure 7.16 Flux and torque components of stator current.

The other ('torque') component (which is proportional to the rotor current) can be thought of as being responsible for nullifying the opposing effect of the rotor current that results when the rotor conductors are 'cut' by the traveling flux wave. This current component is therefore seen as the counterpart of the armature or work current in the d.c. motor.

Looking back to the left-hand diagram in Figure 7.15, we see that at small slips (light load) the stator current is small and practically in phase with the flux; this is what we referred to as the magnetizing current in previous chapters. At higher slips, the stator current is larger, reflecting that it now has a torque or 'work' component in addition to its magnetizing component, which again accords with our findings in previous chapters.

6.2 Establishing the flux

In the previous discussion we assumed that steady-state conditions prevailed, with the rotor flux wave remaining of constant magnitude and rotating relative to the rotor at the slip speed. We now look at how the flux wave was established, and we will see that the reaction of the rotor is very different from its subsequent steady-state behavior.

We start with the rotor at rest, no current in any of the windings, and hence no flux. With reference to Figure 7.9, we suppose that we supply a step (d.c.) current into phase A, which will split with half exiting from each of phases B and C, and producing a stationary sinusoidally distributed m.m.f. that, ultimately, will produce the flux pattern labeled 'final state' in Figure 7.17.

But of course the rotor windings are short-circuited, with no flux through them, and closed electrical circuits behave like many things in the physical world in that they react to change by opposing it. In this context, if the flux linking a winding changes, Faraday's law tells us there will be an induced e.m.f. The direction of the e.m.f. is such that if it acts in a closed circuit and produces a current, the m.m.f. produced by that current will be in opposition to the 'incoming' m.m.f./flux. (Formally this is expressed by Lenz's law, and sometimes by the use of a negative sign preceding the e.m.f. equation that quantifies Faraday's law.)

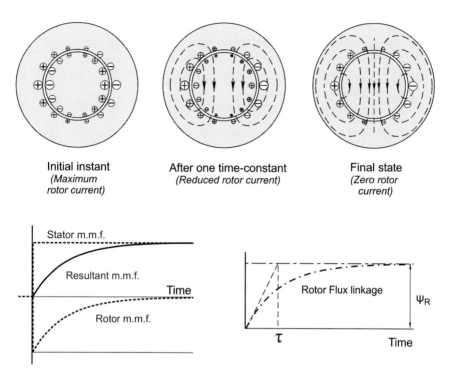

Figure 7.17 Diagrams illustrating the build-up of rotor flux when a step of stator m.m.f. occurs.

So when the stator m.m.f. phasor suddenly comes into existence, the immediate reaction of the rotor is the production of a negative stationary rotor m.m.f. pattern, i.e. in direct opposition to the stator m.m.f.: this is labeled 'initial instant' in Figure 7.17. Instantaneously, the magnitude of the rotor m.m.f. is such as to keep the rotor flux linkage at zero, as it was previously.

However, because of the rotor resistance, the rotor current needs a voltage to sustain it, and the voltage can only be induced if the flux changes. So the rotor flux begins to increase, rising rapidly at first (high e.m.f.) then with ever-decreasing gradient leading to lower current and lower rotor m.m.f. The response is a first-order one, governed by the rotor time-constant, so after one time-constant (middle sketch) the flux linking the rotor reaches about 63% of its final value, while the rotor current has fallen to 37% of its initial value. Finally, the rotor's struggle to prevent the flux changing comes to an end and the rotor flux linkage reaches a steady value determined by the stator current. If the resultant rotor flux linkage is the target value for steady-state running (ψ_R), the corresponding stator current is what we previously referred to as the 'flux component'.

The physical reason why it takes time to build the flux is that energy is stored in a magnetic field, so we cannot suddenly produce a field because that would require an impulse of infinite power. If we want to build up the flux more rapidly, we can put in a bigger step of stator current at first, so that the flux heads for a higher final value than we really need, and then reduce the current when we get close to the flux we are seeking.

We began this section with d.c. current in the stator, which in effect corresponds to zero slip frequency, all the field patterns being stationary in space. Because there is no relative motion involved, there is no motional e.m.f. and hence no torque. The 'torque component' only comes into play when there is relative motion between the rotor and the rotor flux wave, i.e. when there is slip. Obviously, to cause rotation the frequency must be increased, and as we have seen in the previous section the stator current then has to be adjusted with slip and torque to keep the rotor flux linkage constant.

Finally, it is worth revisiting Figure 7.16 briefly to reconcile what we have discussed in this section with our picture of steady-state operation, where the rotor currents are at slip frequency. On the left we have the fictitious 'flux component' of stator current, which remains constant in magnitude and aligned with the rotor flux linkage, ψ_R, along the so-called direct axis. When we first established this flux, the rotor reacted as we have discussed above, but after a few time-constants the flux settled to a constant value along the direct axis in the direction of the flux component of the stator m.m.f. This is why the arrows on ψ_R point in the same direction as the stator flux producing component.

However, we note from Figure 7.16 that the rotor flux linkage phasor $(L_R I_R)$ is always equal in magnitude to the torque component of the stator mutual flux linkage phasor, but, as shown by the arrows, it is in the opposite direction. There is therefore no resultant m.m.f. or tendency for flux to develop along this, the so-called quadrature axis. This is what we would expect in the light of the previous discussion, where we saw that the reaction of mutually coupled windings to any suggestion of change is for currents to spring up so as to oppose the change. In the literature when, as here, the 'torque' current does not affect the flux, the axes are said to be 'decoupled'.

7. DYNAMIC TORQUE CONTROL

If we want to obtain a step increase in torque, we have to change the rotor flux or the rotor current instantaneously, so as to jump instantaneously from one steady-state operating condition to another. We have seen above that because a magnetic field has stored energy associated with it, it is not possible to change the rotor flux linkage instantaneously. In the case of the induction motor, any change in the rotor flux is governed by the rotor time-constant, which will be as much as 0.25 seconds for even a modest motor of a few kW rating, and much longer for large motors. This is not acceptable when we are seeking instantaneous changes in torque.

The alternative is to keep the flux constant, and make the rotor current change as quickly as possible.

In the previous section, our aim was to grow the rotor flux, which, because of its stored energy, took a while to reach the steady state. However, if we keep the rotor flux linkage constant (by ensuring that the flux component of the stator current is constant and aligned with the flux) we can cause sudden changes to the motionally induced rotor current by making sudden changes in the torque component of the stator current.

We achieve sudden step changes in the stator currents by means of a fast-acting closed-loop current controller. Fortunately, under transient conditions the effective inductance looking in at the stator is quite small (it is equal to the leakage inductance), so it is possible to obtain very rapid changes in the stator currents by applying high, short-duration impulsive voltages to the stator windings. In this respect the stator current controller closely resembles the armature current controller used in the d.c. drive.

When a step change in torque is required the magnitude, frequency, and phase of the stator currents are changed (almost) instantaneously in such a way that the rotor current jumps suddenly from one steady state to another. But in this transition it is only the torque component of stator current that is changed, leaving the flux component aligned with the rotor flux. There is therefore no change in the magnitude of the rotor flux wave and no change in the stored energy in the field, so the change can be accomplished almost instantaneously.

We can picture what happens by asking what we would see if we were able to observe the stator m.m.f. wave at the instant that a step increase in torque was demanded. For the sake of simplicity, we will assume that the rotor speed remains constant, and consider an increase in torque by a factor of three (as between the middle and right-hand sketches in Figure 7.15), in which case we would find that:

- the stator m.m.f. wave suddenly increases its amplitude;
- it suddenly accelerates to a new synchronous speed, so the slip increases by a factor of three;
- it jumps forward to retain its correct relative phase with respect to the rotor flux; i.e. the angle between the stator m.m.f. and the rotor flux increases from ϕ_2 to ϕ_3.

Thereafter the stator m.m.f. retains its new amplitude, and rotates at its new speed.

The rotor current experiences a step change from a steady state at its initial slip frequency to a new steady state with three times the amplitude and frequency, and there is a step increase in torque by a factor of three, as shown in Figure 7.18(a). The new current is maintained by the new (higher) stator currents and slip frequency.

We should note particularly that it is the jump in the *angular position* (i.e. *space phase angle*) that accompanies the step changes in magnitude and frequency of the stator m.m.f. phasor that allows the very rapid and transient-free control of torque. Given that the definition of a vector quantity is one which has magnitude and

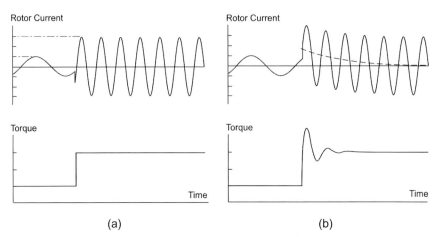

Figure 7.18 Step changes in rotor current. (a) Transient-free transition with correct changes to magnitude, frequency and instantaneous position of stator m.m.f. wave (i.e. vector control). (b) Same changes to magnitude and frequency, but not phase.

direction, and that the angular position of the phasor defines the direction in which it is pointing, it is clear why this technique gained the name 'vector control'.[2]

To underline the importance of the sudden change in *phase* of the stator current (i.e. the sudden jump in angular position of the stator m.m.f. in achieving a step of torque), Figure 7.18(b) shows what happens typically if only the magnitude and frequency, but not the position, are suddenly changed. The steady-state conditions are ultimately reached, but only after an undesirable transient governed by the (long) rotor time-constant, which may persist for several cycles at the slip frequency. The fundamental reason for the transient is that if the magnitude of the stator current is suddenly increased without a change of position, the flux and torque components both increase proportionately. The change in the flux component portends a change in the rotor flux (and associated stored energy), which in turn is resisted by induced rotor currents until they decay and the steady state is reached.

7.1 Summary

This section has described the underlying principles by which very rapid and precise torque control can be achieved from an induction motor, but we should remember that until sophisticated power-electronic control became possible the approach outlined here was only of academic interest. The fact that the modern inverter-fed drive is able to implement torque control and achieve such outstandingly impressive

---

[2] The term vector control has sometimes been misused to refer to drives that do not include field orientation. However, the term is so ubiquitous that we cannot avoid it, so when we refer to 'vector control' we mean a proper field-oriented system.

performance from a motor whose inherent transient behavior is poor, represents a major milestone in the already impressive history of the induction motor. The way in which such drives achieve field-oriented control is discussed next.

8. IMPLEMENTATION OF FIELD-ORIENTED CONTROL

An essential requirement if we are to unravel the workings of the overall scheme for field-oriented control is an understanding of the pulse-width modulation (PWM) vector modulator/inverter combination that is a feature of all such schemes, so this is covered first.

8.1 PWM controller/vector modulator

In the inverters we have looked at so far (see section 4 of Chapter 2) we have supposed that the periodic switching required to approximate a sinusoidal output was provided from a master oscillator. The frequency of the oscillator determined the frequency of the a.c. voltage applied to the motor, and the amplitude was controlled separately. In terms of space phasors this allows control of the amplitude and frequency, but not the instantaneous angular position of the voltage and current phasors. As we have seen, it is the additional ability to make instantaneous changes to the *angular position* of the output phasor that is the key to dynamic torque control, and this is the key feature provided by the 'vector modulator'.

We now explore what the inverter can produce in terms of its output voltage phasor. We recall that there are six devices (switches) in three legs (see Figure 2.17), and to avoid a short-circuit both switches in one leg must not be turned on at the same time. If we make the further restriction that each phase winding must at all times be connected to one or other of the d.c. link terminals, there are only eight possible combinations, as shown in Figure 7.19.

The six switching combinations labeled 1–6 each produce an output voltage phasor of equal amplitude but displaced in phase by 60° as shown in the lower part of each diagram, while the final two combinations have all three terminals joined together so the voltage is zero. The six unit vectors are shown with their correct relative phase, but rotated so as to bring U6 horizontal, in Figure 7.20.

Having only six states of the voltage phasor at our disposal is clearly not satisfactory, because we need to exert precise control over the magnitude and position of the voltage phasor at any instant, so this is where the 'time modulation' aspect comes into play. For example, if we switch rapidly between states U1 and U6, spending the same amount of time with each, we will effectively have synthesized a voltage phasor lying half way between them, and of magnitude U1 cos 30° (or 86.7% of U1), as shown by the vector U(x) in the upper-right part of Figure 7.20. If we spend a higher proportion of the time on U1 and the remainder on U2, we could produce the vector U(y). As long as we spend

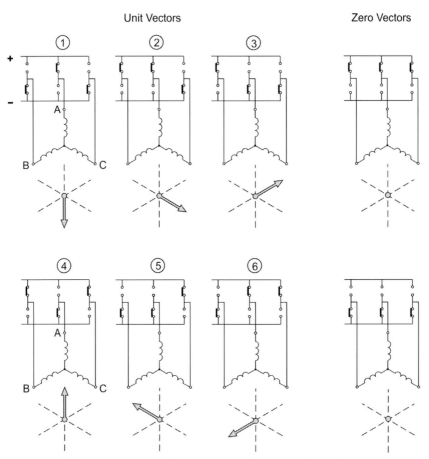

Figure 7.19 Voltage phasors for all acceptable combinations of switching for a 3-phase inverter.

the whole of the sample time on either U1 or U6, we will end up somewhere along the line joining U1 to U6.

We have used the terms 'switch rapidly' and 'the time' without specifying what they mean. In practice, we would expect the modulating frequency to be perhaps a few kHz up to the low tens of kHz, so 'the time' means one cycle at this frequency, say 100 microseconds at 10 kHz. So for as long as we wished the voltage phasor to remain at U(x), we would spend 50 µs of each sample period alternately connected to U1 and U6.

Recalling that ideally we want to be able to choose both magnitude and position it is clearly not satisfactory to be constrained to the outer edges of the hexagon. So now we bring the zero vector into play. For example, suppose we wish the voltage phasor to be U(z), as in the lower sketch. This is composed of (0.5)U5

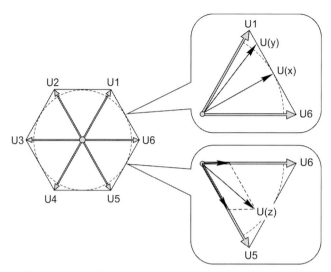

Figure 7.20 Synthesis of intermediate voltage phasors in vector modulator.

plus (0.3)U6. Hence in each modulating cycle of 100 μs, we will spend times of 50 μs on U5, 30 μs on U6 and 20 μs with one of the zero states.

The precise way in which these periods are divided within one cycle of the modulating frequency is a matter of important detail in relation to the distribution and minimization of losses between the six switching devices, but need not concern us here. Suffice it to say that it is a straightforward matter to arrange for digital software/hardware that has input signals representing the magnitude and instantaneous position of the output voltage phasor, and which selects and modulates the six switches appropriately, to create the desired output until told to move to a new location.

When we introduced the idea of space phasors earlier in this chapter, we saw that if we begin with balanced 3-phase sinusoidal voltages, the voltage phasor is of constant length and rotates at a uniform rate. Looking at it the other way round, it should be clear that if we arrange for the output of the inverter to be a voltage phasor of constant length, rotating at a constant rate, then the corresponding phase voltages must form a balanced sinusoidal set, which is what we want for steady-state running.

We conclude that in the steady state, the magnitude of the input signal to the vector modulator would have a constant amplitude and its angle would increase at a linear rate corresponding to the desired angular velocity of the output. Clearly in order to avoid having to deal with ever-increasing angles, the input signal to the modulator will reset each time a full cycle of 360° is reached, as shown in Figure 7.21.

Away from the steady-state condition, for example during acceleration, we should recall that to preserve the linear relation between torque and the stator current component (I_T), the flux component of the stator current phasor (I_F) must

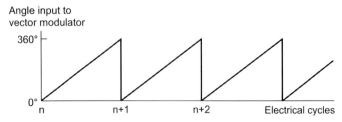

Figure 7.21 Angle reference to vector modulator corresponding to constant frequency operation of inverter. Note that the angle resets to zero at the end of each cycle.

remain aligned with the rotor flux. As we will see in the next section, this is achieved by deriving the angle input to the vector modulator directly from the absolute angular position of the rotor flux.

8.2 Torque control scheme

A simplified block diagram of a typical field-oriented torque control system is shown in Figure 7.22.

The first and most important fact to bear in mind in the discussion that follows is that Figure 7.22 represents a *torque* control scheme, and that for applications that require speed control, it will form the 'inner loop' of a closed-loop speed control scheme. The torque and flux inputs will therefore be outputs from the speed controller, as indicated in Figure 7.23.

Returning to Figure 7.22 it has to be acknowledged that it looks rather daunting, and getting to grips with it is not for the faint-hearted. However, if we examine it a bit at a time, it should be possible to grasp the essential features of its

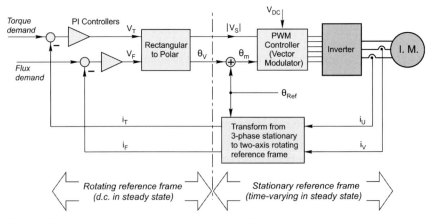

Figure 7.22 Simplified block diagram of a typical field-oriented torque control system.

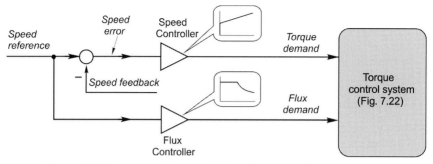

Figure 7.23 Schematic diagram of closed-loop speed control system.

operation. To simplify matters, we will focus on steady-state conditions, despite the fact that the real merit of the system lies in its ability to provide precise torque control even under transient conditions.

Taking the broad overview first, we can see that there are similarities with the d.c. drive with its inner current (torque) control loop (see Chapter 4), notably the stator current feedback and the use of proportional and integral (PI) controllers to control the torque and flux components of the stator current. It would be good if we could measure the flux and torque components directly, but of course the current components do not have separate existences: they are merely components of the stator current, which is what we can measure. The motor has three phases, but because we are assuming that there is no neutral connection, it suffices to measure only two of the line currents (because the sum of the three is zero). The information from these two currents allows us to keep track of the angular position of the stator current phasor with respect to the stationary reference frame (θ_S) as shown in Figure 7.24.

However, the stator current feedback signals are alternating at the frequency supplied by the inverter, and the corresponding stator space phasor is rotating at the

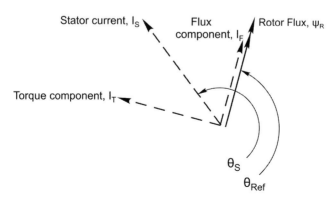

Figure 7.24 Stator current and rotor flux reference angles.

supply frequency with respect to a stationary reference frame. Before the flux and torque components of these signals (I_F and I_T) can be identified (and subsequently fed back to the PI controllers) they must first be transformed (see section 4) into a reference frame that rotates with the rotor flux. As explained previously, the rotor flux angle θ_{Ref} is therefore an essential input to the transformation algorithm, as shown in Figure 7.22.

The broken line in the middle of Figure 7.22 separates quantities defined in the stationary reference frame (on the right) from those in the rotating reference frame (on the left). In the steady state, all those on the left are d.c., while all those on the right are time-varying.

The reader might wonder why, when we follow the signal path of the current control loops, beginning on the right with the phase current transducers, there is no matching 'inverse transform' to get us back from the rotating reference frame on the left to the stationary reference frame on the right. The answer lies in the nature of the input signal to the PWM/vector modulator and inverter, which we discussed above. Let us suppose that the motor is running in the steady state, so that the output voltage phasor rotates at a constant rate with angular frequency ω. Under these conditions the rotor flux phasor also rotates with constant angular velocity ω, so the angle of the flux vector with respect to the stationary reference frame (θ_{Ref}) increases linearly with time. Also, in the steady state, the output from the PI controllers is constant, so the angle θ_V (Figure 7.22) is constant. Hence the input angle to the modulator (θ_m in Figure 7.22), which is the sum of θ_{Ref} and θ_V, is also a ramp in time, and this is what provides the rotation of the output voltage phasor. In effect, the system is self-sustaining: the primary time-varying input angle to the modulator comes from the flux position signal (which is already in the stationary reference frame), and the PI controller provides the required magnitude signal ($|V_S|$) and the additional angle (θ_V).

Turning now to the action of the PI controllers, we see from Figure 7.22 that the outputs are voltage commands in response to the differences between the feedback (actual) values of the transformed currents and their demanded values. The flux demand will usually be constant up to base speed, while the torque demand will usually be the output from the speed or position controller, as shown in Figure 7.23. The proportional term gives an immediate response to an error, while the integral term ensures that the steady-state error is zero. The outputs from the two PI controllers (which are in the form of quadrature voltage demands, V_F and V_T) are then converted from rectangular to polar form, to produce amplitude and phase signals, $|V_S|$ and θ_V, where

$$|V_S| = \sqrt{V_F^2 + V_T^2}, \text{ and } \theta_V = \tan^{-1} \frac{V_T}{V_F},$$

as shown in Figure 7.25.

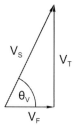

Figure 7.25 Derivation of voltage phasor from flux and torque components.

The amplitude term specifies the magnitude of the output voltage phasor (and thus the three phase voltages applied to the motor), with any variation of the d.c. link voltage (V_{dc}) being compensated in the PWM controller. The phase angle (θ_V) represents the desired angle between the stator voltage phasor and the rotor flux phasor, both of which are measured in the stationary reference frame. The angle of the rotor flux phasor is θ_{Ref}, so θ_V is added at the input to the vector modulator to yield the stator voltage phasor angle θ_m, as shown in Figure 7.22.

We can usefully conclude our look at the steady state by adding the stator voltage phasor to Figure 7.24 to produce Figure 7.26, to provide reassurance that, in the steady state, the rather different approach we have taken in this section is consistent with the classical approach taken earlier.

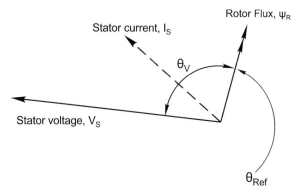

Figure 7.26 Time phasor diagram showing stator voltage and current under steady-state conditions.

8.3 Transient operation

We concluded earlier that for the motor torque to be directly proportional to the torque component of stator current, it is necessary to keep the magnitude of the rotor flux constant and to ensure that the flux component of stator current is aligned with the rotor flux. This is achieved automatically because the principal angle input to the vector modulator comes directly from the rotor flux angle (θ_{Ref}), as shown in Figure 7.22. So during acceleration, for example, the instantaneous angular velocity of the rotor flux wave will remain in step with that of the stator current phasor, so that there is no possibility of the two waves falling out of synchronism with one another.

In section 7 we discussed a specific example of how to obtain a step change in torque by making near-instantaneous changes to the magnitude, speed and position of the stator m.m.f. wave, and we are now in a position to see how this particular strategy is effected using the control scheme shown in Figure 7.22.

A step demand for torque causes a step increase in $|V_S|$ and θ_V at the output of the rectangular to polar converter in order to effect a very rapid increase in the magnitude and instantaneous position of the stator current phasor. At the same time, the algorithm that calculates the slip velocity of the flux wave (see later, equation (7.3)) yields a step increase because of the sudden increase in the torque component of stator current. The principal angular input to the vector modulator (the flux angle (θ_{Ref})) therefore changes gradient abruptly, as shown in Figure 7.27.

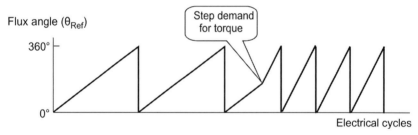

Figure 7.27 Flux angle reference showing response to a sudden step demand for increased torque.

Recalling that the steady-state stator frequency is governed by the angular velocity of the flux (i.e. $d\theta_{\mathrm{Ref}}/dt$), this lines up with our expectation that (assuming the rotor velocity is constant) the stator frequency will increase in order to increase the slip and provide the new higher torque.

8.4 Acceleration from rest

Measured results showing how the real and transformed currents behave during acceleration from rest to base speed, are shown in Figure 7.28. These relate to a motor whose rotor time–constant is approximately 0.1 s, and cover a time of 0.5 s.

The upper diagram shows the demanded values for the transformed flux and torque components of stator current; the middle diagram shows the measured (actual) flux and torque components; and the lower shows the three phase currents.

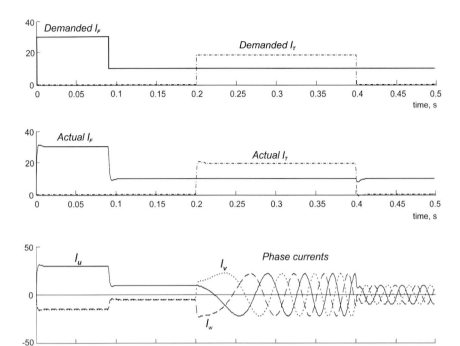

Figure 7.28 Experimental results showing build-up of flux followed by sudden demand for step increase in torque until motor reaches its target speed. *(Courtesy of Emerson – Control Techniques)*

At $t = 0$, the flux demand signal is set to its highest possible level of 30 A in order to raise the flux from zero as quickly as possible to its target value (10 A). Note how the actual transformed flux component of stator current (I_T) follows the demand signal very rapidly, with only slight overshoot. The signal I_T is the transformed version of the phase winding currents, so the fact that I_T is on target is of course indicative that the phase currents are established rapidly and held while the flux builds up, as we can see in the lower figure. During this period phase U carries a positive current of 30 A while phases V and W each carry a negative current of 15 A.

After about one time-constant the demand is reduced to 10 A, and thereafter held constant. This 'rapid forcing' ensures that by 0.2 s, the rotor flux has been fully established.

At 0.2 s, a torque producing demand is applied to accelerate the motor, and maintained until 0.4 s, when the torque producing reference is reduced to zero, and the motor stops accelerating.

We note the almost immediate and transient-free transition of the 3-phase currents from their initial steady (d.c.) values immediately prior to 0.2 s, into constant amplitude, 'smoothly increasing frequency' a.c. currents over the next 0.2 s. And then there is a similarly near-perfect transition to reduced amplitude steady-state conditions (at about 40 Hz) after 0.4 s. In the steady state, the torque component is negligible because the motor is unloaded, and the stator current consists only of the flux component, which traditionally would be referred to as the magnetizing current.

Younger readers will doubtless not require convincing of the validity of these remarkable results, but they might find it salutary to know that until the 1970s it was widely believed that such performance would never be possible.

To conclude this section we can draw a further parallel between the field-oriented induction motor and the d.c. motor. We see from Figure 7.28 that as the motor accelerates, the frequency of the stator currents increases with the speed. If we stationed ourselves on the rotor of a d.c. motor as it accelerated, the rate at which the current in each rotor coil reversed as it was commutated would also increase in proportion to the speed, though of course we are not aware of it when we are in the stationary reference frame.

8.5 Deriving the rotor flux angle

By now, the key role played by the rotor flux angle should have become clear, so finally we look at how it is obtained. It is not practical or economic to fit a flux sensor to the motor, so industrial control schemes invariably estimate the position of the flux.

We will first establish an expression for absolute rotor flux angle (θ_{Ref}) in the stationary reference frame in terms of quantities that can either be measured or estimated. Readers who find the derivation indigestible need not worry as it is the conclusions that are important, not the analytical detail.

If we let the angle of the rotor body with respect to the stationary reference frame be θ, then the instantaneous angular velocities of the rotor flux wave and the rotor itself are given by

$$\omega_{flux} = \frac{d\theta_{Ref}}{dt}$$

$$\omega_{rotor} = \frac{d\theta}{dt}$$

The rotor motional e.m.f. is directly proportional to the rotor flux linkage and the slip velocity, i.e.

$$V_R = \psi_R \left(\omega_{flux} - \omega_{rotor}\right),$$

and the rotor current is therefore given by

$$I_R = \frac{\psi_R \left(\omega_{flux} - \omega_{rotor}\right)}{R_R}$$

The corresponding component of stator current is given (see Figure 7.16) by

$$I_{ST} = \frac{L_R}{M} I_R$$

Combining these equations and rearranging gives

$$\frac{d\theta_{Ref}}{dt} = \frac{MR_R}{\psi_R L_R} I_{ST} + \omega_{rotor} = \left(\frac{M}{\tau \psi_R}\right) I_{ST} + \omega_{rotor} \qquad (7.3)$$

where τ is the rotor time-constant. Hence to find the rotor flux angle at time t we must integrate the expression above.

The mutual inductance M is a constant, and although the time-constant will vary because the rotor resistance varies with temperature, it will change relatively slowly, so we can treat it as constant, in which case the rotor flux angle is given by

$$\theta_{Ref} = \int_0^t \omega_{rotor}\,dt + \frac{M}{\tau} \int_0^t \frac{I_{ST}}{\psi_R}\,dt = \theta + \frac{M}{\tau} \int_0^t \frac{I_{ST}}{\psi_R}\,dt$$

Note that because of the symmetry of the rotor, we only need the time-varying element of the rotor body angle (θ), not the absolute position, so the constant of integration is not required. (In contrast, for vector control of permanent magnet motors, the absolute position is important, because the rotor has saliency.)

The various methods that are used to keep track of the flux angle are what differentiate the various practical and commercial implementations of field-oriented control, as we will now see.

If we have a shaft encoder we can measure the rotor position (θ), or if we have a measured speed signal, we can derive θ by direct integration. This approach involves the fewest estimations, and therefore will normally offer superior performance, especially at low speeds, but is more costly because it requires extra transducers. We will refer to systems that use shaft feedback as 'closed-loop', but in the literature they may be also referred to as 'direct vector control'. In common with all schemes, the second term has to be estimated.

Many different methods of estimating the instantaneous parameter values are employed, but all employ a digital simulation or mathematical model of the motor/inverter system. The model runs in real time and is subjected to the same inputs as the actual motor, the model then being continuously fine-tuned so that the predicted and actual outputs match. Modern drives measure the circuit parameters automatically at the commissioning stage, and can even refine them on a near-continual basis to capture parameter variations.

The majority of vector control schemes eliminate the need for measurement of rotor position, and instead the rotor position term in equation (7.3) is also estimated from a motor model, based on the known motor voltage and currents. Rather confusingly, in order to differentiate them from schemes that do have shaft transducers, these systems are known as 'open-loop' or 'indirect' vector control. The term 'open loop' is a misleading one because at its heart is the closed-loop torque control shown in Figure 7.22, but it is widely used: what it really means is 'no shaft feedback'.

The main problems of the open-loop approach occur at low speeds where motor voltages become very low and measurement noise can render the algorithms unreliable. Techniques such as high-frequency injection of diagnostic signals exist, but are yet to find acceptance in the market on standard motors. Open-loop inverter-fed induction motors are usually unsuitable for continuous operation at frequencies below 0.75 Hz, and struggle to produce full torque in this region.

An additional difficulty is that the significant variation of rotor resistance with temperature is reflected in the value of the all-important rotor time-constant τ. Any difference between the real rotor time constant and the value used by the model causes an error in the calculation of the flux position and so the reference frame becomes misaligned. If this happens, the flux and torque control are no longer completely decoupled, which results in suboptimum performance and possible instability. To avoid this, routines are included in the drive to provide ongoing estimates of the rotor time constant.

9. DIRECT TORQUE CONTROL

Direct torque control is an alternative high-performance strategy for vector/field orientation, and warrants a brief discussion to conclude our look at contemporary schemes. Developed from work first published in 1985 it theoretically provides the fastest possible torque response by employing a 'bang-bang' approach to maintain flux and torque within defined hysteresis bands. Like field-oriented control, it only became practicable with the emergence of relatively cheap and powerful digital signal processing.

Direct torque control avoids coordinate transformations because all the control actions take place in the stator reference frame. In addition there are no PI controllers, and a switching table determines the switching of devices in the inverter in place of the PWM approach favored for field-oriented control. These apparent advantages are offset by the need for a higher sampling rate (up to 40 kHz as compared with 6–15 kHz) leading to higher switching loss in the inverter; a more complex motor model; and inferior torque ripple. Because a hysteresis method is used the inverter has a continuously variable switching frequency which may be seen as an advantage in spreading the spectrum of acoustic noise from the motor.

We saw in the previous sections that in field-oriented control, the torque was obtained from the product of the rotor flux and the torque component of stator current. But there are many other ways in which the torque can be derived, for example in terms of the product of the rotor and stator fluxes and the sine of the angle between, or the stator flux and current and the sine of the angle between them. The latter is the approach discussed in the next section, but first a word about hysteresis control.

A good example of hysteresis control is discussed later in this book, in relation to 'chopper drives' for stepping motors in Chapter 9. Another more familiar example is the control of temperature in a domestic oven. Both are characterized by a simple approach in which full corrective action is applied whenever the quantity to be controlled falls below a set threshold, and when the target is reached, the power is switched off until the controlled quantity again drops below the threshold. The frequency of the switching depends on the time-constant of the process and the width of the hysteresis band: the narrower the band and the shorter the time-constant, the higher the switching frequency.

In the domestic oven, for example, the 'on' and 'off' temperatures can be a few degrees apart because the cooking process is not that critical and the time-constant is many minutes. As a result the switching on and off is not so frequent as to be irritating and wear out the relay contacts. If the hysteresis band were to be narrowed to a fraction of a degree to get tighter control of the cooking temperature, the price to be paid would be incessant clicking on and off, and shortened life of the relay.

9.1 Outline of operation

The block diagram of a typical direct torque control scheme is shown in Figure 7.29. There are several similarities with the scheme shown in Figure 7.22, notably the inverter, the phase current feedback, and the separate flux and torque demands, which may be generated by the speed controller, as in Figure 7.23.

However, there are substantial differences. Earlier we discovered that the inverter output voltage space phasor has only six active positions, and two zero states

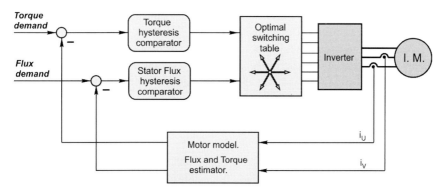

Figure 7.29 Block diagram of typical direct torque control scheme.

(see Figure 7.19), corresponding to the eight possible combinations of the six switching devices. This means that at every instant there are only eight options in regard to the voltage that we can apply to the motor terminals. In the field-oriented approach, PWM techniques are employed to alternate between adjacent unit vectors to produce an effective voltage phasor of any desired magnitude and instantaneous position. However, with direct torque control, only one of the eight intrinsic vectors is used for the duration of each sample, during which the estimated stator flux and torque are monitored.

The motor model is exposed to the same inputs as the real motor, and from it the software continuously provides updated estimates of the stator flux and torque. These are compared with the demanded values and as soon as either strays outside its target hysteresis band, a logical decision is taken as to which of the six voltage phasors is best placed to drive the flux and/or torque back onto target. At that instant the switching is changed to bring the desired voltage phasor into play. The duration of each sample therefore varies according to the rate of change of the two parameters being monitored: if they vary slowly it will take a long time before they hit the upper or lower hysteresis limit and the sample will be relatively long, whereas if they change rapidly, the sample time will be shortened and the sample frequency will increase. Occasionally, the best bet will be to apply zero voltage, so one of the two zero states then takes over.

9.2 Control of stator flux and torque

We will restrict ourselves to operation below base speed, so we should always bear in mind that although we will talk about controlling the stator flux, what we really mean is keeping its magnitude close to its normal (rated) value at which the magnetic circuit is fully utilized. We should also recall that when the stator flux is at its rated value and in the steady state, so is the rotor flux.

It is probably easiest to grasp the essence of the direct torque method by focusing on the stator flux linkage, and in particular on how (a) the magnitude of the stator flux is kept within its target limits and (b) how its phase angle with respect to the current is used to control the torque.

The reason for using stator flux linkage as a reference quantity is primarily the ease with which it can be controlled. When we discussed the basic operation of the induction motor in Chapter 5, we concluded that the stator voltage and frequency determined the flux, and we can remind ourselves why this is by writing the voltage equation for the stator as

$$V_S = I_S R_S + \frac{d\psi_S}{dt}$$

(We are being rather loose here, by treating space phasor quantities as real variables, but there is nothing to be gained by being pedantic when the message we take away will be valid.) In the interests of clarity we will make a further simplification by ignoring the resistance voltage term, which will usually be small compared with V_S. This yields

$$V_S = \frac{d\psi_S}{dt} \text{ or, in integral form, } \psi_s = \int V_S \, dt$$

The differential form shows us that the rate of change of stator flux is determined by the stator voltage, while the integral form reminds us that to build the flux (e.g. from zero) we have to apply a fixed volt-second product, with either a high voltage for a short time, or a low voltage for a long time. We will limit ourselves to the fine-tuning of the flux after it has been established, so we will only be talking about very short sample intervals of time (Δt) during which the change in flux linkage that results ($\Delta\psi_S$) is given by

$$\Delta\psi_S = V_S\Delta t$$

As far as we are concerned, ψ_S represents the stator flux linkage space phasor, which has magnitude and direction relative to the stator reference frame, and V_S represents one of the six possible stator voltage space phasors that the inverter can deliver. So if we consider an initial flux linkage vector ψ_S as shown in Figure 7.30(a), and assume that we apply, over time, Δt, each of the six possible options, we will produce six new flux-linkage vectors. The tips of the new vectors are labeled ψ_1 to ψ_6 in the figure, but only one (ψ_4) is fully drawn (dashed) to avoid congestion. There is also the option of applying zero voltage, which would of course leave the initial flux-linkage unchanged.

In (a) option 4 results in a reduction in amplitude and (assuming anticlockwise rotation) a retardation in phase of the original flux, but if the original flux linkage

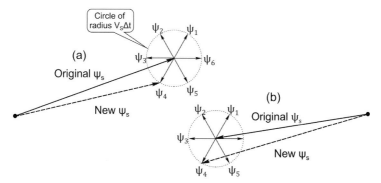

Figure 7.30 Space phasor diagram of stator flux linkage showing how the outcome of applying a given volt-second product depends on the original phase angle. In (a) the magnitude is reduced and the phase is retarded, while in (b) the magnitude is increased and the phase is advanced.

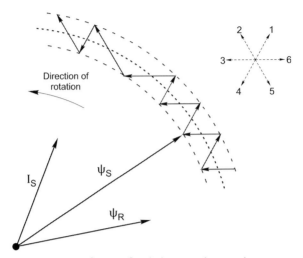

Figure 7.31 Trajectory of stator flux linkage under steady-state conditions.

had a different phase, as shown in diagram (b), option 4 results in an increase in magnitude and an advance in phase. It should be clear that outcomes vary according to initial conditions, and therefore an extensive look-up table will be needed to store all this information.

Having seen how we can alter the magnitude and phase of the stator flux, we now consider the flux linkage phasor during steady-state operation with constant speed and torque, in which case we know that ideally all the space phasors will be rotating at a constant angular velocity.

The locus of the stator flux linkage space phasor (ψ_S) is shown in Figure 7.31. In this diagram the spacing of hysteresis bands indicated by the innermost and outermost dashed lines have been greatly exaggerated in order to show the trajectory of the flux linkage phasor more clearly. Ideally, the phasor should rotate smoothly along the center dotted line.

In this example, the initial position shown has the flux linkage at the lower bound, so the first switching brings voltage vector 1 into play to drive the amplitude up and the phase forward. When the upper bound is reached, vector 3 is used, followed by vector 1 again and then vector 3. Recalling that the change in the flux linkage depends on the time for which the voltage is applied, we can see from the diagram that the second application of vector 3 lasts longer than the first. (We should also reiterate that in this example only a few switchings take place while the flux rotates through 60°: in practice the hysteresis band is very much narrower, and there may be many hundreds of transitions.)

We are considering steady-state operation, and so we would wish to keep the torque constant. Given that the flux is practically constant, this means we need to keep the angle between the flux and the stator current constant. This is where the torque hysteresis controller shown in Figure 7.29 comes in. It has to decide what switching will best keep the phase on target, so it runs in parallel with the magnitude controller we have looked at here. Each controller will output a signal for either an increase or decrease in its respective variable (i.e. magnitude or phase) and these are then passed to the optimal switching table to determine the best switching strategy in the prevailing circumstances (see Figure 7.29).

As we saw when discussing field-oriented control, it is not possible to make very rapid changes to the rotor flux because of the associated stored energy. Because the rotor and stator are tightly coupled it follows that the magnitude of the stator flux linkage cannot change very rapidly either. However, just as with field-oriented control, sudden changes in torque can be achieved by making sudden changes to the phase of the flux linkage, i.e. to the tangential component of the phasor shown in Figure 7.31.

CHAPTER EIGHT

Inverter-fed Induction Motor Drives

1. INTRODUCTION

We have seen in Chapter 6 why the relatively simple and low-cost induction motor is the preferred choice for many fixed speed applications, and in Chapter 7 we saw that by using field orientation/vector or direct torque control, it is possible to achieve not only steady-state speed control but also dynamic performance superior to that of a thyristor d.c. drive. We have also seen that achieving such performance is dependent on the ability to perform very fast/real-time modeling of the motor and very rapid control of the motor voltage magnitude and phase.

In this chapter we look at some of the practical aspects of inverter-fed induction motor drives and consider the impressive performance of commercially available drives. We will also briefly revisit the subject of control and, for the sake of balance, look at an application where field orientation struggles and direct torque control simply doesn't work.

While it comes as no surprise that the inverter-fed induction motor is now the best-selling industrial drive, the adoption of the standard induction motor in a variable-speed drive is not without potential problems, so it is important to be aware of their existence and learn something of the methods of mitigation. We will therefore consider some of the important practical issues which result from operating standard (utility supply) motors from a variable frequency inverter.

Finally, we will look at the most popular inverter circuits and highlight their most important characteristics.

2. PULSE-WIDTH MODULATED (PWM) VOLTAGE SOURCE INVERTER (VSI)

Several alternative drive topologies are applied to induction motors and the most relevant of these are discussed later in this chapter. By far the most important for most industrial applications has a diode bridge rectifier (which only allows energy flow from the supply to the d.c. link) and a PWM VSI, as shown in Figure 8.1, and this arrangement will now be the focus of our attention.

Most low-power inverters use MOSFET switching devices in the inverter bridge, and may switch at ultrasonic frequencies, which naturally results in quiet

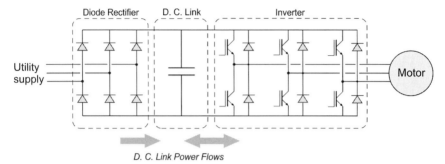

Figure 8.1 Pulse-width modulated (PWM) voltage source inverter (VSI).

operation. Medium- and larger-power inverters use insulated-gate bipolar transistors (IGBTs) which can be switched at high enough frequencies to be ultrasonic for most of the population. It should be remembered, however, that the higher the switching frequency the higher the inverter losses, and hence the lower the efficiency, and so a compromise must be reached.

Variable-frequency inverter-fed induction motor drives are used in ratings up to many megawatts. Standard 50 or 60 Hz motors are often used, though the use of a variable-frequency inverter means that motors of almost any rated frequency can be employed. By 'rated' frequency we mean the frequency at which the maximum possible output voltage of the power converter is achieved. We saw in Chapter 7 that operation above rated frequency limits performance and so this needs to be carefully considered when specifying a drive system. Commercially available inverters operate with output frequencies typically from 0 Hz up to perhaps several hundred Hz, and, in some cases, much higher frequencies. The low-frequency limit is generally determined by the form of control, while the higher frequency depends on the control and the physical dimensioning of the power electronic circuits (where stray inductance can be a problem if interconnections become too long).

The majority of inverters are 3-phase input and 3-phase output, but single-phase input versions are available up to about 7.5 kW. Some inverters (usually less than 3 kW) are specifically designed for use with single-phase motors, but these are unusual and will not be considered further. The upper operating frequency is generally limited by the mechanical stresses in the rotor. Very-high-speed motors for applications such as centrifuges and wood working machines can be designed, with special rotor construction and bearings for speeds up to 40,000 rev/min, or even higher.

A fundamental aspect of any converter which is often overlooked is the instantaneous energy balance. In principle, for any balanced 3-phase load, the total load power remains constant from instant to instant, so if it were possible to build an ideal 3-phase input, 3-phase output converter, there would be no need for the converter to include any energy storage elements. In practice, all converters require some energy storage (in capacitors or inductors), but these are relatively small when the input is 3-phase because the energy balance is good. However, as mentioned above, many low-power (and some high-power, rail traction) converters are supplied from a single-phase source. In this case, the instantaneous input power is zero at least twice per cycle of the mains (because the voltage and current go through zero every half-cycle). If the motor is 3-phase and draws power at a constant rate from the d.c. link, it is obviously necessary to store sufficient energy in the converter to supply the motor during the brief intervals when the load power is greater than the input power. This explains why the most bulky components in many power inverters are electrolytic capacitors in the d.c. link. (Some drive manufacturers are now designing products, for connection to a 3-phase supply, with low values of d.c. capacitance for undemanding applications where the subsequent reduction in control/performance is acceptable, and this is discussed in section 7.)

The output waveform produced by the PWM inverter in an a.c. drive also brings with it challenges for the motor, which we will consider later. When we looked at the converter-fed d.c. motor we saw that the behavior was governed primarily by the mean d.c. voltage, and that for most purposes we could safely ignore the ripple components. A similar approximation is useful when looking at how the inverter-fed induction motor performs. We make use of the fact that although the actual voltage waveform supplied by the inverter will not be sinusoidal, the motor behavior depends principally on the fundamental (sinusoidal) component of the applied voltage. This is a somewhat surprising but extremely welcome simplification, because it allows us to make use of our knowledge of how the induction motor behaves with a sinusoidal supply to anticipate how it will behave when fed from an inverter.

In essence, the reason why the harmonic components of the applied voltage are much less significant than the fundamental is that the impedance of the motor at the harmonic frequencies is much higher than at the fundamental frequency. This causes the current to be much more sinusoidal than the voltage (as previously shown in principle in Figure 7.1), and this means that we can expect a sinusoidal traveling field to be set up in much the same way as discussed in Chapter 5.

In commercial inverters the switching frequency is high and the measurement and interpretation of the actual waveforms is not straightforward. For example, the voltage and current waveforms in Figure 8.2 relate to an industrial drive with a 3 kHz switching frequency. Note the blurring of the individual voltage pulses (a result of sampling limitations of the oscilloscope), and the near-sinusoidal fundamental component of current. We might be concerned at what appear to be spikes of current, but consideration of the motor leakage inductance and the limited forcing voltage will confirm that such rapid rates of change of current are impossible: the spikes are in fact the result of noise on the signal from the current transducer.

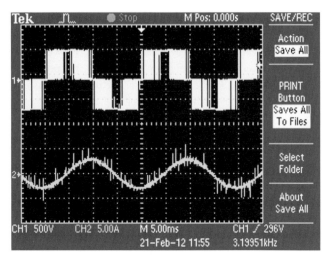

Figure 8.2 Actual voltage and current waveforms for a star-connected, PWM-fed induction motor. Upper trace – voltage across U and V terminals; lower trace – U phase motor current. (See Plate 8.2)

Measurement of almost all quantities associated with power electronic converters is difficult, and great care must be taken in the selection of instrumentation and interpretation of the results. A clear understanding of grounding is also important when reviewing inverter d.c. link and output waveforms since unlike a utility supply there is no clear, or at least simple, ground reference.

3. PERFORMANCE OF INVERTER-FED INDUCTION MOTOR DRIVES

It has often been said that the steady-state performance of the inverter-fed induction motor is broadly comparable with that of an industrial d.c. drive, but in fact the performance of the inverter-fed induction motor is better in almost all respects.

To illustrate this, we can consider how quickly an induction motor drive, with field-oriented control, can change the motor shaft torque. Remarkably, the torque can be stepped from zero to rated value and held there in less than 1 ms, and this can now be achieved by a motor even without a speed/position sensor. For comparison, a thyristor-fed d.c. motor could take up to one-sixth of a 50/60 Hz mains cycle, i.e. around 3 ms before the next firing pulse can even initiate the process of increasing the torque, and clearly considerably longer to complete the task.

The induction motor is also clearly more robust and better suited to hazardous environments, and can run at higher speeds than the d.c. motor, which is limited by the performance of its commutator.

The field-oriented control strategy, coupled with the ability, through a PWM inverter, to change the stator voltage phasor in magnitude, phase and frequency very rapidly, is at the heart of this exceptional motor shaft performance. The majority of commercial inverter systems embody such control strategies, but the quantification of shaft performance is subject to a large number of variables and manufacturers' data in this respect needs to be interpreted with care. Users are interested in how quickly the speed of the motor shaft can be changed, and often manufacturers quote the speed loop response, but many other factors contribute significantly to the overall performance. Some of the most important, considering a spectrum of applications, are:

- Torque response: The time needed for the system to respond to a step change in torque demand and settle to the new demanded level.
- Speed recovery time: The time needed for the system to respond to a step change in the load torque and recover to the demanded speed.
- Minimum supply frequency at which 100% torque can be achieved.
- Maximum torque at 1 Hz.
- Speed loop response: This is defined in a number of different ways but a useful measure is determined by running the drive at a non-zero speed and applying a square wave speed reference and looking at the overshoot of speed on the leading edge of the square wave: an overshoot of 15% is – for most applications – considered practically acceptable. For the user, it is always a good idea to seek clarification from the manufacturer whenever figures are quoted for the speed (or current/torque) loop bandwidth of a digital drive, because this can be defined in a number of ways (often to the advantage of the supplier and not to the benefit of the application).

Note that the above measures of system performance should be obtained under conditions which avoid the drive hitting current limits, as this obviously limits the performance. Tests should typically be undertaken on a representative motor with load inertia approximately equal to the motor inertia.

Indications of the performance of open-loop and closed-loop field-oriented induction motor control schemes are shown in Table 8.1.

The performance of a closed-loop inverter-fed induction motor is comparable to that of a closed-loop permanent magnet motor, which we discuss in Chapter 9. This comes as a great surprise to many people (including some who have spent a lifetime in drives). It is particularly noteworthy given that the majority of induction motor drives use standard motors which were designed for fixed speed operation and broad application. Permanent magnet motors, however, tend to be customized and many have been designed with relatively low inertias (long length and small diameter rotor), which facilitate rapid speed changes, or high inertias (short shaft and large diameter rotor), which promote smooth rotation in the presence of load changes. Special induction motor designs are available, however, and are sometimes the preferred solution.

Table 8.1 Performance of open-loop and closed-loop field-oriented induction motor control schemes.

	Open-loop (without position feedback)	Closed loop (with position feedback)
Torque response (ms)	<0.5	<0.5
Speed recovery time (ms)	<20	<10
Min. speed with 100% torque (Hz)	0.8	Standstill
Max. torque at 1 Hz (%)	>175	>175
Speed loop response (Hz)	75	125

In the remainder of this section we give broad indications of the applicability of the various drive configurations that should prove helpful when looking at specific applications.

3.1 Open-loop (without speed/position feedback) induction motor drives

Open-loop induction motor drives are used in applications that require moderate performance (fans and pumps, conveyors, centrifuges, etc.). The performance characteristics of these drives are summarized below:

- There is moderate transient performance with full torque production down to approximately 2% of rated speed.
- Although a good estimate of stator resistance improves torque production at low speeds, the control system will work with an inaccurate estimate, albeit with reduced torque.
- A good estimate of motor slip improves the ability of the drive to hold the reference speed, but the control system will work with an inaccurate estimate, albeit with poorer speed holding.

The performance of open-loop induction motor drives continues to improve. Techniques for sensing the rotational speed of an induction motor without the need for a shaft-mounted speed or position sensor pervade the technical literature, and will in time find their way into some commercial drives.

3.2 Closed-loop (with speed/position feedback) induction motor drives

Induction motor drives with closed-loop control are used in similar applications to d.c. motor drives (cranes and hoists, winders and unwinders, paper and pulp processing, metal rolling, etc.). These drives are also particularly suited to applications that must operate at very high speeds with a high level of field weakening, for example spindle motors. The performance characteristics of these drives are summarized below.

- There is good dynamic performance at speeds down to standstill when position feedback is used.
- Only incremental position feedback is required. This can be provided with a position sensor or alternatively a sensorless scheme can be used. The transient performance of a sensorless scheme will be lower than when a position sensor is used and lower torque is produced at very low speeds.
- The robustness of the rotor makes induction motors particularly well suited to high-speed applications that require field weakening. The motor current reduces as the speed is increased and the flux is reduced.
- Induction motors are generally less efficient than permanent magnet motors because of their additional rotor losses.

3.3 When field orientation and direct torque control cannot be used

Field orientation and direct torque control both rely upon modeling the flux in the motor. If a single inverter is being used to feed more than one motor as in Figure 8.3, neither control strategy can be used.

In such systems individual motor control is not possible and the only practical form of control is to feed the motor group with an appropriate voltage source, of which the magnitude and frequency can be controlled. In fact, this is exactly the traditional form of V/f control which predominated in early inverters. The output frequency, and hence the no-load speed of the motor, is set by the speed reference signal, which was traditionally applied in the form of an analogue voltage (0–10 V) or current (4–20 mA), but is now a digital signal. In most automated applications the speed demand comes from a remote controller such as a PLC (programmable logic controller), while in simpler applications there will be a digital user interface on a control panel, or on the drive itself. The drive would

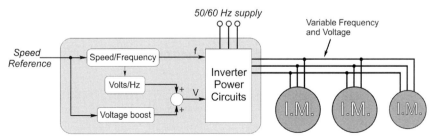

Figure 8.3 Simple V/f induction motor control strategy applied to a multi-motor system.

have the facility to adjust the V/f ratio. It would also have the facility to boost the voltage at low frequencies to compensate for the dominating influence of the stator winding resistance.

The fact that field orientation schemes use a vector modulator/PWM controller (see Figure 7.22) indicates that to adapt the field-oriented scheme for multi-motor drives is relatively simple: all that is required is to provide a sawtooth waveform of the appropriate frequency as input to the vector modulator (see Figure 7.21). Unfortunately, for direct torque control there is no such controller available and in order to provide multi-motor operation manufacturers employing this control strategy must also provide a PWM controller specifically to cater for such a load.

4. EFFECT OF INVERTER WAVEFORM AND VARIABLE SPEED ON THE INDUCTION MOTOR

It is often stated that standard 'off-the-shelf' a.c. motors can be used without problem on modern PWM inverters. While such claims may be largely justified, inverters do have some impact and limitations are inevitable. In particular, the harmonic components of the voltages and currents create acoustic noise; they always give rise to additional iron and copper losses; and they have other effects which are perhaps less obvious. In addition, the operation of a standard motor – with its cooling system designed to suit fixed-speed operation – can be a significant limitation, and this will also be considered here.

4.1 Acoustic noise

Acoustic noise can usually be reduced by selecting a higher switching frequency (at the cost of higher inverter losses). It is interesting to note that not all motors exhibit the same motor characteristic when connected to identical inverters. The differences are usually small, and relate primarily to the clamping of the iron in the stator of the motor. Certain switching frequencies may excite resonances in some motors often related to the tie bars between the end frames: these can be alarming but easily remedied by changing the switching frequency, or, if the tie bar is external to the motor frame, by adding a wedge to change the natural frequency of the bar.

4.2 Motor insulation and the impact of long inverter-motor cables

The PWM waveform has another very significant, but perhaps less obvious, effect, related to the very high rates of change of voltage ($\mathrm{d}V/\mathrm{d}t$), which results in transiently uneven voltage distribution across the motor winding, as well as short duration voltage overshoots because of reflection effects in the motor cable. This might damage the motor insulation.

In a modern 400 V power converter the d.c. link voltage is around 540 V, the voltage switches in a time of the order of 100 ns, and so *at the terminal of the drive* there is a $\mathrm{d}V/\mathrm{d}t$ of over 5000 V/µs. The motor insulation will usually cope with this high $\mathrm{d}V/\mathrm{d}t$, but in practical installations the motor may be some distance, perhaps hundreds of meters, away from the drive, and in this case we also have to recognize that at these very high rates of change of voltage the cable behaves as a transmission line. Hence when the voltage edge reaches the motor terminals, a reflection occurs because the motor impedance is higher than that of the cable. Consequently, the motor terminal voltage sees an overshoot of theoretically up to twice the step voltage. Fortunately the inductance and the high-frequency resistance of the cable mitigate these effects somewhat, causing the actual rise time to increase so that in practice the voltage overshoot usually has little detrimental effect on the main motor insulation systems between phases and from phase to earth, which are traditionally designed to withstand overvoltage pulses. However, some low-cost (or very old) motors may have relatively poor inter-phase insulation, which can lead to premature insulation failure. Further, at each pulse edge the drive has to provide a pulse of current to charge the capacitance of the converter-motor cable, and in small drives, with very long motor cables, this charging current may, in extreme cases, exceed the rated current of the motor, and determine the rating of the required drive!

In case all this sounds alarming, the fact is that such problems are extremely unusual and usually associated with systems employing very old or very-low-cost motors with poor insulation systems, and/or with drive systems with rated voltages over 690 V. Naturally enough, the problem is more pronounced on medium voltage drives where it is not uncommon for $\mathrm{d}V/\mathrm{d}t$ filters to be fitted between the inverter and the motor.

This phenomenon is now very well understood by reputable motor manufacturers. International standards on appropriate insulation systems have also been published, notably IEC 34-17 and NEMA MG1pt31.

4.3 Losses and impact on motor rating

Operation of induction motors on an inverter supply inevitably results in additional losses in the machine as compared with a sinusoidal utility supply. These losses fall into three main categories:

- Stator copper loss: This is proportional to the square of the r.m.s. current although additional losses due to skin effect associated with the high-frequency components also contribute. We have seen in Figure 8.2 that the motor current is reasonably sinusoidal and hence, as we would expect, the increase in copper loss is seldom significant.

- Rotor copper loss: The rotor resistance is different for each harmonic current present in the rotor due to skin effect (and is particularly pronounced in deep bar rotors). Since the rotor resistance is a function of frequency, the rotor copper loss must be calculated independently for each harmonic. While these additional losses used to be significant in the early days of PWM inverters with low switching frequencies, in modern drives with switching frequencies above 3 kHz the additional losses are minimal.

• Iron loss: This is increased by the harmonic components in the motor voltage.

For PWM voltage source inverters using sinusoidal modulation and switching frequencies of 3 kHz or higher, the additional losses are therefore primarily iron losses and are generally small, resulting in a loss of motor efficiency by 1–2%. Motors designed for enhanced efficiency, e.g. to meet the IEC IE2/IE3 requirements or NEMA EPACT and premium efficiency requirements, also experience a proportionately lower increase in losses with inverter supplies because of the use of reduced-loss magnetic steels.

However, the increase in losses does not directly relate to a derating factor for standard machines since the harmonic losses are not evenly distributed through the machine. The harmonic losses mostly occur in the rotor and have the effect of raising the rotor temperature. Whether or not the machine was designed to be stator critical (stator temperature defining the thermal limit) or rotor critical clearly has a significant impact of the need for, or magnitude of, any derating. The cooling system (see below) is at least as important, however, and in practice it emerges that a standard motor may have to be derated by 5 or even 10% for use on an inverter supply.

Whereas a d.c. motor was invariably supplied with through ventilation provided by an auxiliary blower, to allow it to operate continuously at low speeds without overheating, the standard induction motor has no such provision. Having been designed primarily for fixed-frequency/full-speed operation, most induction motors tend to be totally enclosed (IP44 or IP54) with a shaft-mounted fan at the non-drive end running within a cowl to duct the cooling air over a finned motor body as shown in Figure 8.4. Note also the cast 'paddles' on the rotor endrings which provide internal air circulation and turbulence to assist with transmitting the heat from the rotor to the stator housing and from there to the atmosphere.

Thus although the inverter is capable of driving the induction motor with full torque at low speeds, continuous operation *at rated torque* is unlikely to be possible because the standard shaft-mounted cooling fan will be less effective at reduced speed and the motor will overheat. We should say, however, that for applications such as fans and pumps where the load torque is proportional to the cube of the speed, no such problems exist, but for many applications it is a significant consideration.

4.4 Bearing currents

Scare stories periodically appear in the trade press and journals relating to motor failures in inverter-fed a.c. motors. It should be said immediately that such failures are rare, and mainly associated with medium voltage systems.

Figure 8.4 Typical shaft-mounted external cooling fan on an a.c. induction motor. *(Courtesy of Emerson – Leroy Somer.)* (See Plate 8.4)

With a balanced 3-phase sinusoidal supply the sum of the three stator currents in an a.c. motor is zero and there is no further current flow outside the motor. In practice, however, there are conditions which may result in currents flowing through the bearings of a.c. motors even when fed with a sinusoidal 50 or 60 Hz supply, and the risk is further increased when using an inverter supply. Any asymmetric flux distribution within an electrical machine can result in an induced voltage from one end of the rotor shaft to the other. If the bearing 'breakover voltage' is exceeded (the electrical strength of the lubricant film being of the order of 50 V) or if electrical contact is made between the moving and fixed parts of the bearing this will result in a current flowing through both bearings. The current is of low (often slip) frequency and its amplitude is limited only by the resistance of the shaft and bearings, so it can be destructive. In some large machines it is common and good practice to fit an insulated bearing, usually on the non-drive end, to stop such currents flowing.

Any motor may also be subject to bearing currents if its shaft is connected to machinery at a different ground potential from the motor frame. It is therefore important to ensure that the motor frame is connected through a low-inductance route to the structure of the driven machinery. This issue is well understood and with modern motors such problems are rare.

4.5 'Inverter grade' induction motors

Addressing the above potential hazards, induction motors bearing the name 'inverter grade' or similar are readily available. They would typically have reinforced insulation systems and have a thermal capacity for a constant torque operating range, often down to 30% of base speed, without the need for additional external cooling.

Further they would have options to fit thermocouples, a separate cooling fan (for very-low-speed operation) and a speed/position feedback device.

International standards exist to help users and suppliers in this complex area. NEMA MG1-2006, Part 31 gives guidance on operation of squirrel cage induction motors with adjustable-voltage and adjustable-frequency controls. IEC 60034-17 and IEC 60034-25 give guidance on the operation of induction motors with converter supplies, and design of motors specifically intended for converter supplies, respectively.

5. EFFECT OF THE INVERTER-FED INDUCTION MOTOR ON THE UTILITY SUPPLY

It is a common misconception to believe that the harmonic content of the motor current waveform and the motor power-factor are directly reflected on the utility supply, but this is not the case. The presence of the inverter, with its energy-buffering d.c. link capacitor, results in near unity power-factor as seen by the utility regardless of load or speed of operation, which is of course highly desirable. It is not all good news, however, so we now look at the impact of an inverter-fed drive on the utility.

5.1 Harmonic currents

Harmonic current is generated by the input rectifier of an a.c. drive. The essential circuit for a typical a.c. variable-speed drive is shown in Figure 8.1. The utility supply is rectified by the diode bridge, and the resulting d.c. voltage is smoothed by the d.c. link capacitor and, for drives rated typically at over 2.2 kW, the d.c. current is smoothed by an inductor in the d.c. circuit. The d.c. voltage is then chopped up in the inverter stage, which uses PWM to create a sinusoidal output voltage of adjustable voltage and frequency.

While small drive ratings may have a single-phase supply, we will consider a 3-phase supply. We see from Figure 8.5 that current flows into the rectifier as a series of pulses that occur whenever the supply voltage exceeds that of the d.c. link, which is when the diodes start to conduct. The amplitude of these pulses is much larger than the fundamental component, which is shown by the dashed line.

Figure 8.6 shows the spectral analysis of the current waveform in Figure 8.5.

Note that all currents shown in spectra comprise lines at multiples of the 50 Hz utility frequency. Because the waveform is symmetrical in the positive and negative half-cycles, apart from imperfections, even-order harmonics are present only at a very low level. The odd-order harmonics are quite high, but they diminish with increasing harmonic number. For the 3-phase input bridge there are no triplen harmonics, and by the 25th harmonic the level is negligible. The frequency of this harmonic for a 50 Hz supply is 1250 Hz, which is in the audio frequency part of the

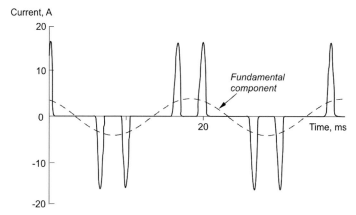

Figure 8.5 Typical current from utility supply for a 1.5 kW 3-phase drive.

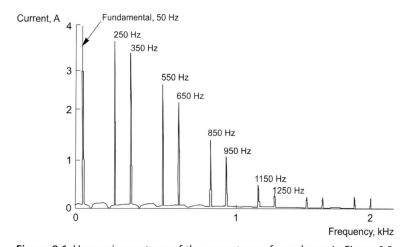

Figure 8.6 Harmonic spectrum of the current waveform shown in Figure 8.5.

electromagnetic spectrum and well below the radio-frequency part, which is generally considered to begin at 150 kHz. This is important, because it shows that supply harmonics are low-frequency effects, which are quite different from radio-frequency EMC effects. They are not sensitive to fine details of layout and screening of circuits, and any remedial measures which are required use conventional electrical power techniques such as tuned power-factor capacitors and phase-shifting transformers. This should not be confused with the various techniques used to control electrical interference from fast switching devices, sparking electrical contacts, etc.

The actual magnitudes of the current harmonics depend on the detailed design of the drive, specifically the values of d.c. link capacitance and, where used, d.c. link inductance, as well as the impedance of the utility system to which it is connected, and the other non-linear loads on the system.

We should make clear that industrial problems due to harmonics are unusual, although with the steady increase in the use of electronic equipment, they may be more common in the future. Problems have occurred most frequently in office buildings with a very high density of personal computers, and in cases where most of the supply capacity is used by electronic equipment such as drives, converters and uninterruptible power supplies (UPS).

As a general rule, if the total rectifier loading (drives, UPS, PCs, etc.) on a power system comprises less than 20% of its current capacity then harmonics are unlikely to be a limiting factor. In many industrial installations the capacity of the supply considerably exceeds the installed load, and a large proportion of the load is not a significant generator of harmonics – uncontrolled (direct-on-line) induction motors and resistive heating elements generate minimal harmonics.

If rectifier loading exceeds 20% then a harmonic control plan should be in place. This requires some experience and guidance can often be sought from equipment suppliers. The good news is that if it is considered that a problem will exist with the estimated level of harmonics then there are a number of options available to reduce the distortion to acceptable levels.

A.C. drives rated over 2.2 kW tend to be designed with inductance built into the d.c. link and/or the a.c. input circuit. This gives the better supply current waveform and dramatically improved spectrum as shown in Figures 8.7 and 8.8, respectively, which are again for a 1.5 kW drive for ease of comparison with the previous illustrations. (In this case the inductance in each line is specified as '2%', which means that when rated fundamental current flows in the line, the volt-drop across the inductor is equal to 2% of the supply voltage.) Note the change of vertical

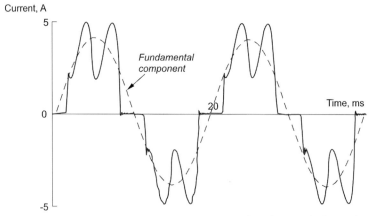

Figure 8.7 Input current waveform for the 3-phase 1.5 kW drive with d.c. and 2% a.c. inductors.

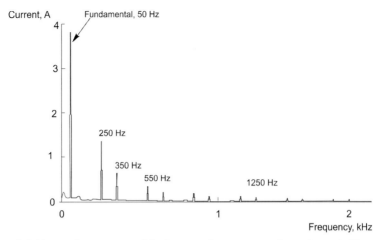

Figure 8.8 Harmonic spectrum of the improved current waveform shown in Figure 8.7.

scale between Figures 8.5 and 8.7, which may tend to obscure the fact that the pulses of current now reach about 5 A, rather than the 17 A or so previously, but the fundamental component remains at 4 A because the load is the same. (Remember that while we have just demonstrated the tremendous improvement in supply harmonics achieved by adding d.c. link inductance to a 1.5 kW drive, standard drives would rarely be manufactured with any inductance because while the harmonic spectrum looks worrying, the currents are at such a low level that they would rarely cause practical problems.)

Standard 3-phase drives rated up to about 200 kW tend to use conventional 6-pulse rectifiers. At higher powers, it may be necessary to increase the pulse number to improve the supply-side waveform, and this involves a special transformer with two separate secondary windings, as shown for a 12-pulse rectifier in Figure 8.9.

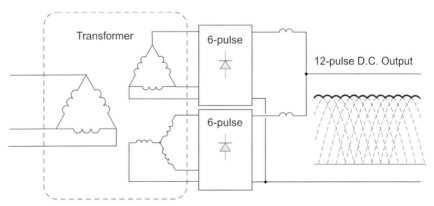

Figure 8.9 Basic 12-pulse rectifier arrangement.

The voltages in the transformer secondary star and delta windings have the same magnitude but a relative phase shift of 30°. Each winding has its own set of six diodes, and each produces a 6-pulse output voltage. The two outputs are generally connected in parallel, and, because of the phase shift, the resultant voltage consists of 12 pulses of 30° per cycle, rather than the six pulses of 60° shown, for example, in Figure 2.13.

The phase shift of 30° is equivalent to 180° at the fifth and seventh harmonics (as well as 17, 19, 29, 31, etc.), so that flux and hence primary current at these harmonics cancels in the transformer, and the resultant primary waveform therefore approximates well to a sinusoid, as shown for the 150 kW drive in Figure 8.10.

The use of drive systems with an input rectifier/converter using PWM which generates negligible harmonic current in the utility supply is becoming increasingly common. This also permits the return of power from the load to the supply, and is discussed later in section 7.

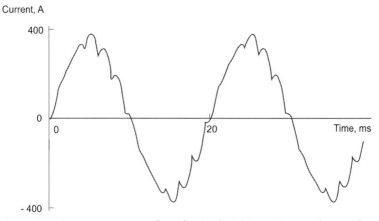

Figure 8.10 Input current waveform for 150 kW drive with 12-pulse rectifier.

5.2 Power-factor

The power-factor of an a.c. load is a measure of the ratio of the average power to the product of r.m.s. current and voltage, and is given by:

$$\text{Power-factor} = \frac{\text{Average power (W)}}{\text{r.m.s. volts} \times \text{r.m.s. amps}}$$

With a sinusoidal supply voltage and a linear load the current will also be sinusoidal, with a phase-shift of ϕ with respect to the voltage. The power is then given by the simple expression

$$W = VI \cos \phi,$$

where V and I are r.m.s. values (which are equal to the peak of the sinusoid divided by $\sqrt{2}$), and so in this case the power-factor is equal to cos ϕ. Clearly the maximum possible power-factor is 1.

Unfortunately, in power electronic circuits either the voltage or the current or both are non-sinusoidal, so there is no simple formula for the r.m.s. values or the mean power, all of which have to be found by integration of the waveforms. There is therefore no simple formula for the power-factor, but frequent use is made of a related quantity known as the fundamental power-factor, given by

$$\text{Fundamental power-factor} = \frac{\text{Average power}}{\text{Fundamental r.m.s. volts} \times \text{Fundamental r.m.s. amps}}$$

The influence of the harmonics in the non-sinusoidal waveforms causes the actual power-factor to be lower than the fundamental power-factor, so users should be aware that when suppliers quote the power-factor of a drive they are usually ignoring the harmonic currents, and quoting cos ϕ, the fundamental power-factor.

It may be worth reminding readers who are not familiar with industrial energy tariffs why maximizing the power-factor is important. All industrial users pay primarily for the energy used, which depends on the integrated total of the product of power and time, but most are also penalized for drawing the power at a low power-factor (because the currents are higher and therefore switchgear and cables have to be larger than would otherwise be necessary). In addition, there may be a penalty related to the maximum volt-ampere product in a specified period, so again a high power-factor is desirable.

Fortunately, for the diode bridge, which is the most common form of rectifier in a commercial a.c. drive, cos ϕ is close to unity for all speed and load conditions. To illustrate this, we can consider a typical 11 kW induction motor operating at full-load, connected either directly to the mains supply or through an a.c. variable-speed drive. Comparative figures are given in Table 8.2.

Table 8.2 Comparative figures for a typical 11 kW induction motor operating at full-load.

At supply terminals	DOL motor	Motor via a.c. drive	Notes on drive parameters
Voltage (V)	400	400	
R.m.s. current (A)	21.1	21.4	No significant change
Fundamental current (A)	21.1	18.8	Reduced because magnetizing current is not drawn directly from the utility supply
Fundamental power factor (cos ϕ)	0.85	0.99	Improved because input rectifier current is in phase with supply voltage
Power (W)	12,440	12,700	Slight increase at full-load due to drive losses

A typical PWM induction motor drive improves the power-factor as compared with a direct-on-line motor because it reduces the requirement of the supply to provide the magnetizing current for the motor, but in return generates harmonics. Power consumption at full-load is slightly increased due to losses of the drive.

6. INVERTER AND MOTOR PROTECTION

We have stressed before that power semiconductors are notoriously intolerant of excess current, and so even in the earliest drives of this type, current was measured in order to trip the drive when a simple current threshold was exceeded and before damage to the inverter could be done. Some protection schemes would also sense high currents and reduce the applied frequency to reduce the current.

The stored energy in the drive and motor inductances and capacitances also needs to be handled without inducing voltages or currents which can damage the system components. As previously mentioned, the basic power circuit is not inherently capable of regenerating energy back into the supply, and, when a braking duty results in energy flow into the d.c link, a 'dump resistor' (see section 7) of the right thermal rating needs to be provided in order to limit the circuit voltages.

Motor protection also requires current measurement, but here it is thermal protection of the motor that is of concern. A very approximate indication of the losses or heating effect in the motor is obtained by monitoring the product of the square of the motor current times time. This so-called i^2t protection is still referred to as motor thermal protection in drives, though many of the thermal algorithms now employed are very much more complex and accurate than their primitive predecessors.

Modern commercial drives include extensive internal protection systems as well as thermal motor modeling systems, but such drives are designed for a multiplicity of applications and motor designs and so must be configured during installation. Where multiple motors are fed from a single inverter (as described in section 3) each motor must have individual thermal trips, because the fault current of an individual motor may not be significant when a large number of motors are connected to the same inverter.

7. ALTERNATIVE CONVERTER TOPOLOGIES

7.1 Braking

The inverter (motor converter) that forms the output stage of the popular voltage source a.c. motor drives described above inherently allows the motor to be controlled in either direction of rotation and also allows power to flow in either direction between the d.c link and the motor, as shown in Figure 8.1.

The simple diode rectifier that is often used between the utility supply and the d.c. terminals of the inverter does not allow power to flow back into the supply. Therefore an a.c. motor drive based on this configuration cannot be used where power is required to flow from the motor to the utility supply. On the face of it, this limitation might be expected to cause a problem in almost all applications when shutdown of a process is involved and the machine being driven by the motor has to be braked by the motor. The kinetic energy has to be dissipated, and during active deceleration power flows from the motor to the d.c. link, thereby causing the voltage across the d.c. link capacitor to rise to reflect the extra stored energy.

To prevent the power circuit from being damaged, d.c. link overvoltage protection is included in most industrial drive control systems. This shuts down the inverter if the d.c. link voltage exceeds a trip threshold, but this is at best a last resort. It may be possible to limit the voltage across the d.c. link capacitor by restricting the energy flow from the motor to the d.c. link by controlling the slowdown ramp, but again this is not always acceptable to the user.

To overcome the limited deceleration possible with the scheme described above, the diode rectifier can be supplemented with a d.c. link braking resistor circuit as shown in Figure 8.11.

In this arrangement, when the d.c. link voltage exceeds the braking threshold voltage, the switching device is turned on. Provided the braking resistor has a low enough resistance to absorb more power than the power flowing from the inverter, the d.c. link voltage will begin to fall and the switching device will be turned off again. In this way the on and off times are automatically set depending on the power from the inverter, and the d.c. link voltage is limited to the braking threshold. To limit the switching frequency of the braking circuit, appropriate hysteresis is included in its control.

A braking resistor circuit is used in many applications where it is practical and acceptable to dissipate stored kinetic energy in a resistor, and hence there is no longer a limit on the dynamic performance of the system.

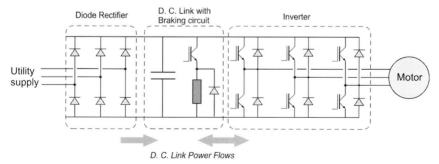

Figure 8.11 Induction motor drive with a braking resistor circuit.

7.2 Active front end

The diode rectifier does not allow power flow from the d.c. link to the supply. There are various circuits that can be used to recover the load energy and return it to the supply, one of which is the active rectifier shown in Figure 8.12, in which the diode rectifier is replaced with an IGBT inverter.

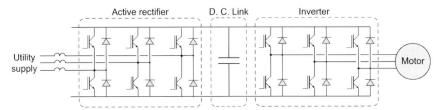

Figure 8.12 Induction motor drive with active front-end rectifier.

The labeling of the two converters shown in Figure 8.12 reflects their functions when the drive is operating in its normal 'motoring' sense, but during active braking (or even continuous generation) the converter on the left will be inverting power from the d.c. link to the utility supply, while that on the right will be rectifying the output from the induction generator.

This arrangement if often referred to as an 'active front end'. Additional input inductors are usually required to limit the unwanted currents generated by the switching action of the inverter, but by using PWM control the front-end converter can be controlled to give near sinusoidal current waveforms with a power-factor close to unity, so that the complete system presents a near perfect load to the utility supply.

An active rectifier is used where full four-quadrant operation and good quality input waveforms are required. Cranes and elevators, engine test rigs and cable laying ships are some applications where an active rectifier may be appropriate. The performance characteristics of this configuration are summarized below:

- Power flow between the motor and the mains supply is possible in both directions, and so this makes the drive more efficient than when a braking resistor is used.
- Good quality input current waveforms, i.e. low harmonic distortion of the utility supply.
- Supply power-factor can be controlled to near unity.

Clearly, however, an active rectifier is more expensive than a simple diode rectifier and a brake resistor.

7.3 Multi-level inverter

The inverter drive circuit described above is widely used in drive systems rated to around 2 MW at voltages from 400 V but is more usually seen at higher voltages including in medium voltage inverters (2–11 kV). At higher powers, switching the current in the devices proves more problematic in terms of the losses, and so switching frequencies have to be reduced. At higher voltages the impact of the rate of change of voltage on the motor insulation causes switching times to be extended and hence losses increase. In addition, while higher voltage power semiconductors are available, they tend to be relatively expensive and so commercial consideration is given to the series connection of devices, but here voltage sharing between devices is a problem due to the disproportionate impact of any small difference between switching characteristics.

Multi-level converters have been developed which address these issues. One example from among many different topologies is shown in Figure 8.13.

In this example four capacitors act as potential dividers to provide four discrete voltage levels, and each arm of the bridge has four series-connected IGBTs with anti-parallel diodes. The four intermediate voltage levels are connected by means of clamping diodes to the link between the series IGBTs. To obtain full voltage between the outgoing lines all four devices in one of the upper arms are switched on, together with all four in a different lower arm, while for say half voltage only the bottom two in an upper arm are switched on. The quarter and three-quarter levels can be selected in a similar fashion, and in this way a good 'stepped' approximation to a sinewave can be achieved.

Clearly there are more devices turned on simultaneously than in a basic inverter, and this gives rise to an unwelcome increase in the total conduction loss. But this is offset by the fact that because the VA rating of each device is smaller than that of the equivalent single device, the switching loss is much reduced.

For the inverter in Figure 8.13 the stepped waveform will have four positive, four negative and a zero level(s). The oscilloscope trace in Figure 8.14 is from a six-level

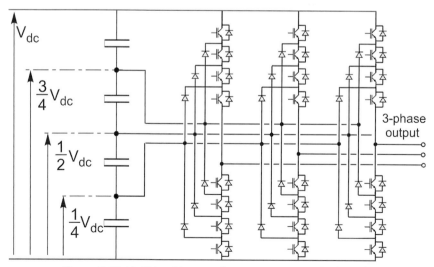

Figure 8.13 Multi-level inverter (motor converter only shown).

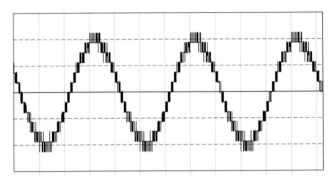

Figure 8.14 Six-level inverter output voltage waveform.

inverter, and clearly displays the six discrete levels. A sophisticated modulation strategy is required in order not only to achieve a close approximation to a sinewave, but also to ensure that the reduction of charge (and voltage) across a capacitor during its periods of discharge period is compensated by a subsequent charging current.

Multi-level inverters have advantages over the conventional PWM inverter:

- Higher effective output switching frequency for a given PWM frequency. Smaller filter components are required.
- Improved EMC due to lower dV/dt at output terminals – less stress on the motor insulation.
- Higher d.c. link voltages achievable for medium voltage applications due to voltage sharing of power devices within each inverter leg.

However, there are drawbacks:

- The number of power devices is increased by at least a factor of two; each IGBT requires a floating gate drive and power supply; and additional voltage clamping diodes are required.
- The number of d.c. bus capacitors may increase, but this is unlikely to be a practical problem as lower voltage capacitors in series are likely to be the most cost-effective solution.
- The balancing of the d.c. link capacitor voltages requires careful management/control.
- Control/modulation schemes are more complicated.

For low-voltage drives (690 V and below) the disadvantages of using a multi-level inverter tend to outweigh the advantages, with even three-level converters being significantly more expensive than conventional topologies. However, multi-level converters are entering this market, so the situation may change.

7.4 Cycloconverter

The operation of the cycloconverter was discussed in Chapter 2, so here – for completeness – we need only to recap the main features and areas of application.

The drive is inherently four quadrant; its maximum output frequency is limited to approximately half the supply frequency by considerations related to harmonics in the motor currents and torque, stability, and dimensions of the drive components. It is thus suited to large low-speed drives, where motors can be made with high pole-number and thus low synchronous speed. The complexity of the drive means that only high power systems ($>>1$ MW), or specialized applications (e.g. conveyor drives for use in hazardous environments) are economic. They are used on large ball mills, mine winders, etc.

Due to the modulation of the converter firing angles, the harmonic content of the utility supply is complex and designs for appropriate harmonic filters become somewhat involved.

7.5 Matrix converter

Recently, much attention has been focused in technical journals on the matrix converter, the principle of which is shown in Figure 8.15.

The matrix converter operates in much the same way as a cycloconverter. For example, if we assume that we wish to synthesize a 3-phase sinusoidal output of a known voltage and frequency, we know at every instant what voltage we want, say between the lines A and B, and we know what the voltages are between the three incoming lines. So we switch on whichever pair of the A and B switches connects us to the two incoming lines whose voltage at the time is closest to the desired output line-to-line voltage, and we stay with it whenever it offers the best approximation to what we want. As soon as a different combination of switches would allow us to hook onto a more appropriate pair of input lines, the switching pattern changes.

Because there are only three different incoming line-to-line voltages to choose from, we cannot expect to synthesize a decent sinusoidal waveform with simple switching, so in order to obtain a better approximation to a sinusoidal waveform we must use chopping. This means that the switches have to be capable of operating at much higher frequencies than fundamental output frequency, so that switching loss becomes an important consideration.

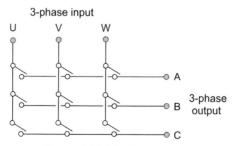

Figure 8.15 Matrix converter.

While the basic circuit is not new, recent advances in power devices offer the potential to overcome many of the drawbacks inherent in the early implementations that used discrete IGBTs due to the lack of suitable packaged modules. On the other hand, the fact that the output voltage is limited to a maximum of 86% of the utility supply voltage means that applications in the commercial industrial market, where standard motors predominate, remain largely problematic. However, there appear to be good prospects in some aerospace applications and possibly in some specific areas of the industrial drives market, notably integrated motors (where the drive is built into the motor, the windings of which can therefore be designed to suit the voltage available).

7.6 PWM voltage source inverter with small d.c. smoothing capacitance

For a 3-phase rectifier, the d.c. link capacitance value can be much reduced provided that the modulation strategy in the inverter is adapted to compensate for the resulting voltage ripple. The input current waveform is then improved, as compared with the very peaky waveform seen in Figure 8.5. This is a useful technique for cost-sensitive applications where harmonic current is a critical factor. However, there are disadvantages resulting from the reduced capacitance, in that the d.c. link voltage becomes more sensitive to transient conditions, both from rapid variations in load and from supply disturbances. This approach is therefore most attractive in applications where the load does not exhibit highly dynamic behavior.

Another practical factor is that the capacitor now has a disproportionately high ripple current, so that a conventional aluminum electrolytic capacitor cannot be used and a more expensive capacitor with a plastic dielectric is required.

Converter circuits with zero d.c. link capacitance exist in the literature. They have no energy storage, so it is evident that this topology is another form of matrix converter, with the output voltage being made up of 'chunks' of the utility supply waveform.

7.7 Current source induction motor drives

The majority of inverters used in motor drives are voltage source inverters (VSIs), in which the output voltage to the motor is controlled to suit the operating conditions of the motor. Current source inverters (CSIs) are still sometimes used, particularly for high-power applications, and warrant a brief mention.

The forced-commutated current-fed induction motor drive, shown in Figure 8.16, was strongly favored for single motor applications for a long period, and was available at power levels in the range 50–3500 kW at voltages normally up to 690 V. High-voltage versions at 3.3/6.6 kV were also developed but they have not proved to be economically attractive. Today it is not seen as having merit and

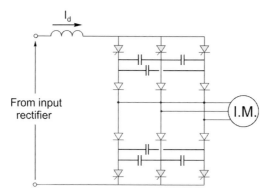

Figure 8.16 Forced-commutated current-fed induction motor drive (motor converter only shown).

has virtually disappeared from the portfolios of most companies. A brief description is included here for interest only.

The d.c. link current I_d, taken from a 'stiff' current source (usually in the form of a thyristor bridge and a series inductor in the d.c. link), is sequentially switched at the required frequency into the stator windings of the induction motor. The capacitors and extra series diodes provide the mechanism for commutating the thyristors by cleverly exploiting the reversal of voltage resulting from resonance between the capacitor and the motor leakage reactance. The resultant motor voltage waveform is, perhaps somewhat surprisingly, approximately sinusoidal apart from the superposition of voltage spikes caused by the rise and fall of machine current at each commutation.

The operating frequency range is typically 5–60 Hz, the upper limit being set by the relatively slow commutation process. Below 5 Hz, torque pulsations can be problematic but PWM control of the current can be used at low frequencies to ease the problem.

This system was most commonly used for single motor applications such as fans, pumps, extruders, and compressors, where very good dynamic performance is not necessary and a supply power-factor which decreases with speed is acceptable.

CHAPTER NINE

Synchronous and Brushless Permanent Magnet Machines and Drives

1. INTRODUCTION

Chapters 6 to 8 have described the virtues of the induction motor and how, when combined with power electronic control, it is capable of meeting the performance and efficiency requirements of many of the most demanding applications. In this chapter another group of a.c. motors is described. In all of them the electrical power that is converted to mechanical power is fed into the stator, so, as with the induction motor, there are no sliding contacts in the main power circuits. All also have stators that are identical (or very similar) to the induction motor, although some new constructional and winding techniques involving segmented construction are being applied at the lower-power end to improve the power density of the machine and/or reduce cost.

It has to be admitted that the industrial and academic communities have served to make life confusing in this area by giving an array of different names to essentially the same machine, so we begin by looking at the terminology. The names we will encounter include:

1. Conventional *synchronous machine* with its rotor field winding (excited-rotor). This is the only machine that may have brushes, but even then they will only carry the rotor excitation current, not the main a.c. power input.
2. *Permanent magnet synchronous machine* with permanent magnets replacing the rotor field winding.
3. *Brushless permanent magnet synchronous motor* (same as (2)). The prefix 'brushless' is superfluous.
4. *Brushless a.c. motor* (same as (2)).
5. *Brushless d.c. motor* (same as (2) except for detailed differences in the field patterns). This name was coined in the 1970s to describe 'inside-out' motors that were intended as direct replacements for conventional d.c. motors, and in this sense it has some justification.
6. *Permanent magnet servo motor* (same as (2)).

We begin by looking at motors that are intended, or at least have the capability, to be operated directly off the utility supply, usually at either 50 or 60 Hz. These are known as 'synchronous' motors, and they operate at a specific and constant speed for a wide range of loads, and therefore can be used in preference to induction motors when precise (within the tolerance of the utility frequency), constant speed operation is essential: there is no load-dependent slip as is unavoidable with the induction motor. These machines are available over a very wide range from tiny single-phase versions in domestic clocks to multi-megawatt machines in large industrial applications such as gas compressors. The clock application means that utility companies have a responsibility to ensure that the average frequency over a 24-hour period always has to be precisely the rated frequency of the supply in order to keep us all on time. Ironically, in order to do this, they control the speed of very large, turbine-driven, synchronous machines that generate the vast majority of the electrical power throughout the world.

To overcome the fixed-speed limitation that results from the constant frequency of the utility supply, inverter-fed synchronous motor drives have been developed. We will see that all forms of this generic technology use a variable-frequency inverter to provide for variation of the synchronous speed, but that in almost all cases, the switching pattern of the inverter (and hence the frequency) is determined by the rotor position and not by an external oscillator. In such so-called self-synchronous drives, the rotor is incapable of losing synchronism and stalling (which is one of the main drawbacks of the utility-fed machine).

We will also see how field-oriented control can be applied to synchronous machines to achieve the highest levels of performance and efficiency with machines which have higher inherent power densities than the equivalent induction motors.

Finally, reluctance motors will be considered briefly. Their simple rotor construction appears to offer great opportunity, but we will see that their low operating power-factor restricts their star rating. It does, however, remain a commercially available product and warrants review. For completeness hysteresis motors also get a mention.

2. SYNCHRONOUS MOTORS

In Chapter 5 we saw that the 3-phase stator windings of an induction motor produce a sinusoidal rotating magnetic field in the air-gap. The speed of rotation of the field (the synchronous speed) was shown to be directly proportional to the supply frequency, and inversely proportional to the pole-number of the winding. We also saw that in the induction motor the rotor is dragged along by the field, but that the higher the load on the shaft, the more the rotor has to slip with respect to the field in order to induce the rotor currents required to produce the torque. Thus although at no-load the speed of the rotor can be close to the synchronous speed, it must always be less; and as the load increases, the speed has to fall.

In the synchronous motor, the stator windings are essentially the same as in the induction motor, so when connected to the 3-phase supply, a rotating magnetic field is produced. However, instead of having a cylindrical rotor with a cage winding, the synchronous motor has a rotor with either a d.c. excited winding (supplied via sliprings, or on larger machines an auxiliary exciter[1]), or permanent magnets, designed to cause the rotor to 'lock on' or 'synchronize with' the rotating magnetic field produced by the stator. Once the rotor is synchronized, it will run at exactly the same speed as the rotating field despite load variation, so under constant-frequency operation the speed will remain constant as long as the supply frequency is stable.

As previously shown, the synchronous speed (in rev/min) is given by the expression

$$N_s = \frac{120f}{p}$$

where f is the supply frequency and p is the pole-number of the winding. Hence for 2-, 4- and 6-pole industrial motors the running speeds on a 50 Hz supply are 3000, 1500 and 1000 rev/min, while on a 60 Hz supply they become 3600, 1800 and 1200 rev/min, respectively. At the other extreme, the little motor in a timeswitch with its cup-shaped rotor with 20 axially projecting fingers and a circular coil in the middle is a 20-pole reluctance (synchronous) motor that will run at 300 rev/min when fed from a 50 Hz supply. Users who want speeds different from these discrete values will be disappointed, unless they are prepared to invest in a variable-frequency inverter.

With the synchronous machine, we find that there is a limit to the maximum (pull-out) torque which can be developed before the rotor is forced out of synchronism with the rotating field. This 'pull-out' torque will typically be 1.5 times the continuous rated torque but can be designed to be as high as 4 or even 6 times higher in the case of high-performance permanent magnet motors where, for example, high accelerating torques are needed for relatively short periods. For all torques below pull-out the steady running speed will be absolutely constant. The torque–speed curve is therefore simply a vertical line at the synchronous speed, as shown in Figure 9.1. We can see from Figure 9.1 that the vertical line extends into quadrant 2, which indicates that if we try to force the speed above the synchronous speed the machine will act as a generator.

Traditionally, utility-fed synchronous motors were used where a constant speed is required, high efficiency desirable, and power-factor controllable. They were also used in some applications where a number of motors were required to run at

---

[1] An auxiliary exciter is simply a second, smaller machine with a 3-phase stator and rotor winding, mounted on the same shaft. A 3-phase supply on the stator is 'transformer coupled' to the rotor winding. The induced e.m.f. in the rotor is rectified and fed to the main motor field winding. The phase rotation of the supply to the auxiliary stator is opposite to that of the main motor so that when the motor comes to high speed the induced rotor e.m.f. remains high.

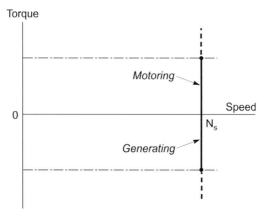

Figure 9.1 Steady-state torque–speed curve for a synchronous motor supplied at constant frequency.

precisely the same speed. However, a group of utility-fed synchronous motors could not always replace mechanical shafting[2] because while their rotational speed would always be matched, the precise relative rotor angle of each motor would vary depending on the load on the individual motor shafts.

We will now look briefly at the various types of synchronous motor, mentioning the advantages and disadvantages of each. The excited-rotor type is considered first because of its importance in large sizes but also because its behavior can be readily analyzed, and its mechanism of operation illuminated, by means of a relatively simple equivalent circuit. We then consider the permanent magnet type that is the most numerous in the drives arena, from which we can also benefit by dint of the use of the same equivalent circuit. Parallels are drawn with both the d.c. motor and the induction motor to emphasize that despite their obvious differences, most electrical machines also have striking similarities.

2.1 Excited-rotor motors

The rotor of a conventional synchronous machine carries a 'field' winding which is supplied with direct current either via a pair of sliprings on the shaft, or via an auxiliary brushless exciter on the same shaft. The field winding is designed to produce an air-gap field of the same pole–number and spatial distribution (usually sinusoidal) as that produced by the stator winding. The rotor may be more or less cylindrical, with the field winding distributed in slots (Figure 9.2(a)), or it may have

---

[2] Drive shafts were used in early textile factories and the like. A single mechanical shaft would be fed through the factory or machine and various functions would be connected to the same shaft via belts. The connected equipment would then run up and down together. With toothed belts, position synchronism could be achieved.

(a)   (b)

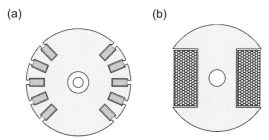

Figure 9.2 Rotors for synchronous motors. (a) 2-pole cylindrical, with field coils distributed in slots; (b) 2-pole salient pole with concentrated field winding.

projecting ('salient') poles around which the winding is concentrated (Figure 9.2(b)). A cylindrical-rotor motor has little or no reluctance (self-aligning) torque, so it can only produce torque when current is fed into the rotor. On the other hand, the salient-pole type also produces some reluctance torque even when the rotor winding has no current. In both cases, however, the rotor 'excitation' power is relatively small, since all the mechanical output power is supplied from the stator side.

Excited-rotor motors are used in sizes ranging from a few kW up to many MW. The large ones are effectively alternators (as used for power generation) but used as motors. Wound-rotor induction motors (see Chapter 6) can also be made to operate synchronously by supplying the rotor with d.c. through the sliprings.

The simplest way to visualize the mechanism of torque production is to focus on a static picture, and consider the alignment force between the stator and rotor field patterns. When the two are aligned with N facing S, the torque is zero and the system is in stable equilibrium, with any displacement to right or left causing a restoring torque to come into play. If the fields are distributed sinusoidally in space, the restoring torque will reach a maximum when the poles are misaligned by half a pole-pitch, or 90°. Beyond 90° the torque reduces with angle, giving an unstable region, zero torque being reached again when N is opposite N.

When the motor is running synchronously, we can use much the same mental picture because the field produced by the 3-phase alternating currents in the stator windings rotates at precisely the same speed as the field produced by the d.c. current in the rotor. At no-load there is little or no angular displacement between the field patterns, because the torque required to overcome friction is small. But each time the load increases, the rotor slows momentarily before settling at the original speed but with a displacement between the two field patterns that is sufficient to furnish the torque needed for steady-state running. This angle is known as the 'load angle', and we can actually see it when we illuminate the shaft of the motor with a supply-frequency stroboscope: a reference mark on the shaft is seen to drop back by a few degrees each time the load is increased. (Note that this is a characteristic of a synchronous machine when connected directly to the utility supply. We will see later in this chapter that when under inverter control with position feedback, precise angular position can be maintained regardless of load, which is important when coordinated motion is required, for example in X–Y tables where a precise path/contour must be followed.)

2.2 Permanent magnet motors

The synchronous machines considered so far require two electrical supplies, the first to feed the field/excitation and the second to supply the stator. Brushless permanent magnet machines have magnets attached to the rotor to provide the field, and so only a stator supply is required. The principle is illustrated for 2-pole and 4-pole surface-mounted versions in Figure 9.3, the direction in which the magnets have been magnetized being represented by the arrows. Motors of this sort have typical output ranging from about 100 W up to perhaps 500 kW, though substantially higher ratings have been made. (We should also mention that the magnets are sometimes buried within the rotor iron, which gives rise to an additional reluctance torque component. Although we discuss reluctance motors briefly later in this chapter, we will only deal with surface-mounted permanent magnet motors here.)

The advantages of the permanent magnet type are that no supply is needed for the rotor and the rotor construction can be robust and reliable. The disadvantage is that the excitation is inherently fixed, so the designer must either choose the shape and disposition of the magnets to match the requirements of one specific load, or seek a general-purpose compromise. Control of power-factor via excitation is of course no longer possible. Within these constraints the brushless permanent magnet synchronous motor behaves in very much the same way as its excited-rotor sister.

We will defer further discussion of some important constructional details of permanent magnet motors to a later section, and concentrate now on the tools available to help us understand the operating characteristics of synchronous motors.

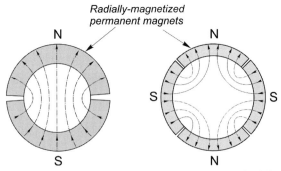

Figure 9.3 Permanent magnet synchronous motor rotors: 2-pole (left); 4-pole (right).

3. EQUIVALENT CIRCUITS OF SYNCHRONOUS MOTORS

Predicting the current and power-factor drawn from the supply by a cylindrical-rotor or permanent magnet synchronous motor is possible by means of the per-phase a.c. equivalent circuit shown in Figure 9.4. To arrive at such a simple circuit inevitably means that approximations have to be made, but we are seeking only a broad-brush picture, so the circuit is perfectly adequate.

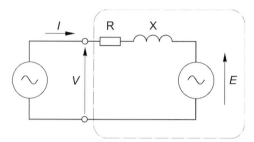

Figure 9.4 Equivalent circuit for synchronous machine.

In this circuit X (known as the synchronous reactance, or simply the reactance) represents the effective inductive reactance of the stator phase winding; R is the stator winding resistance; V is the applied voltage; and E is the e.m.f. induced in the stator winding by the rotating field produced either by the d.c. current on the rotor or by the permanent magnet. (For the benefit of readers who are familiar with the parameters of the induction motor, it should be pointed out that X is equal to the sum of the magnetizing and leakage reactances, but because the effective air-gap in synchronous machines is usually larger than in induction motors, their per-unit synchronous reactance is usually lower than that of an induction machine with the same stator winding.)

The similarity between this circuit and that of the d.c. machine (Figure 3.6) and the induction motor (Figure 5.8) is clear, and it stems from the fact that these machines all produce torque by the interaction of a magnetic field and current-carrying conductors (the so-called 'BIl' effect). In the case of the d.c. machine the inductance was seen not to be important under steady-state conditions because the current was steady (i.e. d.c.), and the resistance emerged as the dominant parameter. In the case of the synchronous motor the current is alternating, so not surprisingly we find that the reactance is the dominant impedance and resistance plays only a minor role, except in the case of small motors.

4. OPERATION FROM CONSTANT-VOLTAGE, CONSTANT-FREQUENCY (UTILITY) SUPPLY

At this point, readers who are not familiar with a.c. circuit theory can be reassured that they will not be seriously disadvantaged by skipping the rest of this section. But although no seminal truths are to follow, discussion of the equivalent circuit and the associated phasor diagram greatly assists the understanding of motor behavior, especially its ability to operate over a range of power-factors.

4.1 Excited-rotor motor

Our aim is to find what determines the current drawn from the supply, which from Figure 9.4 clearly depends on all the parameters therein. But for a given machine operating from a constant-voltage, constant-frequency supply, the only independent variables are the load on the shaft and the d.c. current (the excitation) fed into the rotor, so we will look at the influence of both, beginning with the effect of the load on the shaft.

The speed is constant and therefore the mechanical output power (torque times speed) is proportional to the torque being produced, which in the steady state is equal and opposite to the load torque. Hence if we neglect the losses in the motor, the electrical input power is also determined by the load on the shaft. The input power per phase is given by $VI \cos \phi$ where I is the current and the power-factor angle is ϕ. But V is fixed, so the in-phase (or real) component of input current $I \cos \phi$ is determined by the mechanical load on the shaft. We recall that, in the same way, the current in the d.c. motor (Figure 3.6) was determined by the load. This discussion reminds us that although the equivalent circuits in Figures 9.4 and 3.6 are very informative, they should perhaps carry a 'health warning' to the effect that the single most important determinant of the current (the load torque) does not actually appear explicitly on the diagrams.

Turning now to the influence of the d.c. excitation current, at a given supply frequency (i.e. speed) the utility-frequency e.m.f. (E) induced in the stator is proportional to the d.c. field current fed into the rotor. If we wanted to measure this e.m.f. we could disconnect the stator windings from the supply, drive the rotor at synchronous speed by an external means, and measure the voltage at the stator terminals, performing the so-called open-circuit test. If we were to vary the speed at which we drove the rotor, keeping the field current constant, we would of course find that E was proportional to the speed. We discovered a very similar state of affairs when we studied the d.c. machine (Chapter 3): its induced motional or 'back' e.m.f. (E) turned out to be proportional to the field current, and to the speed of rotation of the armature. The main difference between the d.c. machine and the synchronous machine is that in the d.c. machine the field is stationary and the armature rotates, whereas in the synchronous machine the field system rotates while the stator windings are at rest: in other words, one could describe the synchronous machine, loosely, as an 'inside-out' d.c. machine.

We also saw in Chapter 3 that when the unloaded d.c. machine was connected to a constant-voltage d.c. supply, it ran at a speed such that the induced e.m.f. was (almost) equal to the supply voltage, so that the no-load current was almost zero.

When a load was applied to the shaft, the speed fell, thereby reducing E and increasing the current drawn from the supply until the motoring torque produced was equal to the load torque. Conversely if we applied a driving torque to the shaft of the machine, the speed rose, E became greater than V, current flowed out to the supply and the machine acted as a generator. These findings are based on the assumption that the field current remains constant, so that changes in E are a reflection of changes in speed. Our overall conclusion was the simple statement that if E is less than V, the d.c. machine acts as a motor, while if E is greater than V, it acts as a generator.

The situation with the synchronous motor is similar, but now the speed is constant and we can control E independently via control of the d.c. excitation current fed to the rotor. We might again expect that if E was less than V the machine would draw in current and act as a motor, and vice versa if E was greater than V. But we are no longer dealing with simple d.c. circuits in which phrases such as 'draw in current' have a clear meaning in terms of what it tells us about power flow. In the synchronous machine equivalent circuit the voltages and currents are a.c., so we have to be more careful with our language and pay due respect to the phase of the current, as well as its magnitude. Things turn out to be rather different from what we found in the d.c. motor, but there are also similarities.

4.2 Phasor diagram and power-factor control

To see how the magnitude of the e.m.f. influences behavior we can examine the phasor diagrams of a synchronous machine operating as a motor, as shown in Figure 9.5.

The first point to clarify is that our sign convention is such that motoring corresponds to positive electrical input power to the machine. The power is given by $VI \cos \phi$, so when the machine is motoring (positive power) the angle ϕ lies in

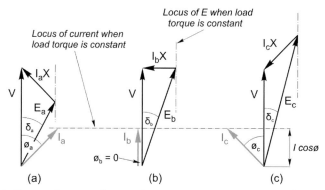

Figure 9.5 Phasor diagrams for synchronous motor operating with constant load torque, for three different values of the rotor (excitation) current.

the range $\pm 90°$. If the current lags or leads the voltage by more than $90°$ the machine will be generating.

Figure 9.5 shows three phasor diagrams corresponding to low, medium and high values of the induced e.m.f. (E), the shaft load (i.e. mechanical power) being constant. As discussed above, if the mechanical power is constant, so is $I \cos \phi$, and the locus of the current is therefore shown by the horizontal dashed line. The load angle (δ), discussed earlier, is the angle between V and E in the phasor diagram. In Figure 9.5, the voltage phasor diagram embodies Kirchhoff's law as applied to the equivalent circuit in Figure 9.4, i.e. $V = E + IR + jIX$, but for the sake of simplicity R is neglected so the phasor diagram simply consists of the volt-drop IX (which leads the current I by $90°$) added to E to yield V.

Figure 9.5(a) represents a condition where the field current has been set so that the magnitude of the induced e.m.f. (E) is less than V. This is called an 'under-excited' condition, and as can be seen the current is lagging the terminal voltage and the power-factor is $\cos \phi_a$, lagging. When the field current is increased (increasing the magnitude of E) the magnitude of the input current reduces and it moves more into phase with V: the special case shown in Figure 9.5(b) shows that the motor can be operated at unity power-factor if the field current is suitably chosen. Finally, in Figure 9.5(c), the field current is considerably higher (the 'overexcited' case), which causes the current to increase again but this time the current leads the voltage and the power-factor is $\cos \phi_c$, leading. We see that we can obtain any desired power-factor by appropriate choice of rotor excitation: this is a freedom not afforded to users of induction motors, and arises because in the synchronous machine there is an additional mechanism for providing excitation, as we will now discuss.

When we studied the induction motor we discovered that the magnitude and frequency of the supply voltage V governed the magnitude of the resultant flux density wave in the machine, and that the current drawn by the motor could be considered to consist of two components. The real (in-phase) component represented the real power being converted from electrical to mechanical form, so this component varied with the load. On the other hand, the lagging reactive (quadrature) component represented the 'magnetizing' current that was responsible for producing the flux, and it remained constant regardless of load.

The stator winding of the synchronous motor is the same as the induction motor, so it is to be expected that the resultant flux will be determined by the magnitude and frequency of the applied voltage. This flux will therefore remain constant regardless of the load, and there will be an associated requirement for magnetizing m.m.f. But now we have two possible means of providing the excitation m.m.f, namely the d.c. current fed into the rotor and the lagging component of current in the stator.

When the rotor is underexcited, i.e. the induced e.m.f. E is less than V (Figure 9.5(a)), the stator current then has a lagging component to make up for the shortfall in excitation needed to yield the resultant field that must be present as determined by the terminal voltage V. With more field current (Figure 9.5(b)), however, the rotor excitation alone is sufficient and no lagging current is drawn by the stator. And in the overexcited case (Figure 9.5(c)), there is so much rotor excitation that there is effectively some reactive power to spare and the leading power-factor represents the export of lagging reactive power that could be used to provide excitation for induction motors elsewhere on the same system.

To conclude our look at the excited-rotor motor we can now quantify the qualitative picture of torque production we talked about earlier, by noting from the phasor diagrams that if the mechanical power (i.e. load torque) is constant, the variation of the load angle (δ) with E is such that E sin δ remains constant. As the rotor excitation is reduced, and E becomes smaller, the load angle increases until it eventually reaches its maximum of $90°$, at which point the rotor will lose synchronism and stall. This means that there will always be a lower limit to the excitation required for the machine to be able to transmit the specified torque. This is just what our simple mental picture of torque being developed between two magnetic fields, one of which becomes very weak, would lead us to expect.

4.3 Permanent magnet motor

Although the majority of permanent magnet motors are supplied from variable-frequency inverters, some are directly connected to the utility supply, and we can explore their behavior using the equivalent circuit shown in Figure 9.4. Because the permanent magnet acts as a source of constant excitation, we no longer have control over the magnitude of the induced e.m.f. E, which now depends on the magnet strength and the speed, the latter being fixed by the utility frequency. So now we only have the load torque as an independent variable, and, as we saw earlier, because the supply voltage is constant, the load torque determines the in-phase or work component of the stator current $I \cos \phi$, as indicated in the phasor diagrams in Figure 9.5.

In order to identify which of the three diagrams in Figure 9.5 applies to a particular motor we need to know the motional e.m.f. (E) with the rotor spinning at synchronous speed and the stator open-circuited. If E is less than the utility voltage, diagram (a) applies; the motor is said to be underexcited; and it will have a lagging power-factor that worsens with load. Conversely, if E is greater than V (the overexcited case), diagram (b) or (c) is typical, and the power-factor will be leading.

4.4 Starting

It should be clear from the discussion of how torque is produced that unless the rotor is running at the same speed as the rotating field, no steady torque can be produced. If the rotor is running at a different speed, the two fields will be sliding past each other, giving rise to a pulsating torque with an average value of zero. Hence a basic synchronous machine is not self-starting, and some alternative method of producing a run-up torque is required.

Most synchronous motors, designed for direct connection to the utility supply, are therefore equipped with some form of rotor cage, similar to that of an induction motor, in addition to the main field winding. When the motor is switched onto the supply, it operates as an induction motor during the run-up phase, until the speed is just below synchronous. The excitation is then switched on, and as long as the load is not too high, the rotor is able to make the final acceleration and 'pull in' to synchronism with the rotating field. Because the cage is only required during starting, it can be short-time rated, and therefore comparatively small. Once the rotor is synchronized, and the load is steady, no currents are induced in the cage, because the slip is zero. The cage does, however, come into play when the load changes, when it provides an effective method for damping out the oscillations of the rotor as it settles at its new steady-state load angle.

Large motors will tend to draw a very heavy current during run-up, perhaps six or more times the rated current, for many tens of seconds, or longer, so some form of reduced voltage starter is often required (see Chapter 6). Sometimes, a separate small or 'pony' motor is used simply to run up the main motor prior to synchronization, but this is only feasible where the load is not applied until after the main motor has been synchronized.

5. VARIABLE-FREQUENCY OPERATION

Just as we have seen in Chapters 7 and 8 for the induction motor, once we interpose a power electronic converter between the utility supply and the machine we introduce new levels of performance and lose most of the inherent drawbacks which we find when the motor is directly connected to the utility supply.

Most obviously, a variable-frequency converter frees the synchronous machine from the fixed-speed constraint imposed by utility-frequency operation. The obvious advantage over the inverter-fed induction motor is that the speed of the synchronous motor is exactly determined by the supply frequency whereas the induction motor always has to run with a finite slip. On the down side, we lose the ability of the excited-rotor motor to vary the power-factor as seen by the utility supply.

In principle, a precision frequency source (oscillator) controlling the inverter switching is all that is necessary to give precise speed control with a synchronous motor, while speed feedback is essential to achieve accuracy with an induction motor. In practice, however, we seldom use open-loop control, where the voltage and frequency are generated within the inverter and are independent of what the motor does. Instead, field-oriented control, almost identical to that described for the inverter-fed induction motor, predominates. The principal advantage of field-oriented control is that it allows us to control the torque and flux components of the stator current independently, and in the case of the synchronous motor it prevents the motor from losing synchronism with the traveling field by locking the supply frequency to the speed of the rotor.

However, in the steady state, an observer looking at the stator voltage and current would see steady-state sinusoidal waveforms, and would be unaware of the underlying control mechanism. We can therefore study the steady-state behavior using the equivalent circuit in much the same way as we did with the utility-fed permanent magnet motor. We will continue to ignore resistance because this makes the phasor diagrams much simpler to understand without seriously compromising our conclusions.

The approach that we take differs somewhat from our previous discussion in this chapter by laying more emphasis on the relation between the torque and the fluxes in the motor. We imagine the flux produced by the magnet and the flux produced by the stator as if they existed independently, although in reality there is only one resultant flux. Intuitively we can see that because the fluxes rotate in synchronism, the magnitude of the torque will depend on the product of the two field strengths and the angle between them: when aligned, the torque is zero, and when perpendicular, it is maximum. This is equivalent to saying that the torque is maximum when the stator current wave is aligned with the magnet flux wave, which is the traditional 'BIl' picture.

Because both flux distributions are sinusoidal, the torque depends on the sine of the angle between them (λ). The stator field strength depends on the current, so maximum torque will be obtained when the current is perpendicular to the magnet flux in the phasor diagram, as shown in later figures (9.7) and (9.8).

When we discussed the utility-connected excited-rotor motor in section 4, we were reminded that with the voltage and frequency fixed, the resultant (stator) flux was constant, and that if the rotor excitation was low, extra magnetizing current would be drawn from the utility supply, and the power-factor would be lagging. When the rotor excitation was high there was a surplus of excitation and a leading current would be exported to the utility. By suitable adjustment of the rotor current we could achieve a power-factor of unity for any value of the load torque. We then saw that for a utility-fed permanent magnet motor, in which the rotor excitation is constant, the stator current adjusted itself to satisfy the requirement for the resultant flux to be constant, but as a result we had no control over the power-factor.

With an inverter-fed motor we gain control of both the stator voltage and frequency, so that together with the load torque we now have three independent variables in the case of the permanent magnet motor, or four for the excited-rotor machine. The majority of inverter-fed synchronous motor drives, and almost all below 200 kW, employ permanent magnet motors, so we will concentrate on their behavior for the remainder of this section.

5.1 Phasor diagram – nomenclature and basic relationships

The general diagram (Figure 9.6) is for an underexcited case; i.e. at the speed in question, the open-circuit e.m.f. (E) is less than the terminal voltage. We will discuss what each phasor represents first, and then turn to the relationships that allow the diagram to be produced.

E is the open-circuit e.m.f. produced by the magnet flux (ϕ_{mag}): it is proportional to the magnet flux and the speed, which is proportional to the stator frequency ω. It is convenient for us to use this as our reference phasor because once the frequency is specified, the magnitude of E is known, so we can start the phasor diagram with the known E.

Recalling that phasors rotate anticlockwise, and that the projection of any phasor onto the vertical axis represents the instantaneous value, we can deduce that at the instant shown in Figure 9.6, the induced e.m.f. is at its maximum. We also know from Faraday's law that the induced e.m.f. depends on the rate of change of the flux, which is greatest when the flux is passing through zero, so in Figure 9.6 the instantaneous flux is zero, as shown. In general, therefore, the flux lags the corresponding induced e.m.f. by 90°, as shown.

ϕ_{arm} is the flux that would be set up if the armature (stator) current existed alone. We are assuming no saliency, so the magnitude and direction of this flux depend only on the magnitude and phase of the stator current.

ϕ_{res} is the resultant ('the') flux. With resistance ignored, this flux is effectively determined by the applied voltage and frequency, as we have seen several times earlier in this book, so in the phasor diagram, the magnitude of this flux is proportional to, and in quadrature with, V.

The self-induced e.m.f. from the stator flux is modeled in the equivalent circuit by means of the voltage across the stator inductive reactance X, equal to ωL, where L is the self inductance of the winding. In the phasor diagram the voltage IX leads

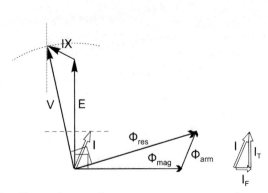

Figure 9.6 Phasor diagram for permanent magnet synchronous motor.

the current by 90°. The e.m.f. could equally well have been represented by an e.m.f. source similar to the E that represents the magnet-induced e.m.f., but historically the circuit representation is preferred because it is a compact way of representing both the magnitude and time-phase of the induced voltage with respect to the current.

V is the applied stator voltage, here regarded as an independent variable, and I is the resulting stator current. The stator phasor equation is $V = E + IX$ (or $V = E + jIX$ for those who prefer the complex notation).

The other independent variable is the load torque, which is our next consideration in developing the diagram.

In the steady state, the motor torque must equal the load torque. The motor torque is proportional to the product of the current, the magnet flux ϕ_{mag}, and the sine of the angle between them, so, knowing the magnet strength, we can find the in-phase or torque component of the stator current (I_T). The locus of the current I is then the horizontal dashed line, and the locus of V then becomes the vertical dotted line as shown in Figure 9.6. So when we finally specify the magnitude of V (shown by the dotted arc), the intersection of the arc and vertical line fixes the phase of the stator current and finalizes the diagram.

The current has two components as shown on the right in Figure 9.6. The component in phase with E is the useful or torque component I_T, while the component in phase with the magnet flux is the flux component I_F. The area of the flux triangle is proportional to the product of I_T and ϕ_{mag}, and thus the area provides an immediate visual indication of the torque.

We note that by control of V, we can alter the flux component (I_F), but the steady-state torque component is determined by the load. In particular, we can adjust V so that the stator current is in phase with E (i.e. the flux component is zero), so that we get the maximum value of torque per ampere of stator current. In this condition the armature (stator) flux is perpendicular to the magnet flux, as discussed above, and is explored in the next section.

Before we move on, we should note that because we specified an applied voltage just a bit bigger than E, the current turned out to be of modest amplitude and at a reasonable power-factor angle. However, if we had specified a much higher V, we would find that the current would have had a much larger lagging flux component (with the same torque component, determined by the load), and a much worse power-factor. And conversely, a much smaller applied voltage would result in a large leading power-factor current. Neither of these conditions is desirable because of the increased copper loss.

5.2 Field-oriented control

In a later section we will look at the torque–speed capabilities of the inverter-fed permanent magnet motor, and we will find that, as with the d.c. drive and the induction motor drive, there is a so-called constant torque region that extends up to base speed, within which full rated torque is available on a continuous basis. And in common with the other drives, at higher speeds there is a field weakening region where the maximum available torque is reduced. However, the usual constraints imposed by the maximum supply voltage and allowable continuous stator current may result in more serious restrictions on operation in this region than we have seen with other drives, including the likelihood that only intermittent operation will be possible.

Given the number of parameters involved and their variation between motor designs, it is only possible for us paint a broad picture, so we will look at one hypothetical machine and use it to provide an insight into some of the issues involved. We will focus on three conditions, two in the constant-torque region and one in field weakening. All three phasor diagrams are to the same scale to allow easy comparisons to be made.

We discussed field-oriented control in the context of the induction motor in Chapter 7. In principle for both induction and permanent magnet motors the strategy involves independent control over the torque and flux components of the stator current. But whereas the induction motor has to have both components (because it has no other source of flux), the permanent magnet motor only needs the torque component because its flux is provided by the magnets on the rotor. At speeds up to base speed, therefore, field control of the permanent magnet motor involves control of the stator current (via control of the stator voltage) to keep the flux component zero, while at the same time the torque component matches the input torque demand to the controller (see later Figure 9.10). During acceleration or deceleration, the torque demand will be high, and the controller will maintain the correct torque component while the speed is changing, until the target speed is reached when the torque component will be reduced to the appropriate steady-state value.

For convenience in calculations in relation to the phasor diagrams we will take the open-circuit e.m.f. E at the base speed (ω_B) to be 1 p.u. (per-unit). (In practice, the rated applied voltage is usually taken as 1 p.u., but there is no reason why any other value should not be chosen.) We will assume that the reactance of the winding (X) at the base speed is 0.3 p.u. (which means that at rated current (and rated frequency) the voltage across the reactance is 0.3 times the rated voltage).

5.3 Full-load

By full-load we mean that the machine is running at base speed and delivering its full rated torque. We saw above that the current depended on the applied voltage, and that, in particular, if we apply the right voltage, we can minimize the current. This is what we do in field-oriented control, so the diagram (Figure 9.7) represents this condition.

We should note that the stator current has zero flux component, and so the armature flux is perpendicular to the magnet flux, i.e. in the optimum torque-producing position. We will define the current in this situation as 1 p.u., so the

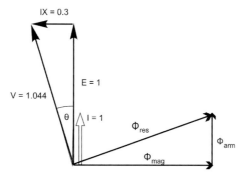

Base speed; freq. ω_B; Rated torque; max. effy.;
Torque angle = 90°; Rated power.

Figure 9.7 Permanent magnet motor phasor diagram – field-oriented control – full (rated) torque at base speed.

volt-drop across the reactance is $1 \times 0.3 = 0.3$ p.u. From the property of the right-angled triangle, V turns out to be 1.044 p.u., and because we have defined this as the rated power condition, we will regard 1.044 V as the maximum voltage that the inverter can produce. This condition is the most efficient state for the given current and torque, because the stator copper loss is minimized.

5.4 Full torque at half base speed (half power)

The phasor diagram is shown in Figure 9.8, again with the voltage optimized to give the most efficient stator current. The magnet flux is the same as in Figure 9.7 of course, and so for the same (rated) torque the stator current has to be 1 p.u., as before. However, the frequency is now only half ($\omega_B/2$) so the open-circuit e.m.f. is reduced to 0.5. The stator reactance is proportional to frequency, so it has also

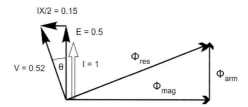

Half speed; freq. $\omega_B/2$; Rated torque; max. effy;
Torque angle = 90°; Half power.

Figure 9.8 Permanent magnet motor phasor diagram – field-oriented control – full (rated) torque at half base speed.

halved, to 0.15 p.u., and the volt-drop IX becomes 0.15 p.u. also. In view of the similarity between the diagrams it is no surprise that the applied voltage turns out to be half, i.e. 0.52 p.u., and given the emphasis we have previously placed on the fact that the flux in an a.c. machine is determined by the V/f ratio, we will not be surprised to see that this is again confirmed by these results.

The angle θ between V and I (the power-factor angle) is the same at full and half speed, and since the input power is given by $VI\cos\theta$, it is clear that the input power at half speed is half of that at full speed. This is what we would expect because with resistance neglected the input power equals the mechanical power, which is half in Figure 9.8 because the torques are the same but the speed is half of that in Figure 9.7.

The two previous examples have shown how the motor can be operated to produce full rated torque up to a speed at which the full available voltage is applied; i.e. this is what we have referred to previously as the 'constant torque region'. In this region, the permanent magnet motor with field-oriented control behaves, in the steady state, in a very similar manner to a d.c. motor drive, in that the applied voltage (and frequency) is proportional to the speed and the stator current is proportional to the torque.

We saw with the d.c. drive and the induction motor drive that once we had reached full voltage and current, any further increase in speed could only be achieved at the expense of a corresponding reduction in torque, because the input power was already at its maximum or rated value. In both cases, higher speeds were obtained by entering the aptly named 'field weakening' region.

For the d.c. motor, the field flux is under our direct control, so we reduce the current in the field circuit. In the induction motor, the field is determined indirectly by the V/f ratio, so if f increases while V remains constant, the field flux reduces. The permanent magnet motor behaves in a somewhat similar fashion to the induction motor, in that if the voltage is constant at speeds above base speed, the V/f ratio reduces as the frequency (speed) is increased, so the resultant flux also reduces. However, whereas the only source of excitation in the induction motor is the stator winding, the magnet in the permanent magnet motor remains a constant (and potent!) source of excitation at all times, and so we can anticipate that in order to arrive at a reduced flux to satisfy the V/f condition, the stator current will have to nullify some of the influence of the magnet. We must therefore expect less than ideal behavior in the field weakening region, which we now examine.

5.5 Field weakening – operation at half torque, twice base speed (full power)

We will consider a situation well into the field weakening region; i.e. we will assume that in line with other drives we can expect full power, and so settle for operation at twice base speed and half rated torque. We will see that while this is achievable on an intermittent basis, it is not possible without exceeding the rated current, a conclusion that we must expect to apply to the whole of the field weakening region with a permanent magnet drive.

The condition is represented by the phasor diagram shown in Figure 9.9. The frequency is twice the base value, i.e. $2\omega_B$, and so the open-circuit e.m.f. is 2 p.u. As in other drives, above base speed we apply the maximum possible stator voltage, in this case 1.04 p.u. Because the voltage cannot be greater, it is not possible for us to arrange for the current to be in phase with E, and we are obliged to settle for a rather poor second-best.

The load torque is half rated, which means that the torque component is 0.5 p.u. However, we note that the very large difference in voltage between E and V results in a very large stator volt-drop of $IX = 1.044$ p.u. The reactance is twice as large as at base speed because the frequency is doubled, so the current is given by $1.044/0.6 = 1.74$ p.u., which is 74% above its continuously rated value. The copper loss will therefore be increased by a factor of 1.74^2, i.e. to three times the rated value. Clearly continuous operation will not be possible because the stator will

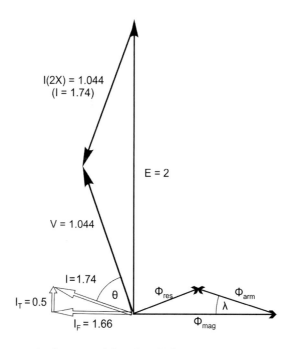

2 x Base speed; freq. $2\omega_B$; Half rated torque; Rated power;
Current = 1.74 p.u.; Torque angle λ = 16.8°.

Figure 9.9 Permanent magnet motor phasor diagram – twice base speed, half rated torque, full power.

overheat, so in this field weakening condition only intermittent operation at half torque will be possible.

By comparing the flux triangles in Figures 9.7 and 9.9 we can see why the latter is referred to as field weakening. In Figure 9.7 our freedom to adjust V allowed us to ensure that the stator or armature flux is in quadrature with the magnet flux (i.e. the torque angle λ is $90°$), leading to a slightly higher resultant flux, and maximizing efficiency by minimizing the current for the given torque.

In contrast the constraint on V means that the armature flux has a large component which is in opposition to the magnet flux, leading to a resultant flux that is much less than the magnet flux, and a very low torque angle (λ) of only $16.8°$. In a sense, therefore, most of the stator current is 'wasted' in being used to oppose the magnet flux. Clearly this is not a desirable operating condition, but it is the best that we can get with the limited voltage at our disposal.

As a final check we can calculate the input power. The angle θ is given by $90 - 2\lambda = 56.4°$, so the input power ($VI \cos \theta$) is $1.044 \times 1.74 \times 0.553 = 1$ p.u., as expected since we assumed that the mechanical power was at rated value (twice base speed, half rated torque) and we ignored resistance in our calculations.

At this point we should recall that the aim throughout section 5 has been to discover how the motor parameters determine the steady-state behavior when the voltage, frequency and load vary over a range that is representative of a typical inverter-fed drive. Few readers will find it necessary to retain all of the detail (although a general awareness of the broad picture is always helpful) so those who have found it hard going can take comfort from the fact that in practice the drive will take care of everything for them. This is discussed in the next section.

6. SYNCHRONOUS MOTOR DRIVES

Inverter-fed operation of synchronous machines plays a very important and growing role in the overall drives market, as customers seek higher efficiency and higher power density than can be achieved with induction motors. We will also see later when we consider some of the available permanent magnet motor designs and control strategies that very high dynamic performance can be achieved, making this the motor of choice in many of the most demanding applications.

Inverter-fed synchronous motor drives are well established in two distinct guises. At one extreme, large (multi-MW) excited-rotor synchronous motors are used, particularly where high speeds are required or when the motor must operate in a hazardous atmosphere (e.g. in a large gas compressor). At the other (100 W to 500 kW), permanent magnet synchronous motors are used in a wide range of applications. Each of these groupings has distinct converter and control strategies and so we will look at each, beginning with the latter.

6.1 Permanent magnet motor drives

Permanent magnet motor drives are a very important and rapidly growing sector of the drives market, which is why we have devoted a large section of this chapter to a theoretical study aimed at understanding what determines their steady-state behavior. In terms of their application in drives we have little new to introduce because the converter circuits used are exactly as we have already discussed in Chapters 7 and 8 for the induction motor, and the control is also very similar.

Our discussion of field-oriented control in Chapter 7 focused heavily on the ability to control the torque-producing and field-producing components of the stator current (i_T and i_F) independently and rapidly, in order to provide virtually instantaneous control of torque. We do exactly the same for the permanent magnet synchronous machine except we normally set the demand for i_F to zero, as there is no need to provide the working flux from the stator side because it is provided by the rotor magnets.

The arrangement of a typical field-oriented control system is shown in Figure 9.10.

As with the induction motor drive the motor is supplied from a voltage source inverter, and again we are therefore controlling the stator currents via the voltage. The control strategy and the inverter are essentially the same as for an induction motor – indeed some manufacturers make life easier for users by offering a single inverter product suitable for both types of motor.

As with the induction motor control strategy, the critical issue is determining the flux position (θ_{Ref}), but in this case the task is very much easier as the flux is aligned to the rotor position. In high-performance systems a high-resolution absolute encoder would be fitted to the motor shaft to provide an accurate rotor position signal so that the flux position is known precisely at all times. However, if a position sensor is not used then the reference frame angle, θ_{Ref}, can be derived from computed motor voltages and currents in the same way as for the induction motor, but in this case we no longer have the complication of the temperature-dependent rotor time-constant.

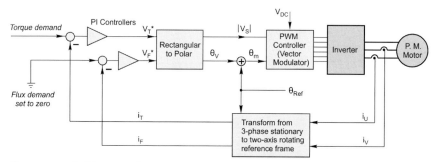

Figure 9.10 Field-oriented control scheme for permanent motor synchronous motor.

As we have discussed before, model-based schemes break down at very low frequencies because the voltage components become very small. For low-speed operation it is therefore necessary to supplement the approach with an alternative such as a position-sensing scheme using injected high-frequency currents. Or, in systems where the load is reasonably predictable, such as in a domestic washing machine, the motor can be brought up to a particular speed using open-loop switching in much the same fashion as a stepping motor (see Chapter 10) until the motional e.m.f. is large enough to be used as the control signal.

We have said with some satisfaction that because we have magnets on the rotor, we do not need to provide current to develop the field, and consequently the field current demand in Figure 9.10 is set to zero. However, if the field current demand were not set to zero, the control system would provide a flux component of stator current that could either increase or decrease the magnet's flux, depending on the polarity of the reference signal. Clearly there will be an upper limit on the flux because of saturation of the magnetic circuit, and in practice, reducing the flux is a more attractive proposition because it allows us to operate in a field weakening mode and so extend the speed range into a constant power region, as discussed at the end of section 5.

6.2 Converter-fed synchronous machine (multi-MW, excited-rotor) drive

The basic components of the drive system are shown in Figure 9.11. Current in the top of the d.c. link always flows from left to right (as shown by the direction of the thyristors), but the link voltage can be positive or negative so that energy can flow in either direction. The labels 'rectifier' and 'inverter' in Figure 9.11 indicate how each converter operates when the machine is operating as a motor, their roles being reversed when the machine is generating. An additional attraction of this set-up is that the direction of rotation of the machine is determined electronically by the switching sequence within the machine converter, and hence full four-quadrant operation is available without extra hardware.

In view of what we have read so far, we might have expected the switching devices in the motor inverter to be IGBTs, and to see PWM control of the output

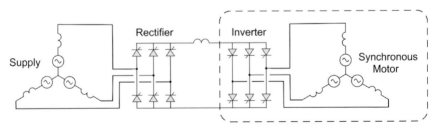

Figure 9.11 Converter-fed large synchronous machine drive.

voltage waveform, but in these large sizes, IGBT devices are prohibitively expensive, and this explains why the motor converter is exactly like that used on the supply side of a d.c. motor drive. The reason why we can use this cheaper inverter lies in the fact that, once rotating, a synchronous machine generates a.c. voltages, which can be used for the natural commutation of a converter connected to its terminals. This is why it is frequently referred to as a load commutated inverter (LCI).

In effect the motor converter behaves in the same way as it does when connected to the utility supply, and as such it is inherently capable of power flow in both directions. On the down side, we have no means of profiling the output voltage waveform, and as a result we have to accept waveforms for the motor current that are less than the perfect sinusoids that we would like.

We have already likened the combination of an inverter and synchronous motor (enclosed by the dashed line in Figure 9.11) to an 'inside-out' d.c. machine, and this view is instructive to help us to understand how this particular drive operates. We can begin by focusing on the motional-induced e.m.f.

In the synchronous motor the moving rotor carries the d.c. field winding. The rotating flux induces a sinusoidal e.m.f. in the three stator windings that is proportional to speed and field current. The firing of the devices in the motor converter is locked to the angular position of the rotor, so that each phase is connected in sequence to the d.c. link. The d.c. link therefore 'sees' a rectified 3-phase voltage that, although unidirectional, is not smooth d.c.

Conversely, the d.c. motor field is stationary, and its flux induces an alternating e.m.f. in the (many) armature coils on the rotor. The mechanical commutator and its sliding brushes rectify the e.m.f. so that at the armature terminals we get a smooth d.c.-induced voltage.

So when viewed from the d.c. link, the two are essentially the same. It is therefore not surprising that, as with a d.c. motor, the no-load speed of the synchronous motor depends on the d.c. link voltage provided by the supply-side converter, while when load is applied and speed tends to fall, the d.c. link current automatically increases until the steady state is reached. The effects of varying the d.c. excitation of the synchronous machine also mirror those in the d.c. motor, so that field weakening will lead to higher speed but reduced torque.

Of course, the ripple frequency of the rectified d.c. on the motor side of the d.c. link depends on the speed of the motor, while that on the supply side of the d.c. link will be 300 or 360 Hz depending on whether the supply is 50 or 60 Hz. So in order to prevent unwanted harmonic currents and to smooth the d.c. link current, a series inductor is included, as shown in Figure 9.11. The reactance of the inductor (ωL) is high at the ripple frequencies, but zero at d.c.

The switching strategy of the motor converter is synchronized to the back e.m.f. of the machine, each switch conducting for 120° (elect), so that the current waveform in each phase of the motor is roughly constant for 120°, and zero for the next 60°, and this pattern is repeated in the negative half-cycle. Ideally, the waveform should be sinusoidal, but the presence notably of the fifth and seventh harmonics (which leads to some unwelcome torque pulsations at six times the fundamental frequency) is the price to be paid for having a relatively unsophisticated inverter. Happily, resonances excited by the torque ripple are rare and can usually be overcome by preventing steady/continuous operation at particular speeds associated with the resonance.

It is necessary to maintain an approximately constant angular relationship between the rotor and stator m.m.f.s and hence automatically maintain the correct inverter frequency. This is an important point: the inverter does not impose a frequency upon the machine, rather the machine itself determines the frequency, and so the motor cannot lose synchronism (pole-slip). As already mentioned, the drive is accelerated by increasing the current fed to the motor, increasing the motor torque and acceleration, which thereby increases the frequency.

Once rotating, the motor terminal voltage (or better the calculated back e.m.f.) can be used to determine the rotor position, but at start-up and at very low speeds the magnitude of the voltage signals is too small to be used for control purposes or to commutate the current. A shaft-mounted absolute position sensor is therefore used to provide position information to determine when the current should be commutated from one switch to the next. The commutation itself is usually achieved by momentarily switching off the d.c. link current (by phase control of the utility-side converter) every sixth of a cycle. This allows the thyristors in the inverter to turn off so that the next pair can be fired. Above approximately 5% of rated speed the machine generates sufficient voltage for natural commutation, and subsequent control is undertaken in a similar manner to that of a d.c. drive.

As we saw earlier with the permanent magnet drive, where the load is predictable some systems actually impose a predetermined sequence of current pulses (applied sequentially to the motor phases) to 'crank' the motor up to a speed at which the back e.m.f. becomes of sufficient magnitude to be used for position sensing and commutation.

As in the d.c. drive, the a.c. supply power-factor is poor at low speeds, but on the plus side, full four-quadrant operation is possible without additional mechanical switching.

Applications for this type of drive fall into two main categories: first, as a starting mechanism for very large synchronous machines, the converter then being rated only for a fraction of the machine rating – the main motor is started off load, synchronized to the utility supply and then the load is applied; and secondly, as large high-power (and sometimes high-speed) variable-speed drives for a variety of applications. Power ratings, typically from 2 to 100 MW at speeds up to 8000 r.p.m. are available, though at powers up to 5 MW voltage source inverters are now proving to be more popular. It is also important to note that with the advent of very high-voltage thyristors, high-voltage drives can be readily designed. Supply voltages up to 12 kV are typical, but systems over 25 kV are in service where the high voltage converter technology is similar to that used for HVDC power-system converters. As powers increase, in order to keep both motor windings and interconnecting power cabling manageable it is important to increase operating voltages and thereby moderate the current levels.

Some manufacturers of synchronous machines of much more modest ratings (e.g. a few tens of kW) also offer this type of converter technology, but they tend to be niche products.

The detailed design of this type of high-power drive is highly specialized and well outside the scope of this book. The impact on the supply of such high-power converters is also an area where a great deal of detailed consideration is necessary.

7. PERFORMANCE OF BRUSHLESS MOTORS

Throughout this section we will use the terms 'brushless' and 'permanent magnet' to be synonymous when they are applied to an electric motor. We will also avoid the confusing term 'brushless d.c. motor' because, as already mentioned, such motors are always supplied with alternating currents.

However, we should mention that the inherent electromagnetic properties of a permanent magnet motor can be quantified by its 'motor constant' in much the same way as a d.c. motor (Chapter 3). If we spin the rotor of a permanent magnet machine at angular velocity ω, the r.m.s. value of the sinusoidal e.m.f. induced in each phase is given by $E = k\omega$, and if we supply balanced currents of r.m.s value I_a to the 3-phase windings in quadrature to the field, the torque is given by $T = kI_a$, where k is the machine constant, expressed in the SI units of volts per radian per second or the equivalent newton-meters per ampere. (In practice, manufacturers usually quote k in terms of volts per thousand revs per min.) These relationships are identical to those we discussed earlier for the d.c. machine, and once again they underline the unity of machines that operate on the 'BIl' principle.

We have hinted previously that permanent magnet motors offer outstanding performance in terms of power density and performance in comparison with induction and d.c. machines, and in this section we look briefly at the underlying reasons. We then discuss the limitations that govern performance and finally give an example that illustrates the impressive results that can be obtained.

7.1 Advantages of permanent magnet motors

The stator windings of permanent magnet motors do not have to carry the excitation or magnetizing current required by the induction motor, so a given winding can carry a higher work current without generating more heat, thereby increasing the electric loading and the specific power output (as discussed in Chapter 1).

Cooling the rotor is difficult in any enclosed machine because ultimately the heat has to get to the stator, so the absence of current on the rotor not only improves efficiency by reducing the total copper loss, but also eases the cooling problem.

Historically, brushless permanent magnet motors only became practicable with the advent of power electronics, so it became normal for them to be supplied via power electronics with the associated expectation that they would operate in a speed-controlled drive. The majority were therefore not expected to operate directly off the utility supply, and as a result their designers had much greater freedom to produce bespoke designs, tailored to a specific purpose.

For example, suppose we require a motor that can accelerate very quickly, which implies that the ratio of torque to inertia should be maximized. We saw in Chapter 1 that with given values of the specific magnetic and electric loadings, the torque is broadly dependent on the volume of the rotor, so we are free to choose long and thin or short and fat. The inertia of a homogeneous rotor is proportional to the fourth power of its radius, so clearly for this application we want to minimize the rotor radius, so a long thin design is required. Fortunately, there is considerable flexibility in regard to the shape and size of the rotor magnets, so no serious constraint applies in relation to rotor diameter. Many so-called servo motors (see later) have this profile, as illustrated in Figure 9.12.

A different application with the same continuous torque and power requirements would require the same rotor volume, but if, for example, the application requires that unwanted changes in speed caused by step changes in load torque must be minimized, the inertia should clearly be maximized, with a short rotor of larger diameter.

A section through a high inertia permanent magnet motor typically rated up to 100 Nm and 3000 rev/min is shown in Figure 9.13 and Plate 9.1. It is interesting to

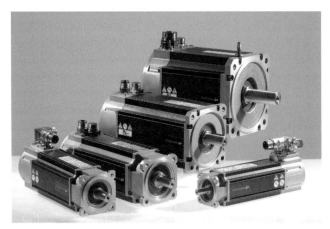

Figure 9.12 Permanent magnet servo motors. *(Courtesy of Emerson – Control Techniques Dynamics.)* (See Plate 9.12)

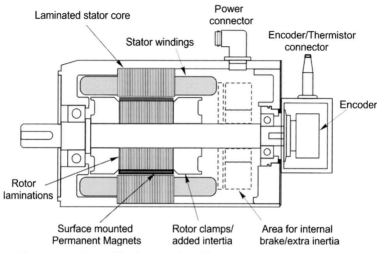

Figure 9.13 Typical high inertia brushless permanent magnet motor.

note how little of the volume of the motor is actually taken up with active material, namely the stator and rotor laminations and the surface-mounted magnets on the rotor. The end regions of the stator windings are seen to contribute significantly to the overall volume of the motor illustrated. (The move to using segmented stator windings reduces the impact of the end winding and can lead to significant reductions in total motor volume.) Most permanent magnet motors employ rare-earth magnets which have much higher energy product (in effect a measure of their magnetizing 'power') than traditional materials such as alnico, so they are very small, as shown in Figure 9.13.

As in other motors the heat dissipated in the stator diffuses into the air through the frame to the finned case and hence to the surrounding air. However, the design of some motors of this type is based on the requirement that a substantial proportion (perhaps 40%) of the loss is conducted through the mounting flange to a suitable heatsink, so this is an area where great care needs to be taken with the thermal properties of the mounting.

7.2 Industrial permanent magnet motors

In the preceding section we mentioned that bespoke design of permanent magnet motors has long been considered unexceptional, but recent years have seen the emergence of permanent magnet motors packaged in the same industrial motor (IEC or NEMA) housings that are being used for induction motors, as shown in Figure 9.14.

This type of motor is now marketed as a direct competitor of the induction motor in variable-speed applications. They are targeted at general applications

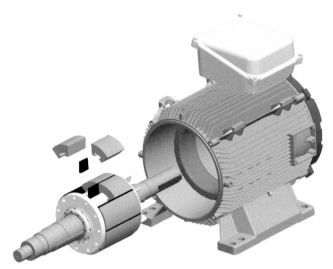

Figure 9.14 Permanent magnet industrial motor in standard IEC frame. *(Courtesy of Emerson Leroy Somer.)* (See Plate 9.14)

where the higher initial cost is offset by their higher efficiency and relatively high power density. The heat loss in the permanent magnet rotor is much less than in the corresponding induction motor, so the rotor runs cooler, which may also be an advantage in aspects such as bearing life.

7.3 Summary of performance characteristics

Permanent magnet motors with low inertia rotors are used in high-performance servo applications such as machine tools or pick and place applications where fast, precise movements are required, and motors with high inertia rotors (and high pole-numbers) suit low-speed applications such as gearless lift systems.

The performance characteristics of these drives are summarized below:

- They have excellent dynamic performance at speeds down to standstill when position feedback is used.
- For precision positioning the position feedback must define the absolute position uniquely within an electrical revolution of the motor. This can be provided with a position sensor or alternatively a sensorless scheme can be used. The performance of a sensorless scheme will be lower than when a position sensor is used.
- Field weakening of permanent magnet motors is possible to extend their speed range, but (as shown in section 5) this requires additional motor current, and so the motor becomes less efficient in the field weakening range. This form of control also increases the rotor losses and raises the temperature of the magnet material, thereby increasing the risk of demagnetization. Care is also needed in

such applications to avoid overvoltage at the motor and drive terminals in the event of a loss of control: at high speeds, the open-circuit voltage will exceed the rated value (see Figure 9.9).

- Permanent magnet motors exhibit an effect called cogging that results in torque ripple. It is caused by magnetic reluctance forces acting mainly in the teeth of the stator, and can be minimized by good motor design, but can still be a problem in sensitive applications.
- Permanent magnet motors can be very efficient as the rotor losses are very small.

To give an impression of the outstanding performance that can be achieved by a brushless permanent magnet motor, Figure 9.15 shows the results from a bench test in which the speed reference begins with a linear ramp from zero to 6000 rev/min in 0.06 s, followed shortly by a demand for the speed to reverse to 6000 rev/min, then back to full forward speed and finally to rest, the whole process lasting less than one second.

The motor was coupled to a high-inertia load of 78 times the rotor inertia, which makes the fact that the speed reversal is accomplished in only 120 ms even more remarkable. It takes only three revolutions to come to rest and a further three to accelerate in reverse. In common with many high-performance applications, the drive control is actually implemented in the form of position control, with the shaft angle being incremented at a rate equivalent to the required speed. The dotted trace in Figure 9.15 shows that the maximum position error of the motor shaft throughout the speed reversal is less than 0.05°, so by any standard this is truly impressive. It is no wonder that the brushless permanent magnet motor is frequently chosen for applications where closely coordinated motion control is called for.

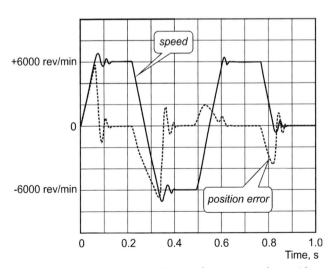

Figure 9.15 Permanent magnet motor drive performance under rapid reversal test.

Finally, on a matter of terminology, it is worth pointing out that brushless permanent magnet motors are sometimes referred to as 'servo' motors. The name 'servo' originates from 'servomechanism', defined as a mechanical or electrical system for control of speed or position. The term tends to be used loosely, but broadly speaking when it is applied to a motor it implies superior levels of performance.

7.4 Limits of operation of a brushless permanent magnet motor

We have previously talked about the limits of operation that determine the rating and operating envelope of other types of electrical machine, so we will conclude this section by taking a closer look at the limits of operation for a brushless permanent magnet servo motor (which usually has no external cooling fins or fan). A typical torque–speed characteristic is shown in Figure 9.16.

The individual limits shown in Figure 9.16 are discussed below, but the most striking feature is clearly the very large area where operation above rated torque is possible (albeit on an intermittent basis). This provision clearly reflects the potential application areas, such as rapid positioning systems, where high acceleration is needed for relatively short periods.

The continuously rated region is, as usual, limited by the allowable temperature rise of the motor. At standstill the predominant source of loss is the stator copper loss (shown as I^2R limit in Figure 9.16), but at higher speeds the iron loss becomes significant and the full-load torque at rated speed is therefore less than the standstill torque.

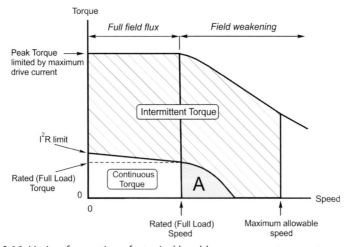

Figure 9.16 Limits of operation of a typical brushless permanent magnet servo motor.

The upper boundary on the intermittent torque region is usually determined by the maximum current that the drive converter can supply: only intermittent operation is possible otherwise the motor overheats.

Continuous operation is not possible when field weakening either, because, as explained in section 5, the stator current is large in order to reduce the magnet flux. For this thermal reason region A shown in Figure 9.16 can be somewhat truncated compared with the characteristic field weakening region of the induction motor.

Operation of permanent magnet motors at excessive temperatures can lead to demagnetization of the rare earth magnets in the motor, as well as the hazards common to all other electrical machines such as loss of robustness of insulation. Good thermal protection is therefore necessary. Below base speed, for applications involving few excursions outside the continuous operating region, a relatively simple motor thermal model in the drive control scheme may be adequate. For applications involving significant operation in the intermittent torque region, and certainly where field weakening operation is used, more complex thermal models would be needed and these are usually supplemented by thermistors embedded in the stator windings.

One final note of warning in relation to operating permanent magnet motors in the field weakening region: the maximum motor speed must be selected with due consideration of what may happen if the drive were to trip. In such a case the stator voltage could rise to a dangerously high level, as discussed previously, causing damage to the drive, the motor itself or indeed the cables or connectors. Crowbar circuits can be designed to provide protection against overvoltages, but the motor and drive system must be designed to withstand the short circuit currents that result.

7.5 Brushless permanent magnet generators

As with almost all forms of electrical machines the permanent magnet motor can be operated as a generator, with mechanical energy supplied to the shaft being converted to electrical energy. The advantages of high power density and efficiency offered by the permanent magnet motor of course apply equally when the machine is being used as a generator.

The majority of commercial wind generators up to 75 kW use permanent magnet synchronous machines. Much larger wind generators are also in service with some multi-pole, multi-MW motors being applied in utility-scale turbines with direct drive systems, i.e. systems which do not employ a gearbox between the wind turbine and the generator.

8. RELUCTANCE AND HYSTERESIS MOTORS

There are a number of synchronous motors which have niche applications and so a brief introduction is appropriate.

8.1 Reluctance motors

The reluctance motor is arguably the simplest synchronous motor of all, the rotor consisting simply of a set of laminations shaped so that it tends to align itself with the field produced by the stator. This 'reluctance torque' action is discussed in Chapter 10, which deals with stepping and switched reluctance motors.

Here we are concerned with utility-frequency reluctance motors, which differ from steppers in that they only have saliency on the rotor, the stator being identical with that of a 3-phase induction motor. In fact, since induction motor action is required in order to get the rotor up to synchronous speed, a reluctance-type rotor resembles a cage induction motor, with parts of the periphery cut away in order to force the flux from the stator to enter the rotor in the remaining regions where the air-gap is small, as shown in Figure 9.17(a). Alternatively, the 'preferred flux paths' can be imposed by removing iron inside the rotor so that the flux is guided along the desired path, as shown in Figure 9.17(b).

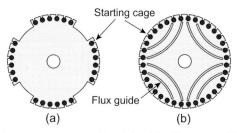

Figure 9.17 Reluctance motor rotors (4-pole) – (a) salient type, (b) flux-guided type.

The rotor will tend to align itself with the field, and hence is able to remain synchronized with the traveling field set up by the 3-phase winding on the stator in much the same way as a permanent magnet rotor. Early reluctance motors were invariably one or two frame sizes bigger than an induction motor for a given power and speed, and had low power-factor and poor pull-in performance. As a result they fell from favor except for some special applications such as textile machinery where cheap constant speed motors were required. Understanding of reluctance motors is now much more advanced, though their fundamental performance still lags the induction motor as regards power output, power-factor and efficiency.

Historically, reluctance motors were frequently used in large numbers connected in parallel to large 'bulk' power electronic inverters. The fact that the motors were locked in synchronism was a great advantage over the induction motor for synchronized/coordinated applications, notably in the textile industry. However, the falling cost of power electronics and advances in control saw such applications being addressed by means of individual inverters with dedicated induction motors and overarching control. Recently, however, some drives companies are once again promoting reluctance motor drives utilizing off-the-shelf converter and control hardware, perhaps stretched to the limit, to obtain performance comparable to that of an induction motor.

8.2 Hysteresis motors

Whereas most motors can be readily identified by inspection when they are dismantled, the hysteresis motor is likely to baffle anyone who has not come across it before. The rotor consists simply of a thin-walled cylinder of what looks like steel, while the stator has a conventional single-phase or 3-phase winding. Evidence of very weak magnetism may just be detectable on the rotor, but there is no hint of any hidden magnets as such, and certainly no sign of a cage. Yet the motor runs up to speed very sweetly and settles at exactly synchronous speed with no sign of a sudden transition from induction to synchronous operation.

These motors (the operation of which is quite complex) rely mainly on the special properties of the rotor sleeve, which is made from a hard steel which exhibits pronounced magnetic hysteresis. Normally in machines we aim to minimize hysteresis in the magnetic materials, but in these motors the effect (which arises from the fact that the magnetic flux density B depends on the previous 'history' of the m.m.f.) is deliberately accentuated to produce torque. There is actually also some induction motor action during the run-up phase, and the net result is that the torque remains roughly constant at all speeds.

Small hysteresis motors were once used extensively in office equipment, fans, etc. The near constant torque during run-up and the very modest starting current (of perhaps 1.5 times rated current) mean that they are also suited to high inertia loads such as gyro compasses and small centrifuges.

CHAPTER TEN

Stepping and Switched-reluctance Motors

1. INTRODUCTION

In previous editions of this book, switched-reluctance motors were included with one of their direct industrial competitors, brushless permanent magnet motors, which appear in Chapter 9 in this edition. In this edition we have decided that because the fundamental mechanism of torque production is the same in both switched-reluctance and stepping motors, there is a strong argument for grouping them together. We will see that there are substantial differences between the modes of operation (and the power ratings), but both have projecting 'iron' poles or teeth on both stator and rotor, and produce torque by reluctance action, rather than by the '*BIl*' mechanism we have encountered hitherto. Stepping motors are considered first as they provide useful groundwork for the switched-reluctance material that follows.

2. STEPPING MOTORS

Stepping (or stepper) motors became attractive because they can be controlled directly by computers or microcontrollers.[1] Their unique feature is that the output shaft rotates in a series of discrete angular intervals, or steps, one step being taken each time a command pulse is received. When a definite number of pulses has been supplied, the shaft will have turned through a known angle, and this makes the motor well suited for open-loop position control.

The idea of a shaft progressing in a series of steps might conjure up visions of a ponderous device laboriously indexing until the target number of steps has been reached, but this would be quite wrong. Each step is completed very quickly, usually in less than a millisecond; and when a large number of steps is called for the step command pulses can be delivered rapidly, sometimes as fast as several thousand steps per second. At these high stepping rates the shaft rotation becomes smooth, and the behavior resembles that of an ordinary motor. Typical applications include disc head drives, and small numerically controlled machine tool slides, where the

---

[1] Many servo drives now have digital pulse train inputs, so the uniqueness of the stepper has somewhat diminished.

motor would drive a lead screw; and print feeds, where the motor might drive directly, or via a belt.

Most stepping motors look much like conventional motors, and as a general guide we can assume that the torque and power of a stepping motor will be similar to the torque and power of a conventional totally enclosed motor of the same dimensions and speed range. Step angles are mostly in the range 1.8°–90°, with torques ranging from 1 μNm (in a tiny wristwatch motor of 3 mm diameter) up to perhaps 40 Nm in a motor of 15 cm diameter suitable for a machine tool application where speeds of 500 rev/min might be called for, but the majority of applications use motors which can be held comfortably in the hand.

2.1 Open-loop position control

A basic stepping motor system is shown in Figure 10.1.

The drive contains the electronic switching circuits which supply the motor, and is discussed later. The output is the angular position of the motor shaft, while the input consists of two low-power digital signals. Every time a pulse occurs on the step input line, the motor takes one step, the shaft remaining at its new position until the next step pulse is supplied. The state of the direction line ('high' or 'low') determines whether the motor steps clockwise or anticlockwise. A given number of step pulses will therefore cause the output shaft to rotate through a definite angle.

This one-to-one correspondence between pulses and steps is the great attraction of the stepping motor: it provides *position* control, because the output is the angular position of the output shaft. It is a *digital* system, because the total angle turned through is determined by the *number* of pulses supplied; and it is *open-loop* because no feedback need be taken from the output shaft.

2.2 Generation of step pulses and motor response

The step pulses may be produced by a digital controller or microprocessor (or even an oscillator controlled by an analogue voltage). When a given number of

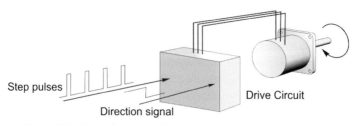

Figure 10.1 Open-loop position control using a stepping motor.

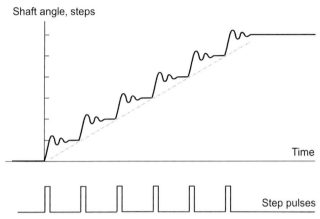

Figure 10.2 Typical step response to low-frequency train of step command pulses.

steps are to be taken, the step pulses are gated to the drive and the pulses are counted, until the required number of steps is reached, when the pulse train is gated off. This is illustrated in Figure 10.2, for a six-step sequence. There are six step command pulses, equally spaced in time, and the motor takes one step following each pulse.

Three important general features can be identified with reference to Figure 10.2. First, although the total angle turned through (six steps) is governed only by the number of pulses, the average speed of the shaft (which is shown by the slope of the broken line in Figure 10.2) depends on the frequency. The higher the frequency, the shorter the time taken to complete the six steps.

Secondly, the stepping action is not perfect. The rotor takes a finite time to move from one position to the next, and then overshoots and oscillates before finally coming to rest at the new position. Overall single-step times vary with motor size, step angle and the nature of the load, but are commonly within the range 5–100 ms. This is often fast enough not to be seen by the unwary newcomer, though individual steps can usually be heard; small motors 'tick' when they step, and larger ones make a satisfying 'click' or 'clunk'.

Thirdly, in order to be sure of the absolute position at the end of a stepping sequence, we must know the absolute position at the beginning. This is because a stepping motor is an incremental device. As long as it is not abused, it will always take one step when a drive pulse is supplied, but in order to keep track of absolute position simply by counting the number of drive pulses (and this is after all the main virtue of the system) we must always start the count from a known datum position. Normally the step counter will be 'zeroed' with the motor shaft at the datum position, and will then count up for clockwise direction and down for anticlockwise rotation. Provided no steps are lost (see later) the number in the step counter will then always indicate the absolute position.

2.3 High-speed running and ramping

The discussion so far has been restricted to operation when the step command pulses are supplied at a constant rate, and with sufficiently long intervals between the pulses to allow the rotor to come to rest between steps. Very large numbers of small stepping motors in watches and clocks do operate continuously in this way, stepping perhaps once every second, but most commercial and industrial applications call for a more exacting and varied performance.

To illustrate the variety of operations which might be involved, and to introduce high-speed running, we can look briefly at a typical industrial application. A stepping motor-driven table feed on a numerically controlled milling machine nicely illustrates both of the key operational features discussed earlier. These are the ability to control position (by supplying the desired number of steps) and the ability to control velocity (by controlling the stepping rate).

The arrangement is shown diagrammatically in Figure 10.3. The motor turns a leadscrew connected to the worktable, so that each motor step causes a precise incremental movement of the workpiece relative to the cutting tool. By making the increment small enough, the fact that the motion is discrete rather than continuous will not cause any difficulties in the machining process in most applications. We will assume that we have selected the step angle, the pitch of the leadscrew, and any necessary gearing so as to give a table movement of 0.01 mm per motor step. We will also assume that the necessary step command pulses will be generated by a digital controller or computer, programmed to supply the right number of pulses, at the right speed for the work in hand.

If the machine is a general-purpose one, many different operations will be required. When taking heavy cuts, or working with hard material, the work will have to be offered to the cutting tool slowly, at say 0.02 mm/s. The stepping rate will then have to be set to 2 steps/s. If we wish to mill out a slot 1 cm long, we will therefore program the controller to put out 1000 steps, at a uniform rate of 2 steps/s, and then stop. On the other hand, the cutting speed in softer material could be much higher, with stepping rates in the range 10–100 steps/s being in order. At the

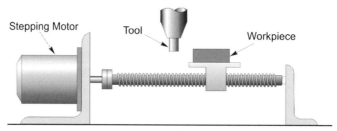

Figure 10.3 Application of stepping motor for open-loop position control.

completion of a cut, it will be necessary to traverse the work back to its original position, before starting another cut. This operation needs to be done as quickly as possible, to minimize unproductive time, and a stepping rate of perhaps 2000 steps/s (or even higher) may be called for.

It was mentioned earlier that a single step (from rest) takes upwards of several milliseconds. It should therefore be clear that if the motor is to run at 2000 steps/s (i.e. 0.5 ms/step), it cannot possibly come to rest between successive steps, as it does at low stepping rates. Instead, we find in practice that at these high stepping rates, the rotor velocity becomes quite smooth, with hardly any outward hint of its stepwise origins. Nevertheless, the vital one-to-one correspondence between step command pulses and steps taken by the motor is maintained throughout, and the open-loop position control feature is preserved. This extraordinary ability to operate at very high stepping rates (up to 20,000 steps/s in some motors), and yet to remain fully in synchronism with the command pulses, is the most striking feature of stepping motor systems.

Operation at high speeds is referred to as 'slewing'. The transition from single-stepping (as shown in Figure 10.2) to high-speed slewing is a gradual one and is indicated by the plots in Figure 10.4. Roughly speaking, the motor will 'slew' if its stepping rate is above the frequency of its single-step oscillations. When motors are in the slewing range, they generally emit an audible whine, with a fundamental frequency equal to the stepping rate.

Naturally, a motor cannot be started from rest and expected to 'lock on' directly to a train of command pulses at say 2000 steps/s, which is well into the slewing range. Instead, it has to be started at a more modest stepping rate, before being accelerated (or 'ramped') up to speed: this is discussed more fully in section 6. In undemanding applications, the ramping can be done slowly, and spread over a large number of steps; but if the high stepping rate has to be reached quickly, the timings of individual step pulses must be very precise.

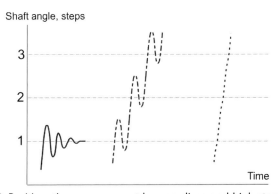

Figure 10.4 Position–time responses at low, medium and high stepping rates.

We may wonder what will happen if the stepping rate is increased too quickly. The answer is simply that the motor will not be able to remain 'in step' and will stall. The step command pulses will still be delivered, and the step counter will be accumulating what it believes are motor steps, but, by then, the system will have failed completely. A similar failure mode will occur if, when the motor is slewing, the train of step pulses is suddenly stopped, instead of being progressively slowed. The stored kinetic energy of the motor (and load) will cause it to overrun, so that the number of motor steps will be greater than the number of command pulses. Failures of this sort are prevented by the use of closed-loop control, as discussed later.

Finally, it is worth mentioning that stepping motors are designed to operate for long periods with their rotor held in a fixed (step) position, and with rated current in the winding (or windings). We can therefore anticipate that overheating after stalling is generally not a problem for a stepping motor.

3. PRINCIPLE OF MOTOR OPERATION

The principle on which stepping motors are based is very simple. When a bar of iron or steel is suspended so that it is free to rotate in a magnetic field, it will align itself with the field. If the direction of the field is changed, the bar will turn until it is again aligned, by the action of the so-called reluctance torque. (The mechanism is similar to that of a compass needle, except that if a compass had an iron needle instead of a permanent magnet it would settle along the earth's magnetic field but it might be rather slow and there would be ambiguity between N and S!)

Before exploring constructional details, it is worth saying a little more about reluctance torque, and its relationship with the torque-producing mechanism we have encountered so far in this book. The alert reader will be aware that, until this chapter, there has been little mention of reluctance torque, and might therefore wonder if it is entirely different from what we have considered so far.

The answer is that in the vast majority of electrical machines torque is produced by the interaction of a magnetic field (produced by either the stator or rotor) with current-carrying conductors on either the rotor or the stator. We based our understanding of how d.c. and induction motors produce torque on the simple formula $F = BIl$ for the force on a conductor of length l carrying a current I perpendicular to a magnetic flux density B (see Chapter 1). There was no mention of reluctance torque because (with very few exceptions) machines which exploit the 'BIl' mechanism do not have reluctance torque.

As mentioned above, reluctance torque originates in the tendency of an iron bar to align itself with magnetic field: if the bar is displaced from its alignment position it experiences a restoring torque. The rotors of machines that produce torque by reluctance action are therefore designed so that the rotor iron has projections or 'poles' (see Figure 10.5) that align with the magnetic field produced by the stator windings. All the torque is then produced by reluctance action, because, with no conductors on the rotor to carry current, there is obviously no 'BIl' torque. In contrast, the iron in the rotors of d.c. and induction motors is (ideally) cylindrical, in which case there is no 'preferred' orientation of the rotor iron, i.e. no reluctance torque.

Because the two torque-producing mechanisms appear to be radically different, the approaches taken to develop theoretical models have also diverged. As we have seen, simple equivalent circuits are available to allow us to understand and predict the behavior of mainstream 'BIl' machines such as d.c. and induction motors, and this is fortunate because of the overwhelming importance of these machines. Unfortunately, no such simple treatments are available for stepping and other reluctance-based machines. Circuit-based numerical models for performance prediction are widely used by manufacturers but they are not really much of use for illuminating behavior, so we will content ourselves with building up a picture of behavior from a study of typical operating characteristics.

The two most important types of stepping motor are the variable-reluctance (VR) type and the hybrid type. Both types utilize the reluctance principle, the difference between them lying in the method by which the magnetic fields are produced. In the VR type the fields are produced solely by sets of stationary current-carrying windings. The hybrid type also has sets of windings, but the addition of a permanent magnet (on the rotor) gives rise to the description 'hybrid' for this type of motor. Although both types of motor work on the same basic principle, it turns out in practice that the VR type is attractive for the larger step angles (e.g. 15°, 30°, 45°), while the hybrid tends to be best suited when small angles (e.g. 1.8°, 2.5°) are required.

3.1 Variable-reluctance motor

A simplified diagram of a 30°/step VR stepping motor is shown in Figure 10.5. The stator is made from a stack of steel laminations, and has six equally spaced projecting poles, or teeth, each carrying a separate coil. The rotor, which may be solid or laminated, has four projecting teeth, of the same width as the stator teeth. There is a very small air-gap – typically between 0.02 and 0.2 mm – between rotor and stator teeth. When no current is flowing in any of the stator coils, the rotor will therefore be completely free to rotate.

Diametrically opposite pairs of stator coils are connected in series, such that when one of them acts as a north pole, the other acts as a south pole. There are thus three independent stator circuits, or phases, and each one can be supplied with direct current from the drive circuit (not shown in Figure 10.5).

When phase A is energized (as indicated by the thick lines in Figure 10.5(a)), a magnetic field with its axis along the stator poles of phase A is created. The rotor is

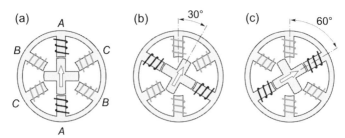

Figure 10.5 Principle of operation of 30°/step variable-reluctance stepping motor.

therefore attracted into a position where the pair of rotor poles distinguished by the marker arrow line up with the field, i.e. in line with the phase A pole, as shown in Figure 10.5(a). When phase A is switched off, and phase B is switched on instead, the second pair of rotor poles will be pulled into alignment with the stator poles of phase B, the rotor moving through 30° clockwise to its new step position, as shown in Figure 10.5(b). A further clockwise step of 30° will occur when phase B is switched off and phase C is switched on. At this stage the original pair of rotor poles comes into play again, but this time they are attracted to stator poles C, as shown in Figure 10.5(c). By repetitively switching on the stator phases in the sequence ABCA, etc. the rotor will rotate clockwise in 30° steps, while if the sequence is ACBA, etc. it will rotate anticlockwise. This mode of operation is known as '1-phase-on', and is the simplest way of making the motor step. Note that the polarity of the energizing current is not significant: the motor will be aligned equally well regardless of the direction of current.

3.2 Hybrid motor

A cross-sectional view of a typical 1.8° hybrid motor is shown in Figure 10.6. The stator has eight main poles, each with five teeth, and each main pole carries a simple coil. The rotor has two steel end-caps, each with 50 teeth, and separated by a permanent magnet. The rotor teeth have the same pitch as the teeth on the stator poles, and are offset so that the centerline of a tooth at one end-cap coincides with a slot at the other end-cap. The permanent magnet is axially magnetized, so that one set of rotor teeth is given a north polarity, and the other a south polarity. Extra torque is obtained by adding more stacks, as shown in Plate 10.2.

When no current is flowing in the windings, the only source of magnetic flux across the air-gap is the permanent magnet. The magnet flux crosses the air-gap from the N end-cap into the stator poles, flows axially along the body of the stator, and returns to the magnet by crossing the air-gap to the S end-cap. If there were no offset between the two sets of rotor teeth, there would be a strong periodic alignment torque when the rotor was turned, and every time a set of stator teeth was in line with the rotor teeth we would obtain a stable equilibrium position.

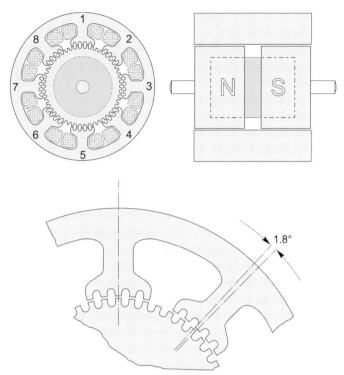

Figure 10.6 Hybrid (200 step/rev) stepping motor. The detail shows the rotor and stator tooth alignments, and indicates the step angle of 1.8°.

However, there is an offset, and this causes the alignment torque due to the magnet to be almost eliminated. In practice a small 'detent' torque remains, and this can be felt if the shaft is turned when the motor is de-energized: the motor tends to be held in its step positions by the detent torque. This is sometimes very useful: for example, it is usually enough to hold the rotor stationary when the power is switched off, so the motor can be left without fear of it being accidentally nudged into a new position.

The eight coils are connected to form two phase-windings. The coils on poles 1, 3, 5 and 7 form phase A, while those on 2, 4, 6 and 8 form phase B. When phase A carries positive current stator poles 1 and 5 are magnetized as south, and poles 3 and 7 become north. The teeth on the north end of the rotor are attracted to poles 1 and 5 while the offset teeth at the south end of the rotor are attracted into line with the teeth on poles 3 and 7. To make the rotor step, phase A is switched off, and phase B is energized with either positive current or negative current, depending on the sense of rotation required. This will cause the rotor to move by one-quarter of a tooth pitch (1.8°) to a new equilibrium (step) position.

The motor is continuously stepped by energizing the phases in the sequence +A, −B, −A, +B, +A (clockwise) or +A, +B, −A, −B, +A (anticlockwise). It will be clear from this that a bipolar supply is needed (i.e. one which can furnish +ve or −ve current). When the motor is operated in this way it is referred to as '2-phase, with bipolar supply'.

If a bipolar supply is not available, the same pattern of pole energization may be achieved in a different way, as long as the motor windings consist of two identical ('bifilar wound') coils. To magnetize pole 1 north, a positive current is fed into one set of phase A coils. But to magnetize pole 1 south, the same positive current is fed into the other set of phase A coils, which have the opposite winding sense. In total, there are then four separate windings, and when the motor is operated in this way it is referred to as '4-phase, with unipolar supply'. Since each winding only occupies half of the space, the m.m.f. of each winding is only half of that of the full coil, so the thermally rated output is clearly reduced as compared with bipolar operation (for which the whole winding is used).

We round off this section on hybrid motors with a comment on identifying windings, and a warning. If the motor details are not known, it is usually possible to identify bifilar windings by measuring the resistance from the common to the two ends. If the motor is intended for unipolar drive only, one end of each winding may be commoned inside the casing; for example, a 4-phase unipolar motor may have only five leads, one for each phase and one common. Wires are also usually color coded to indicate the location of the windings; for example, a bifilar winding on one set of poles will have one end red, the other end red and white, and the common white. Finally, it is not advisable to remove the rotor of a hybrid motor because rotors are magnetized in situ: removal typically causes a 5–10% reduction in magnet flux, with a corresponding reduction in static torque at rated current.

3.3 Summary

The construction of stepping motors is simple and robust, the only moving part being the rotor, which has no windings, commutator or brushes. The rotor is held at its step position solely by the action of the magnetic flux between stator and rotor. The step angle is a property of the tooth geometry and the arrangement of the stator windings, and accurate punching and assembly of the stator and rotor laminations is therefore necessary to ensure that adjacent step positions are exactly equally spaced. Any errors due to inaccurate punching will be non-cumulative, however.

The step angle is obtained from the expression

$$\text{Step angle} = \frac{360°}{(\text{rotor teeth}) \times (\text{stator phases})}$$

The VR motor in Figure 10.5 has four rotor teeth, three stator phase-windings, and the step angle is therefore 30°, as already shown. It should also be clear from the equation why small angle motors always have to have a large number of rotor teeth. The most widely used motor is the 200 step/rev hybrid type (see Figure 10.6). This has a 50-tooth rotor, 4-phase stator, and hence a step angle of 1.8° (=360°/(50 × 4).

The magnitude of the aligning torque clearly depends on the magnitude of the current in the phase-winding. However, the equilibrium position itself does not depend on the magnitude of the current, because it is simply the position where the rotor and stator teeth are in line. This property underlines the digital nature of the stepping motor.

4. MOTOR CHARACTERISTICS

4.1 Static torque–displacement curves

From the previous discussion, it should be clear that the shape of the torque–displacement curve, and the peak static torque, will depend on the internal electromagnetic design of the rotor. In particular the shapes of the rotor and stator teeth, and the disposition of the stator windings (and permanent magnet(s)), all have to be optimized to obtain the maximum static torque.

We now turn to a typical static torque–displacement curve, and look at how it determines motor behavior. Several aspects will be discussed, including the explanation of basic stepping (which has already been looked at in a qualitative way); the influence of load torque on step position accuracy; the effect of the amplitude of the winding current; and half-step and mini-stepping operation. For the sake of simplicity, the discussion will be based on the 30°/step 3-phase VR motor introduced earlier, but the conclusions reached apply to any stepping motor.

Typical static torque–displacement curves for a 3-phase 30°/step VR motor are shown in Figure 10.7. These show the torque that is developed by the motor when the rotor is displaced from its aligned position. Because of the rotor/stator symmetry, the magnitude of the restoring torque when the rotor is displaced by a given angle in one direction is the same as the magnitude of the restoring torque when it is displaced by the same angle in the other direction, but of opposite sign.

There are three curves in Figure 10.7, one for each of the three phases, and for each curve we assume that the relevant phase-winding carries its full (rated) current. If the current is less than rated, the peak torque will be reduced, and the shape of the curve is likely to be somewhat different. The convention used in Figure 10.7 is that a clockwise displacement of the rotor corresponds to a movement to the right, while a positive torque tends to move the rotor anticlockwise.

When only one phase, say A, is energized, the other two phases exert no torque, so their curves can be ignored and we can focus attention on the solid line in Figure 10.7. Stable equilibrium positions (for phase A excited) exist at $\theta = 0°$, 90°, 180° and 270°. They are stable (step) positions because any attempt to move the

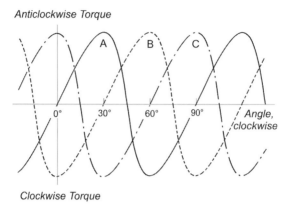

Anticlockwise Torque

Clockwise Torque

Figure 10.7 Static torque–displacement curves for 30°/step variable-reluctance stepping motor.

rotor away from them is resisted by a counteracting or restoring torque. These points correspond to positions where successive rotor poles (which are 90° apart) are aligned with the stator poles of phase A, as shown in Figure 10.5(a). There are also four unstable equilibrium positions (at $\theta = 45°$, 135°, 225° and 315°) at which the torque is also zero. These correspond to rotor positions where the stator poles are mid-way between two rotor poles, and they are unstable because if the rotor is deflected slightly in either direction, it will be accelerated in the same direction until it reaches the next stable position. If the rotor is free to turn, it will therefore always settle in one of the four stable positions.

4.2 Single-stepping

If we assume that phase A is energized, and the rotor is at rest in the position $\theta = 0°$ (Figure 10.7) we know that if we want to step in a clockwise direction, the phases must be energized in the sequence ABCA, etc., so we can now imagine that phase A is switched off, and phase B is energized instead. We will also assume that the decay of current in phase A and the build-up in phase B take place very rapidly, before the rotor moves significantly.

The rotor will find itself at $\theta = 0°$, but it will now experience a clockwise torque (see Figure10.7) produced by phase B. The rotor will therefore accelerate clockwise, and will continue to experience clockwise torque, until it has turned through 30°. The rotor will be accelerating all the time, and it will therefore overshoot the 30° position, which is of course its target (step) position for phase B. As soon as it overshoots, however, the torque reverses, and the rotor experiences a braking torque, which brings it to rest before it accelerates back towards the 30° position. If there was no friction or other cause of damping, the rotor would continue to oscillate; but in practice it comes to rest at its new position quite quickly in much the same way as a damped second-order system. The next 30° step is achieved in the same way, by switching off the current in phase B, and switching on phase C.

In the discussion above, we have recognized that the rotor is acted on sequentially by each of the three separate torque curves shown in Figure 10.7. Alternatively, since the three curves have the same shape, we can think of the rotor being influenced by a single torque curve which 'jumps' by one step (30° in this case) each time the current is switched from one phase to the next. This is often the most convenient way of visualizing what is happening in the motor.

4.3 Step position error and holding torque

In the previous discussion the load torque was assumed to be zero, and the rotor was therefore able to come to rest with its poles exactly in line with the excited stator poles. When load torque is present, however, the rotor will not be able to pull fully into alignment, and a 'step position error' will be unavoidable.

The origin and extent of the step position error can be appreciated with the aid of the typical torque–displacement curve shown in Figure 10.8. The true step position is at the origin in the figure, and this is where the rotor would come to rest in the absence of load torque. If we imagine the rotor is initially at this position, and then consider that a clockwise load (T_L) is applied, the rotor will move clockwise, and as it does so it will develop progressively more anticlockwise torque. The equilibrium position will be reached when the motor torque is equal and opposite to the load torque, i.e. at point A in Figure 10.8. The corresponding angular displacement from the step position (θ_e in Figure 10.8) is the step position error.

The existence of a step position error is one of the drawbacks of the stepping motor. The motor designer attempts to combat the problem by aiming to produce

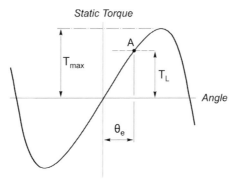

Figure 10.8 Static torque–angle curve showing step position error (θ_e) resulting from load torque T_L.

a steep torque–angle curve around the step position, and the user has to be aware of the problem and choose a motor with a sufficiently steep curve to keep the error within acceptable limits. In some cases this may mean selecting a motor with a higher peak torque than would otherwise be necessary, simply to obtain a steep enough torque curve around the step position.

As long as the load torque is less than T_{max} (see Figure 10.8), a stable rest position is obtained, but if the load torque exceeds T_{max}, the rotor will be unable to hold its step position. T_{max} is therefore known as the 'holding' torque. The value of the holding torque immediately conveys an idea of the overall capability of any motor, and it is – after step angle – the most important single parameter which is looked for in selecting a motor. Often, the adjective 'holding' is dropped altogether: for example, 'a 1 Nm motor' is understood to be one with a peak static torque (holding torque) of 1 Nm.

4.4 Half-stepping

We have already seen how to step the motor in $30°$ increments by energizing the phases one at a time in the sequence ABCA, etc. Although this '1-phase-on' mode is the simplest and most widely used, there are two other modes which are also frequently employed. These are referred to as the '2-phase-on' mode and the 'half-stepping' mode. The 2-phase-on can provide greater holding torque and a much better damped single-step response than the 1-phase-on mode; and the half-stepping mode permits the effective step angle to be halved – thereby doubling the resolution – and produces a smoother shaft rotation.

In the 2-phase-on mode, two phases are excited simultaneously. When phases A and B are energized, for example, the rotor experiences torques from both phases, and comes to rest at a point midway between the two adjacent full-step positions. If the phases are switched in the sequence AB, BC, CA, AB, etc., the motor will take full ($30°$) steps, as in the 1-phase-on mode, but its equilibrium positions will be interleaved between the full-step positions.

To obtain 'half-stepping' the phases are excited in the sequence A, AB, B, BC, etc., i.e. alternately in the 1-phase-on and 2-phase-on modes. This is sometimes known as 'wave' excitation, and it causes the rotor to advance in steps of $15°$, or half the full-step angle. As might be expected, continuous half-stepping usually produces a smoother shaft rotation than full-stepping, and it also doubles the resolution.

We can see what the static torque curve looks like when two phases are excited by superposition of the individual phase curves. An example is shown in Figure 10.9, from which it can be seen that for this machine, the holding torque (i.e. the peak static torque) is higher with two phases excited than with only one excited. The stable equilibrium (half-step) position is at $15°$, as expected. The price to be

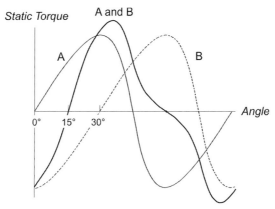

Figure 10.9 Static torque curve (thick line) corresponding to 2-phase-on excitation.

paid for the increased holding torque is the increased power dissipation in the windings, which is doubled as compared with the 1-phase-on mode. The holding torque increases by a factor less than two, so the torque per watt (which is a useful figure of merit) is reduced.

A word of caution is needed in regard to the addition of the two separate 1-phase-on torque curves to obtain the 2-phase-on curve. Strictly, such a procedure is only valid where the two phases are magnetically independent, or the common parts of the magnetic circuits are unsaturated. This is not the case in most motors, in which the phases share a common magnetic circuit which operates under highly saturated conditions. Direct addition of the 1-phase-on curves cannot therefore be expected to give an accurate result for the 2-phase-on curve, but it is easy to do, and provides a reasonable estimate.

Apart from the higher holding torque in the 2-phase-on mode, there is another important difference which distinguishes the static behavior from that of the 1-phase-on mode. In the 1-phase-on mode, the equilibrium or step positions are determined solely by the geometry of the rotor and stator: they are the positions where the rotor and stator are in line. In the 2-phase-on mode, however, the rotor is intended to come to rest at points where the rotor poles are lined up midway between the stator poles. This position is not sharply defined by the 'edges' of opposing poles, as in the 1-phase-on case; and the rest position will only be exactly midway if (a) there is exact geometrical symmetry and, more importantly, (b) the two currents are identical. If one of the phase currents is larger than the other, the rotor will come to rest closer to the phase with the higher current, instead of half-way between the two. The need to balance the currents to obtain precise half-stepping is clearly a drawback to this scheme. Paradoxically, however, the properties of the machine with unequal phase currents can sometimes be turned to good effect, as we now see.

4.5 Step division – mini-stepping

There are some applications where very fine resolution is called for, and a motor with a very small step angle – perhaps only a fraction of a degree – is required. We have already seen that the step angle can only be made small by increasing the number of rotor teeth and/or the number of phases, but in practice it is inconvenient to have more than four or five phases, and it is difficult to manufacture rotors with more than 50–100 teeth. This means it is rare for motors to have step angles below about 1°. When a smaller step angle is required, a technique known as mini-stepping (or step division) is used.

Mini-stepping is a technique based on 2-phase-on operation which provides for the subdivision of each full motor step into a number of 'substeps' of equal size. In contrast with half-stepping, where the two currents have to be kept equal, the currents are deliberately made unequal. By correctly choosing and controlling the relative amplitudes of the currents, the rotor equilibrium position can be made to lie anywhere between the step positions for each of the two separate phases.

Closed-loop current control is needed to prevent the current from changing as a result of temperature changes in the windings, or variations in the supply voltage; and if it is necessary to ensure that the holding torque stays constant for each mini-step both currents must be changed according to a prescribed algorithm. Despite the difficulties referred to above, mini-stepping is used extensively, especially in photographic and printing applications where a high resolution is needed. Schemes involving between three and 10 mini-steps for a 1.8° step motor are numerous, and there are instances where up to 100 mini-steps (20,000 mini-steps/rev) have been achieved.

So far, we have concentrated on those aspects of behavior which depend only on the motor itself, i.e. the static performance. The shape of the static torque curve, the holding torque, and the slope of the torque curve about the step position have all been shown to be important pointers to the way the motor can be expected to perform. All of these characteristics depend on the current(s) in the windings, however, and when the motor is running the instantaneous currents will depend on the type of drive circuit employed.

5. STEADY-STATE CHARACTERISTICS – IDEAL (CONSTANT-CURRENT) DRIVE

In this section we will look at how the motor would perform if it were supplied by an ideal drive circuit, which turns out to be one that is capable of supplying rectangular pulses of current to each winding when required, and regardless of the stepping rate. Because of the inductance of the windings, no real drive circuit will be able to achieve this, but the most sophisticated (and expensive) ones achieve near-ideal operation up to very high stepping rates.

5.1 Requirements of drive

The basic function of the complete drive is to convert the step command input signals into appropriate patterns of currents in the motor windings. This is achieved in two distinct stages, as shown in Figure 10.10, which relates to a 3-phase motor.

The 'translator' stage converts the incoming train of step command pulses into a sequence of on/off commands to each of the three power stages. In the 1-phase-on mode, for example, the first step command pulse will be routed to turn on phase A, the second will turn on phase B, and so on. In a very simple drive, the translator will probably provide for only one mode of operation (e.g. 1-phase-on), but most commercial drives provide the option of 1-phase-on, 2-phase-on and half-stepping. Single-chip ICs with these three operating modes and with both 3-phase and 4-phase outputs are readily available.

The power stages (one per phase) supply the current to the windings. An enormous diversity of types are in use, ranging from simple ones with one switching transistor per phase, to elaborate chopper-type circuits with four transistors per phase, and some of these are discussed in section 6. At this point, however, it is helpful to list the functions required of the 'ideal' power stage. These are first that when the translator calls for a phase to be energized, the full (rated) current should be established immediately; secondly, the current should be maintained constant (at its rated value) for the duration of the 'on' period; and finally, when the translator calls for the current to be turned off, it should be reduced to zero immediately.

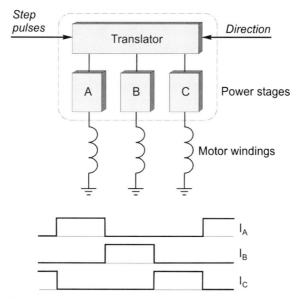

Figure 10.10 General arrangement of drive system for 3-phase motor, and winding currents corresponding to an 'ideal' drive.

The ideal current waveforms for continuous stepping with 1-phase-on operation are shown in the lower part of Figure 10.10. The currents have a square profile because this leads to the optimum value of running torque from the motor. But because of the inductance of the windings, no real drive will achieve the ideal current waveforms, though many drives come close to the ideal, even at quite high stepping rates. Drives which produce such rectangular current waveforms are (not surprisingly) called constant-current drives. We now look at the running torque produced by a motor when operated from an ideal constant-current drive. This will act as a yardstick for assessing the performance of other drives, all of which will be seen to have inferior performance.

5.2 Pull-out torque under constant-current conditions

If the phase currents are taken to be ideal, i.e. they are switched on and off instantaneously, and remain at their full rated value during each 'on' period, we can picture the axis of the magnetic field to be advancing around the machine in a series of steps, the rotor being urged to follow it by the reluctance torque. If we assume that the inertia is high enough for fluctuations in rotor velocity to be very small, the rotor will be rotating at a constant rate which corresponds exactly to the stepping rate.

Now if we consider a situation where the position of the rotor axis is, on average, lagging behind the advancing field axis, it should be clear that, on average, the rotor will experience a driving torque. The more it lags behind, the higher will be the average forward torque acting on it, but only up to a point. We already know that if the rotor axis is displaced too far from the field axis, the torque will begin to diminish, and eventually reverse, so we conclude that although more torque will be developed by increasing the rotor lag angle, there will be a limit to how far this can be taken.

Turning now to a quantitative examination of the torque on the rotor, we will make use of the static torque–displacement curves discussed earlier, and look at what happens when the load on the shaft is varied, the stepping rate being kept constant. Clockwise rotation will be studied, so the phases will be energized in the sequence ABC. The instantaneous torque on the rotor can be arrived at by recognizing that (a) the rotor speed is constant, and it covers one step angle (30°) between step command pulses, and (b) the rotor will be 'acted on' sequentially by each of the set of torque curves.

When the load torque is zero, the net torque developed by the rotor must be zero (apart from a very small torque required to overcome friction). This condition is shown in Figure 10.11(a). The instantaneous torque is shown by the thick line, and it is clear that each phase in turn exerts first a clockwise torque, then an anti-clockwise torque while the rotor angle turns through 30°. The average torque is zero, the same as the load torque, because the average rotor lag angle is zero.

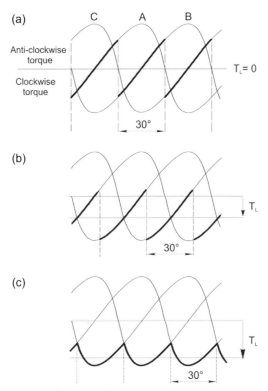

Figure 10.11 Static torque curves indicating how the average steady-state torque (T_L) is developed during constant-frequency operation.

When the load torque on the shaft is increased, the immediate effect is to cause the rotor to fall back in relation to the field. This causes the clockwise torque to increase, and the anticlockwise torque to decrease. Equilibrium is reached when the lag angle has increased sufficiently for the motor torque to equal the load torque. The torque developed at an intermediate load condition like this is shown by the thick line in Figure 10.11(b). The highest average torque that can possibly be developed is shown by the thick line in Figure 10.11(c): if the load torque exceeds this value (which is known as the pull-out torque) the motor loses synchronism and stalls, and the vital one-to-one correspondences between pulses and steps are lost.

Since we have assumed an ideal constant-current drive, the pull-out torque will be independent of the stepping rate, and the pull-out torque–speed curve under ideal conditions is therefore as shown in Figure 10.12. The shaded area represents the permissible operating region: at any particular speed (stepping rate) the load torque can have any value up to the pull-out torque, and the motor will continue to run at the same speed. But if the load torque exceeds the pull-out torque, the motor will suddenly pull out of synchronism and stall.

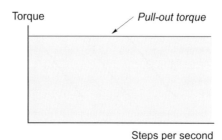

Figure 10.12 Steady-state operating region with ideal constant-current drive. (In such idealized circumstances there would be no limit to the stepping rate, but as shown in Figure 10.16 a real drive circuit imposes an upper limit.)

As mentioned earlier, no real drive will be able to provide the ideal current waveforms, so we now turn to look briefly at the types of drives in common use, and at their pull-out torque–speed characteristics.

6. DRIVE CIRCUITS AND PULL-OUT TORQUE–SPEED CURVES

Users often find difficulty in coming to terms with the fact that the running performance of a stepping motor depends so heavily on the type of drive circuit being used. It is therefore important to emphasize that in order to meet a specification, it will always be necessary to consider the motor and drive together, as a package.

There are three commonly used types of drive. All use transistors which are operated as switches, i.e. they are either turned fully on, or they are cut off. A brief description of each is given below, and the pros and cons of each type are indicated. In order to simplify the discussion, we will consider one phase of a 3-phase VR motor and assume that it can be represented by a simple series $R–L$ circuit in which R and L are the resistance and self-inductance of the winding, respectively. (In practice the inductance will vary with rotor position, giving rise to motional e.m.f. in the windings, which, as we have seen previously in this book, is an inescapable manifestation of an electromechanical energy-conversion process. If we needed to analyze stepping motor behavior fully we would have to include the motional e.m.f. terms. Fortunately, we can gain a pretty good appreciation of how the motor behaves if we model each winding simply in terms of its resistance and self-inductance.)

6.1 Constant-voltage drive

This is the simplest possible drive: the circuit for one of the three phases is shown in the upper part of Figure 10.13, and the current waveforms at low and high

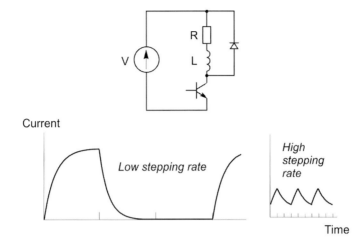

Figure 10.13 Basic constant-voltage drive circuit and typical current waveforms.

stepping rates are shown in the lower part of the figure. The d.c. voltage V is chosen so that when the switching device (usually a MOSFET although a BJT is shown) turns on, the steady current is the rated current as specified by the motor manufacturer.

The current waveforms display the familiar rising exponential shape that characterizes a first-order system: the time-constant is L/R, the current reaching its steady state after several time-constants. When the transistor switches off, the stored energy in the inductance cannot instantaneously reduce to zero, so although the current through the transistor suddenly becomes zero, the current in the winding is diverted into the closed path formed by the winding and the freewheel diode, and it then decays exponentially to zero, again with time-constant L/R: in this period the stored energy in the magnetic field is dissipated as heat in the resistance of the winding and diode.

At low stepping rates (low speed), the drive provides a reasonably good approximation to the ideal rectangular current waveform. (We are considering a 3-phase motor, so ideally one phase should be on for one step pulse and off for the next two, as in Figure 10.10.) But at higher frequencies (right-hand waveform in Figure 10.13), where the 'on' period is short compared with the winding time-constant, the current waveform degenerates, and is nothing like the ideal rectangular shape. In particular the current never gets anywhere near its full value during the 'on' pulse, so the torque over this period is reduced; and even worse, a substantial current persists when the phase is supposed to be off, so during this period the phase will contribute a negative torque to the rotor. Not surprisingly all this results in a very rapid fall-off of pull-out torque with speed, as shown later in Figure 10.16(a).

Curve (a) in Figure 10.16 should be compared with the pull-out torque under ideal constant-current conditions shown in Figure 10.12 in order to appreciate the severely limited performance of the simple constant-voltage drive.

6.2 Current-forced drive

The initial rate of rise of current in a series $R–L$ circuit is directly proportional to the applied voltage, so in order to establish the current more quickly at switch-on, a higher supply voltage (V_f) is needed. But if we simply increased the voltage, the steady-state current (V_f/R) would exceed the rated current and the winding would overheat.

To prevent the current from exceeding the rated value, an additional 'forcing' resistor has to be added in series with the winding. The value of this resistance (R_f) must be chosen so that $V_f/(R + R_f) = I$, where I is the rated current. This is shown in the upper part of Figure 10.14, together with the current waveforms at low and high stepping rates. Because the rates of rise and fall of current are higher, the current waveforms approximate more closely to the ideal rectangular shape, especially at low stepping rates, though at higher rates they are still far from ideal, as shown in Figure 10.14. The low-frequency pull-out torque is therefore maintained to a higher stepping rate, as shown in Figure 10.16(b). Values of R_f from 2 to 10 times the motor resistance (R) are common. Broadly speaking, if $R_f = 10R$, for example, a given pull-out torque will be available at 10 times the stepping rate, compared with an unforced constant-voltage drive.

Some manufacturers call this type of drive an 'R/L' drive, while others call it an 'L/R' drive, or even simply a 'resistor drive'. Sets of pull-out torque–speed curves in

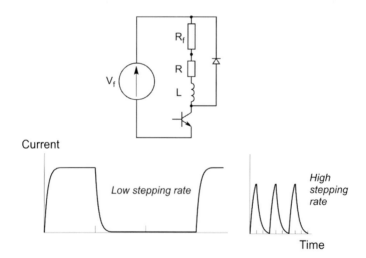

Figure 10.14 Current-forced (*L/R*) drive and typical current waveforms.

catalogues are labeled with values R/L (or L/R) $= 5$, 10, etc. This means that the curves apply to drives where the forcing resistor is five (or 10) times the winding resistance, the implication being that the drive voltage has also been adjusted to keep the static current at its rated value. Obviously, it follows that the higher R_f is made, the higher the power rating of the supply; and it is the higher power rating which is the principal reason for the improved torque–speed performance.

The major disadvantage of this drive is its inefficiency, and the consequent need for a high power-supply rating. Large amounts of heat are dissipated in the forcing resistors, especially when the motor is at rest and the phase current is continuous, and disposing of the heat can lead to awkward problems in the siting of the forcing resistors. These drives are therefore only used at the low-power end of the scale.

It was mentioned earlier that the influence of the motional e.m.f. in the winding would be ignored. In practice, however, the motional e.m.f. always has a pronounced influence on the current, especially at high stepping rates, so it must be borne in mind that the waveforms shown in Figures 10.13 and 10.14 are only approximate. Not surprisingly, it turns out that the motional e.m.f. tends to make the current waveforms worse (and the torque less) than the discussion above suggests. Ideally, therefore, we need a drive which will keep the current constant throughout the 'on' period, regardless of the motional e.m.f. The closed-loop chopper-type drive (below) provides the closest approximation to this, and also avoids the waste of power which is a feature of R/L drives.

6.3 Constant-current (chopper) drive

The basic circuit for one phase of a VR motor is shown in the upper part of Figure 10.15 together with the current waveforms. A high-voltage power supply is used in order to obtain very rapid changes in current when the phase is switched on or off.

The lower transistor is turned on for the whole period during which current is required. The upper transistor turns on whenever the actual current falls below the lower threshold of the hysteresis band (shown dashed in Figure 10.15) and it turns off when the current exceeds the upper threshold. The chopping action leads to a current waveform which is a good approximation to the ideal (see Figure 10.10). At the end of the 'on' period both transistors switch off and the current freewheels through both diodes and back to the supply. During this period the stored energy in the inductance is returned to the supply, and because the winding terminal voltage is then $-V_c$, the current decays as rapidly as it is built up.

Because the current-control system is a closed-loop one, distortion of the current waveform by the motional e.m.f. is minimized, and this means that the ideal (constant-current) torque–speed curve is closely followed up to high stepping rates. Eventually, however, the 'on' period reduces to the point where it is less than the

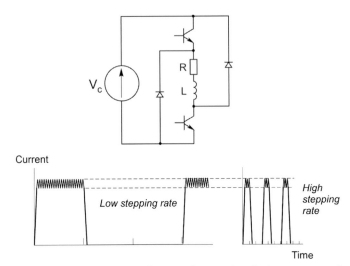

Figure 10.15 Constant-current chopper drive and typical current waveforms.

current rise time, and the full current is never reached. Chopping action then ceases, the drive reverts essentially to a constant-voltage one, and the torque falls rapidly as the stepping rate is raised even higher, as in Figure 10.16(c). There is no doubt of the overall superiority of the chopper-type drive, and it is now the standard drive. Single-chip chopper modules can be bought for small (say 1–2 A) motors, and complete plug-in chopper cards, rated up to 10 A or more, are available for larger motors. (See Plates 10.3 and 10.5)

The discussion in this section relates to a VR motor, for which unipolar current pulses are sufficient. If we have a hybrid or other permanent magnet motor we will need a bipolar current source (i.e. one that can provide positive or negative current), and for this we will find that each phase is supplied from a four-transistor H-bridge, as discussed in Chapter 2.

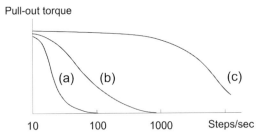

Figure 10.16 Typical pull-out torque–speed curves for a given motor with different types of drive circuit. (a) Constant-voltage drive; (b) current-forced drive; (c) chopper drive.

6.4 Resonances and instability

In practice, measured torque–speed curves frequently display severe dips at or around certain stepping rates: a typical measured characteristic for a hybrid motor with a voltage-forced drive is shown as curve (a) in Figure 10.17. Manufacturers are not keen to stress this feature, so it is important for the user to be aware of the potential difficulty.

The magnitude and location of the torque dips depend in a complex way on the characteristics of the motor, the drive, the operating mode and the load. We will not go into detail here, apart from mentioning the underlying causes and remedies.

There are two distinct mechanisms which cause the dips. The first is a straightforward 'resonance-type' problem which manifests itself at low stepping rates, and originates from the oscillatory nature of the single-step response. Whenever the stepping rate coincides with the natural frequency of rotor oscillations, the oscillations can be enhanced, and this in turn makes it more likely that the rotor will fail to keep in step with the advancing field.

The second phenomenon occurs because at certain stepping rates it is possible for the complete motor/drive system to exhibit positive feedback, and become unstable. This instability usually occurs at relatively high stepping rates, well above the 'resonance' regions discussed above. The resulting dips in the torque–speed curve are extremely sensitive to the degree of viscous damping present (mainly in the bearings), and it is not uncommon to find that a severe dip which is apparent on a warm day (such as that shown at around 1000 steps/s in Figure 10.17) will disappear on a cold day.

The dips are most pronounced during steady-state operation, and it may be that their presence is not serious provided that continuous operation at the relevant speeds is not required. In this case, it is often possible to accelerate through the dips without adverse effect. Various special drive techniques exist for eliminating resonances by smoothing out the stepwise nature of the stator field, or by modulating the supply frequency to damp out the instability, but the simplest

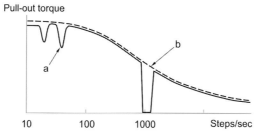

Figure 10.17 Pull-out torque–speed curves for a hybrid stepping motor showing (curve a) low-speed resonance dips and mid-frequency instability at around 1000 steps/s; and improvement brought about by adding an inertial damper (curve b).

remedy in open-loop operation is to fit a damper to the motor shaft. Dampers of the Lanchester type or of the viscously coupled inertia (VCID) type are used. These consist of a lightweight housing which is fixed rigidly to the motor shaft, and an inertia which can rotate relative to the housing. The inertia and the housing are separated either by a viscous fluid (VCID type) or by a friction disc (Lanchester type). Whenever the motor speed is changing, the assembly exerts a damping torque, but once the motor speed is steady, there is no drag torque from the damper. By selecting the appropriate damper, the dips in the torque–speed curve can be eliminated, as shown in Figure 10.17(curve b). Dampers are also often essential to damp the single-step response, particularly with VR motors, many of which have a highly oscillatory step response. Their only real drawback is that they increase the effective inertia of the system, and thus reduce the maximum acceleration.

7. TRANSIENT PERFORMANCE

7.1 Step response

It was pointed out earlier that the single-step response is similar to that of a damped second-order system. We can easily estimate the natural frequency ω_n in rad/s from the equation

$$\omega_n^2 = \frac{\text{Slope of torque} - \text{Angle curve}}{\text{Total inertia}}$$

Knowing ω_n, we can judge what the oscillatory part of the response will look like, by assuming the system is undamped. To refine the estimate, and to obtain the settling time, however, we need to estimate the damping ratio, which is much more difficult to determine as it depends on the type of drive circuit and mode of operation as well as on the mechanical friction. In VR motors the damping ratio can be as low as 0.1, but in hybrid types it is typically 0.3–0.4. These values are too low for many applications where rapid settling is called for.

Two remedies are available, the simplest being to fit a mechanical damper of the type mentioned above. Alternatively, a special sequence of timed command pulses can be used to brake the rotor so that it reaches its new step position with zero velocity and does not overshoot. This procedure is variously referred to as 'electronic damping', 'electronic braking' or 'back phasing'. It involves re-energizing the previous phase for a precise period before the rotor has reached the next step position, in order to exert just the right degree of braking. It can only be used successfully when the load torque and inertia are predictable and not subject to change. Because it is an open-loop scheme it is extremely sensitive to apparently minor changes such as day-to-day variation in friction, which can make it unworkable in many instances.

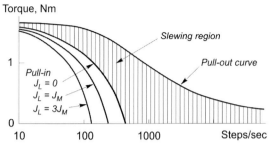

Figure 10.18 Typical pull-in and pull-out curves showing effect of load inertia on the pull-in torque. (J_M = motor inertia; J_L = load inertia.)

7.2 Starting from rest

The rate at which the motor can be started from rest without losing steps is known as the 'starting' or 'pull-in' rate. The starting rate for a given motor depends on the type of drive, and the parameters of the load. This is entirely as expected since the starting rate is a measure of the motor's ability to accelerate its rotor and load and pull into synchronism with the field. The starting rate thus reduces if either the load torque or the load inertia is increased. Typical pull-in torque–speed curves, for various inertias, are shown in Figure 10.18. The pull-out torque–speed curve is also shown, and it can be seen that for a given load torque, the maximum steady (slewing) speed at which the motor can run is much higher than the corresponding starting rate. (Note that only one pull-out torque is usually shown, and is taken to apply for all inertia values. This is because the inertia is not significant when the speed is constant.)

It will normally be necessary to consult the manufacturer's data to obtain the pull-in rate, which will apply only to a particular drive. However, a rough assessment is easily made: we simply assume that the motor is producing its pull-out torque, and calculate the acceleration that this would produce, making due allowance for the load torque and inertia. If, with the acceleration as calculated, the motor is able to reach the steady speed in one step or less, it will be able to pull in; if not, a lower pull-in rate is indicated.

7.3 Optimum acceleration and closed-loop control

There are some applications where the maximum possible accelerations and decelerations are demanded, in order to minimize point-to-point times. If the load parameters are stable and well defined, an open-loop approach is feasible, and this is discussed first. Where the load is unpredictable, however, a closed-loop strategy is essential, and is dealt with later.

To achieve maximum possible acceleration calls for every step command pulse to be delivered at precisely optimized intervals during the acceleration period. For maximum torque, each phase must be on whenever it can produce positive torque, and off when its torque would be negative. Since the torque depends on the rotor position, the optimum switching times have to be calculated from a full dynamic analysis. This can usually be accomplished by making use of the static torque–angle curves (provided appropriate allowance is made for the rise and fall times of the stator currents), together with the torque–speed characteristic and inertia of the load. A series of computations is required to predict the rotor angle–time relationship, from which the switchover points from one phase to the next are deduced. The train of accelerating pulses is then pre-programmed into the controller, for subsequent feeding to the drive in an open-loop fashion. It is obvious that this approach is only practicable if the load parameters do not vary, since any change will invalidate the computed optimum stepping intervals.

When the load is unpredictable, a much more satisfactory arrangement is obtained by employing a closed-loop scheme, using position feedback from a shaft-mounted encoder. The feedback signals indicate the instantaneous position of the rotor, and are used to ensure that the phase-windings are switched at precisely the right rotor position for maximizing the developed torque. Motion is initiated by a single command pulse, and subsequent step command pulses are effectively self-generated by the encoder. The motor continues to accelerate until its load torque equals the load torque, and then runs at this (maximum) speed until the deceleration sequence is initiated. During all this time, the step counter continues to record the number of steps taken.

Closed-loop operation ensures that the optimum acceleration is achieved, but at the expense of more complex control circuitry, and the need to fit a shaft encoder. Relatively cheap encoders are, however, now available for direct fitting to some ranges of motors, and single chip microcontrollers are available which provide all the necessary facilities for closed-loop control.

An appealing approach aimed at eliminating an encoder is to detect the position of the rotor by online analysis of the signals (principally the rates of change of currents) in the motor windings themselves – in other words, to use the motor as its own encoder. A variety of approaches have been tried, including the addition of high-frequency alternating voltages superimposed on the excited phase, so that, as the rotor moves, the variation of inductance results in a modulation of the alternating current component. Some success has been achieved with particular motors, but the approach has not achieved widespread commercial exploitation, perhaps because of competition from the brushless permanent magnet servo drive.

To return finally to encoders, we should note that they are also used in open-loop schemes when an absolute check on the number of steps taken is required. In this context the encoder simply provides a tally of the total steps taken, and normally plays no part in the generation of the step pulses. At some stage, however, the actual number of steps taken will be compared with the number of step command pulses issued by the controller. When a disparity is detected, indicating a loss (or gain of steps), the appropriate additional forward or backward pulses can be added.

8. SWITCHED-RELUCTANCE MOTOR DRIVES

The switched-reluctance drive was developed in the 1980s to offer advantages in terms of efficiency, power per unit weight and volume, robustness and operational flexibility. The motor and its associated power-electronic drive must be designed as an integrated package, and optimized for a particular specification, e.g. for maximum overall efficiency with a specific load, or maximum speed range, peak short-term torque or torque ripple. Despite being relatively new, the technology has been used in a range of applications which can benefit from its performance characteristics including general purpose industrial drives, compressors and domestic appliances.

8.1 Principle of operation

Like the stepping motor, the switched-reluctance motor is 'doubly-salient', i.e. it has projecting poles on both rotor and stator. However, most switched-reluctance motors are of much higher power than the largest stepper, and it turns out that in the higher power ranges (where the winding resistances become much less signif-icant), the doubly-salient arrangement is very effective as far as efficient electro-magnetic energy conversion is concerned.

A typical switched-reluctance motor is shown in Figure 10.19: this example has 12 stator poles and eight rotor poles, and represents a widely used arrangement, but other pole combinations are used to suit different applications. The stator carries

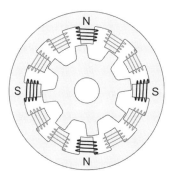

Figure 10.19 Typical switched-reluctance motor. Each of the 12 stator poles carries a concentrated winding, while the eight-pole rotor has no windings or magnets.

coils on each pole, while the rotor, which is made from laminations in the usual way, has no windings or magnets[2] and is therefore cheap to manufacture and extremely robust. (See Plate 10.6)

In Figure 10.19 the 12 coils are grouped to form three phases, which are independently energized from a 3-phase converter.

The motor rotates by exciting the phases sequentially in the sequence A, B, C for anticlockwise rotation or A, C, B for clockwise rotation, the 'nearest' pair of rotor poles being pulled into alignment with the appropriate stator poles by reluctance torque action. In Figure 10.19 the four coils forming phase A are shown in black, the polarities of the coil m.m.f.s being indicated by the letters N and S on the back of the core. Each time a new phase is excited the equilibrium position of the rotor advances by 15°, so after one complete cycle (i.e. each of the three phases has been excited once) the angle turned through is 45°. The machine therefore rotates once for eight fundamental cycles of supply to the stator windings, so in terms of the relationship between the fundamental supply frequency and the speed of rotation, the machine in Figure 10.19 behaves as a 16-pole conventional machine.

The structure is clearly the same as that of the variable-reluctance stepping motor discussed earlier in this chapter, but there are important design differences which reflect the different objectives (continuous rotation for the switched reluctance, stepwise progression for the stepper), but otherwise the mechanisms of torque production are identical. However, while the stepper is designed first and foremost for open-loop operation, the switched-reluctance motor is designed for self-synchronous operation, the phases being switched by signals derived from a shaft-mounted rotor position detector. In terms of performance, at all speeds below the base speed continuous operation at full torque is possible. Above the base speed, the flux can no longer be maintained at full amplitude and the available torque reduces with speed. The operating characteristics are thus very similar to those of the other most important controlled-speed drives, but with the added advantage that overall efficiencies are generally a per cent or two higher.

Given that the mechanism of torque production in the switched-reluctance motor appears to be very different from that in d.c. machines, induction motors and synchronous machines (all of which exploit the 'BIl' force on a conductor in a magnetic field), it might have been expected that one or other type would offer such clear advantages that the other would fade away. In fact, there appears to be little to choose between them overall, and one wonders whether the mechanisms of operation are really so fundamentally different as our ingrained ways of looking at things lead us to believe. Perhaps a visitor from another planet would note the similarity in terms of volume, quantities and disposition of iron and copper, and

---

[2] Recent uncertainties over future supplies of rare-earth magnet materials may renew interest in switched-reluctance technology.

overall performance, and bring some fresh enlightenment to bear so that we emerge recognizing some underlying truth that hitherto has escaped us.

8.2 Torque prediction and control

If the iron in the magnetic circuit is treated as ideal, analytical expressions can be derived to express the torque of a reluctance motor in terms of the rotor position and the current in the windings. In practice, however, this analysis is of little real use, not only because switched-reluctance motors are designed to operate with high levels of magnetic saturation in parts of the magnetic circuit, but also because, except at low speeds, it is not practicable to achieve specified current profiles.

The fact that high levels of saturation are involved makes the problem of predicting torque at the design stage challenging, but despite the highly non-linear relationships it is possible to compute the flux, current and torque as functions of rotor position, so that optimum control strategies can be devised to meet particular performance specifications. Unfortunately this complexity means that there is no simple equivalent circuit available to illuminate behavior.

As we saw when we discussed the stepping motor, to maximize the average torque it would (in principle) be desirable to establish the full current in each phase instantaneously, and to remove it instantaneously at the end of each positive torque period. But, as illustrated in Figure 10.13, this is not possible even with a small stepping motor, and certainly out of the question for switched-reluctance motors (which have much higher inductance) except at low speeds where current chopping (see Figure 10.15) is employed. For most of the speed range, the best that can be done is to apply the full voltage available from the converter at the start of the 'on' period, and (using a circuit such as that shown in Figure 10.15) apply full negative voltage at the end of the pulse by opening both of the switches.

Operation using full positive voltage at the beginning and full negative voltage at the end of the 'on' period is referred to as 'single-pulse' operation. For all but small motors (of less than say 1 kW) the phase resistance is negligible and consequently the magnitude of the phase flux linkage is determined by the applied voltage and frequency, as we have seen many times previously with other types of motor.

The relationship between the flux linkage (ψ) and the voltage is embodied in Faraday's law, i.e. $v = d\psi/dt$, so with the rectangular voltage waveform of single-pulse operation the phase flux linkage waveforms have a very simple triangular shape, as in Figure 10.20 which shows the waveforms for phase A of a 3-phase motor. (The waveforms for phases B are C are identical, but are not shown: they are displaced by one-third and two-thirds of a cycle, as indicated by the arrows.) The upper half of the diagram represents the situation at speed N, while the lower half corresponds to a speed of $2N$. As can be seen, at the higher speed (high frequency) the 'on' period halves, so the amplitude of the flux halves, leading to a reduction in

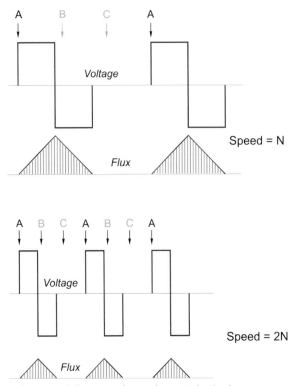

Figure 10.20 Voltage and flux waveforms for switched-reluctance motor in 'single-pulse' mode.

available torque. The same limitation was seen in the case of the inverter-fed induction motor drive, the only difference being that the waveforms in that case were sinusoidal rather than triangular.

It is important to note that these flux waveforms do not depend on the rotor position, but the corresponding current waveforms do because the m.m.f. needed for a given flux depends on the effective reluctance of the magnetic circuit, and this of course varies with the position of the rotor.

To get the most motoring torque for any given phase flux waveform, it is obvious that the rise and fall of the flux must be timed to coincide with the rotor position: ideally, the flux should only be present when it produces positive torque, and be zero whenever it would produce negative torque, but given the delay in build-up of the flux it may be better to switch on early so that the flux reaches a decent level at the point when it can produce the most torque, even if this does lead to some negative torque at the start and finish of the cycle.

The job of the torque control system is to switch each phase on and off at the optimum rotor position in relation to the torque being demanded at the time, and this is done by keeping track of the rotor position, possibly using a rotor position sensor.[3] Just what angles constitute the optimum depends on what is to be optimized (e.g. average torque, overall efficiency), and this in turn is decided by reference to the data stored digitally in the controller 'memory map' that relates current, flux, rotor position and torque for the particular machine. Torque control is thus considerably less straightforward than in d.c. drives, where torque is directly proportional to armature current, or induction motor drives where torque is proportional to slip.

8.3 Power converter and overall drive characteristics

An important difference between the switched-reluctance motor and all other self-synchronous motors is that its full torque capability can be achieved without having to provide for both positive and negative currents in the phases. This is because the torque does not depend on the direction of current in the phase-winding. The advantage of such 'unipolar' drives is that because each of the main switching devices is permanently connected in series with one of the motor windings (as in Figure 10.15), there is no possibility of the 'shoot-through' fault (see Chapter 2) which necessitates the inclusion of 'dead time' in the switching strategies of the conventional inverter.

Overall closed-loop speed control is obtained in the conventional way with the speed error acting as a torque demand to the torque control system described above. However, if a position sensor is fitted the speed can be derived from the position signal.

In common with other self-synchronous drives, a wide range of operating characteristics is available. If the input converter is fully controlled, continuous regeneration and full four-quadrant operation is possible, and the usual constant torque, constant power and series-type characteristic is regarded as standard. Low-speed torque can be uneven unless special measures are taken to profile the current pulses, but the particular merit of the switched-reluctance drive is that continuous low-speed high-torque operation is usually better than for most competing systems in terms of overall efficiency.

---

[3] As in all drives the use of a position sensor is avoided wherever possible because of the significant extra cost and complexity. Sophisticated software models of the motor are therefore employed to deduce rotor position from analysis of the current and voltage waveforms.

CHAPTER ELEVEN

Motor/Drive Selection

1. INTRODUCTION

The selection process often highlights difficulties in three areas. First, as we have discovered in the preceding chapters, there is a good deal of overlap between the characteristics of the major types of motor and drive. This makes it impossible to lay down a set of hard and fast rules to guide the user straight to the best solution for a particular application. Secondly, users tend to underestimate the importance of starting with a comprehensive specification of what they really want, and they seldom realize how much weight attaches to such things as the steady-state torque–speed curve, the dynamic performance required, the inertia of the load, the pattern of operation (continuous or intermittent), the environment and the question of whether or not the drive needs to be capable of regeneration. And thirdly they may be unaware of the existence of standards and legislation, and hence can be baffled by questions from any potential supplier.

The aim in this chapter is to assist the user in taking the first steps by giving some of these matters an airing. We begin by laying out the availability, ratings and speed, and then the main characteristics of the various motor and drive types that we have highlighted earlier in the book. We then move on to the questions which need to be asked about the load and pattern of operation, and finally look briefly at the matter of standards. The whole business of selection is so broad that it really warrants a book to itself, but the cursory treatment here should at least help the user to specify the drive rating and arrive at a shortlist of possibilities.

2. POWER RATINGS AND CAPABILITIES

The four tables in this section represent an attempt to condense the most important aspects that will be of interest to the user into a readily accessible form. It will be evident from the weight given to a.c. drives in the book that they now dominate the scene, and this is reflected in the fact that only the first table deals with d.c. drives. The scope of a.c. drives is too great for a single table, so to aid digestion we have split them into three.

The first (Figure 11.1) covers conventional (brushed) d.c. motor drives, which remain significant despite their diminishing market share. However, the second table (Figure 11.2 – a.c. drives) is by far the most important because it includes the inverter-fed induction motor and brushless permanent magnet motor drives which

D.C. Motor (Separately Excited or Permanent Magnet)			
Drive Type	**Single Quadrant**	**Four Quadrant**	**d.c. Chopper**
Converter	Single or three-phase fully (or half) controlled thyristor bridge.	Dual single/three phase fully controlled thyristor bridge.	d.c. Chopper
Torque/Speed range	Motoring in one direction only (Braking in other direction.)	Motoring and Braking in both directions.	2 or 4 quadrant versions.
Speed Control	Closed loop control of armature voltage with inner current control loop.		
Torque Control	Closed loop control of armature current.		
Ratings	10 W to 5 MW (Fractional HP to 7000 HP)		0.5 to 5 kW (0.7 to 7 HP). Traction > 500 kW.
Max Power	Available to multi-MW ratings, but motor limitations restrict the product of Power and Speed to 3×10^6 kW.rev/min.		
Min Speed	Good control down to standstill.		
Notable features	Separately-excited motors often used above base speed in constant-power mode.		d.c. to d.c. conversion.
		Fast torque reversals.	Smooth torque possible.
Market Position	d.c. drives remain a significant part of the overall drives market. Popularity is diminishing and is focused mainly on applications where a d.c. motor already exists; on very simple drives; and high power drives where the d.c. motor can still be competitive.		
	Popular low-cost solution for low power drives in simple applications or for retrofit to existing motors.	Popular 4 quadrant solution for retrofit to existing motors. Motor and drive can be competitive at higher powers.	Once very popular as servo drive with brushed motor. Gradually losing markets to a.c equivalents. Remains important for some retro-fits/upgrades in traction applications.

Figure 11.1 Conventional (brushed) d.c. motor drives.

are now the most widely used industrial drives. Figure 11.3 deals with (predominantly large) a.c. drives, which have limited and/or niche application and are therefore less common; for completeness the static Scherbius drive is included although it was not discussed. The fourth table (Figure 11.4) contains three unlikely bedfellows whose conjunction reflects the fact that they could not be fitted in elsewhere.

3. DRIVE CHARACTERISTICS

The successful integration of electronic variable-speed drives into a system depends upon knowledge of the key characteristics of the application and the site where the system will be used. The correct drive also has to be selected, but this is often simpler than it used to be because the voltage source pulse-width modulation (PWM) inverter has become universally adopted for almost all applications, and its control characteristics can be adapted to suit most loads with relative ease. It is still important

	Induction Motor			Brushless PM Motor		Excited Rotor Synch. Motor
Drive/ Converter	PWM Inverter	Multi-level PWM	Current source converter	PWM Inverter	Multi-level PWM	3-phase fully-controlled bridges
Torque/Speed Quadrants	Motoring and Braking in both directions with either dual converter on supply side, or d.c. link chopper and brake resistor.		Motoring and Braking in both directions	Motoring and Braking in both directions with either dual converter on supply side, or d.c. link chopper and brake resistor.		Motoring and Braking in both directions
Method of Torque/Speed Control	Field Orientation (Vector) (or Direct Torque Control). Simpler Scalar (V/f) variants available.			Field Orientation (Vector) (or Direct Torque Control). Position feedback is standard, but sensorless options are available.		Same as brushless PM
Typical Power Ratings	Up to 3 MW (400/690 V; >10 MW (6600 V)		4 MW (400/690 V; 10 MW (6600 V)	10 W to 500 kW, with higher powers available as custom designs.		2 to > 20 MW (MV only).
Maximum Speed (Typical)	> 40,000 rev/min.		> 6000 rev/min	> 70,000 Rev/min	>10,000 rev/min	>10,000 rev/min
Minimum Speed (Typical)	Control dependent. With position feedback to standstill with full torque: without, to 1 Hz.		5 Hz standard: lower possible.	To standstill with full torque. (Without feedback to 1 Hz.)		Open loop to 3 Hz; closed loop to standstill with full torque.
Notable Features	Simple, robust, readily available induction motor.			Very high power density. Excellent control dynamics. Motor inertia can be specified from many suppliers. Motors with integral position sensors available.		Simple converter available at very high voltages. High speed motors available.
	Low torque ripple. Excellent control dynamics.					
		Low motor dV/dt (important at MV).	High efficiency		Low motor dV/dt (important at MV).	High efficiency. Good dynamic control.
Market Position	Voltage source PWM inverters are popular because they are readily applied to different types of motor and load.		Once the topology of choice for single motor drives, but has largely been superseded by the PWM inverter.	Most popular high performance/servo drive for precision motion applications where smooth rotation or rapid accln./decln. is required.	Limited application. Most are below 700 V where mutli-level advantages are less clear.	The most popular solution for very high power (> 5 MW) and / or very high speed applications.
	The industrial workhorse. Most widely used drive.	Popular in MV drives.				

Figure 11.2 Induction, brushless permanent magnet and excited rotor synchronous a.c. motor drives.

to understand the general characteristics of the various types of drive, however, and these are listed in summary form in the fifth table (Figure 11.5). (Note that the data on overloads relates to typical industrial equipment, but users will inevitably encounter a range of different figures from the various suppliers.) Finally, a summary of the pros and cons of d.c. and a.c. drives is given in the sixth table (Figure 11.6). The wealth of information in these tables should be of great value in assisting prospective users to narrow down their options.

3.1 Maximum speed and speed range

This is an important consideration in many drives, so a few words are appropriate. We saw in Chapter 1 that, as a general rule, for a given power the higher the base speed the smaller the motor. In practice there are only a few applications where motors with base speeds below a few hundred rev/min are attractive, and it is

	Static Kramer	**Static Scherbius**	**Cycloconverter**	**Matrix Converter**
Motor	Sliping Induction		Synchronous or Induction	
Converter	Diode bridge and 3-phase full controlled bridge	Cycloconverter or back to back PWM inverters	Cycloconverter	Matrix converter
Torque/Speed Quadrants	Motoring in one direction only	Motoring and braking in both directions		
Speed and Torque Control	Field Orientation control (or Direct Torque Control) Simple scalar V/f control variants available.			
Ratings	500 kW to 20 MW	500 kW to 20 MW	500 kW to > 10 MW	50 kW to 2 MW
Typical Maximum Speed	1470 rev/min	2000 rev/min	< 30 Hz	< supply frequency
Min Speed	900 rev/min (wider speed range impacts competitiveness)		0 Hz	
Features	Economic solution at high powers		No DC link components.	
Market Position	Not prominent but still very economic for limited speed range applications.	Has much of the benefit of the Kramer but with ability to operate above synchronous speed.	The topology of choice for very low speed, very high power applications	Has niche areas of application where d.c. link components troublesome. New topology variants may expand market.

Figure 11.3 Slip energy recovery and direct converter drives.

	Soft Starter	**Switched Reluctance Drive**	**Stepper Motor**
Motor	Induction Motor	Switched Reluctance Motor	Stepper Motor
Converter Type	Inverse parallel thyristors in supply lines	Specific topologies	Specific topologies
Torque/Speed Quadrants	Motoring in one direction only	Motoring and Braking in both directions with either a supply-side dual converter or a d.c. link chopper and brake resistor	
Speed Control	Starting duty only	Specific Control Strategies	
Typical Industrial Power Ratings	1 kW to > 1 MW	1 kW to > 1 MW	10 W to > 5 kW
Typical Maximum Speed	Rated DOL speed	> 10,000 rev/min	> 10,000 rev/min
Typical Minimum Speed	Not Applicable	Standstill	Standstill
Notable Features		Very simple and robust motor	Simple motor and control strategy
Market Position	The soft start has very specific application in reducing the starting current and/or controlling the starting torque of a DOL induction motor	High starting torque applications benefit from SR technology	Open loop positioning systems

Figure 11.4 Soft start, switched reluctance and stepping motor drives.

	D.C. Drives (Separately excited)		A.C. Drives		
	Phase Controlled	Chopper	Induction Motor (Field Orientation Control without speed feedback)	Induction Motor (Field Orientation Control with speed feedback)	Brushless PM Motor (Field Orientation Control with position feedback)
Operating Speed Range	0 to Base Speed at Constant Torque. Above base speed at Constant Power. Max approx. 4 x base speed.		3–100% Base speed at constant torque. Above base speed at constant power. Max approx. 20 x base.	0–100% Base speed at constant torque. Above base speed at constant power. Max approx. 20 x base.	0–100% Base speed at constant torque. Above base speed at constant power. Max approx. 8 x base.
Braking Capability	150% (4Q drive)	150% (4Q drive)	150%		> 200%
Speed Loop Response	10 Hz	50 Hz	50 Hz	150 Hz	> 150 Hz
Speed Holding (100% load change)	0.1% – 0.05% with good speed feedback	0.05% – 0.001% with good speed feedback	0.1%	0.001%	0.0005%
Torque / Speed Capability	Constant Torque + Field Weakening	Constant Torque + Field Weakening	Constant Torque + Field Weakening	Constant Torque + Field Weakening	Constant Torque
Starting Torque	150% (60s)	150% (60s)	150% (60s)	150% (60s)	200% (4s)
Min speed with 100% torque	Standstill	Standstill	2% base speed	Standstill	Standstill
Motor IP Rating	IP23	IP54 / IP23	IP54	IP54	IP65
Motor Inertia	High	High (Low avail.)	Medium (Low avail.)	Medium (Low avail.)	Low (High avail.)
Motor size	Large	Large	Medium	Medium	Small
Cooling	Forced ventilated	FV or natural	Natural Cooling	Natural Cooling	Natural Cooling
Typical f/b device	Tachogenerator or Encoder	Tachogenerator or Encoder	Not Applicable	Encoder	Encoder

Figure 11.5 Characteristics of the most popular types of drive.

usually best to obtain low 'maximum' speeds by means of mechanical speed reduction often in the form of a gearbox.

Motor speeds above 10,000 rev/min are also unusual except in small universal motors and special-purpose inverter-fed motors for applications such as aluminum machining. The majority of motors have base speeds between 1500 and 3000 rev/min: this range is attractive as far as motor design is concerned since acceptable power/weight ratios are obtained, and these speeds are also satisfactory as far as mechanical transmission is concerned. Note, however, that as explained in Chapter 1, the higher the rated speed the more compact the motor, so there will be instances where a high-speed motor is preferred when minimizing motor size is of paramount importance.

In controlled-speed applications, the range over which the steady-state speed must be controlled, the required torque–speed characteristic, the duty cycle and the accuracy of the speed holding, are significant factors in the selection process. In Chapter 8 we looked at some of these aspects, notably in relation to motor cooling. For constant torque loads which require operation at all speeds, the inverter-fed

	D.C. Drives (Separately excited)		A.C. Drives		
	Phase Controlled	Chopper	Induction Motor (Field Orientation Control without speed feedback)	Induction Motor (Field Orientation Control with speed feedback)	Brushless PM Motor (Field Orientation Control with position feedback)
Principal Advantages	• Low Cost Controller • Relatively simple technology	• Good Dynamic performance • Relatively simple technology	• Good dynamic performance • Full torque down to very low speeds • Good starting torque	• Very good dynamic performance • Full torque available down to standstill • Standard motor (but with feedback added) • No zero torque dead band	• Excellent dynamic performance • Low (or High) inertia motors • High Motor IP ratings • Very smooth rotation possible
Principal Disadvantages	• Expensive motor <100 kW • Brush gear Maintenance • Note: Low loads reduce brush life • Zero Torque dead band • Possible failure on supply loss/dip • Possible instability on Fan/Pump type loads • Low Motor IP rating • Chopper at modest powers only (<10 kW) • Additional converter required for regeneration		• Additional converter required for regeneration to the supply • Limited Torque and Speed Loop response	• Additional converter required for regeneration to the supply	• Additional converter required for regeneration to the supply • Field weakening range difficult to facilitate

Figure 11.6 Advantages and disadvantages of d.c. and a.c. drives.

induction motor, the inverter-fed synchronous machine and the d.c. drive are possibilities, but only the d.c. drive would come as standard with a force-ventilated motor capable of continuous operation with full torque at very low speeds.

Fan or pump-type loads (see below), with a wide operating speed range, are a somewhat easier proposition because the torque is low at low speeds. In most circumstances, the inverter-fed induction motor (using a standard motor) is the natural choice.

Figures for accuracy of speed holding can sometimes cause confusion, as they are usually given as a percentage of the base speed. Hence with a drive claiming a decent speed holding accuracy of 0.2% and a base speed of 2000 rev/min, the user can expect the actual speed to be between 1996 rev/min and 2004 rev/min when the speed reference is 2000 rev/min. But if the speed reference is set for 100 rev/min, the actual speed can be anywhere between 96 rev/min and 104 rev/min, and still be within the specification.

4. LOAD REQUIREMENTS – TORQUE–SPEED CHARACTERISTICS

The most important things we need to know about the load are the steady-state torque–speed characteristics, the effective inertia as seen by the motor and what dynamic performance is required. At one extreme, for example in a steel rolling mill, it may be necessary for the speed to be set at any value over a wide range, and for the mill to react very quickly when a new target speed is demanded. Having reached the set speed, it may be essential that it is held very precisely even when subjected to sudden load changes. At the other extreme, for example a large ventilating fan, the range of set speed may be quite limited (perhaps from 80 to 100%); it may not be important to hold the set speed very precisely; and the times taken to change speeds, or to run up from rest, are unlikely to be critical.

At full speed both of these examples may demand the same power, and at first sight might therefore be satisfied by the same drive system. But the ventilating fan is obviously an easier proposition, and it would be overkill to use the same system for both. The rolling mill would call for a regenerative drive with speed or position feedback, while the fan could quite happily manage with a simple open-loop, i.e. sensorless, inverter-fed induction motor drive.

Although loads can vary enormously, it is customary to classify them into two major categories, referred to as 'constant-torque' or 'fan or pump' types. We will use the example of a constant-torque load to illustrate in detail what needs to be done to arrive at a specification for the torque–speed curve. An extensive treatment is warranted because this is often the stage at which users come unstuck.

4.1 Constant-torque load

A constant-torque load implies that the torque required to keep the load running is the same at all speeds. A good example is a drum-type hoist, where the torque required varies with the load on the hook, but not with the speed of hoisting. An example is shown in Figure 11.7.

The drum diameter is 0.5 m, so if the maximum load (including the cable) is say 1000 kg, the tension in the cable (mg) will be 9810 N, and the torque applied by the load at the drum will be given by force × radius = $9810 \times 0.25 \approx 2500$ Nm. When the speed is constant (i.e. the load is not accelerating), the torque provided by the motor at the drum must be equal and opposite to that exerted at the drum by the load. (The word 'opposite' in the last sentence is often omitted, it being

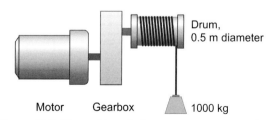

Figure 11.7 Motor driven hoist – a constant-torque load.

understood that steady-state motor and load torque must necessarily act in opposition.)

Suppose that the hoisting speed is to be controllable at any value up to a maximum of 0.5 m/s, and that we want this to correspond with a maximum motor speed of around 1500 rev/min, which is a reasonable speed for a wide range of motors. A hoisting speed of 0.5 m/s corresponds to a drum speed of 19 rev/min, so a suitable gear ratio would be say 80:1, giving a maximum motor speed of 1520 rev/min.

The load torque, as seen at the motor side of the gearbox, will be reduced by a factor of 80, from 2500 to 31 Nm at the motor. We must also allow for friction in the gearbox, equivalent to perhaps 20% of the full-load torque, so the maximum motor torque required for hoisting will be 37 Nm, and this torque must be available at all speeds up to the maximum of 1520 rev/min.

We can now draw the steady-state torque–speed curve of the load as seen by the motor, as shown in Figure 11.8.

The steady-state motor power is obtained from the product of torque (Nm) and angular velocity (rad/s). The maximum continuous motor power for hoisting is therefore given by

$$P_{max} = 37 \times 1520 \times \frac{2\pi}{60} = 5.9\,kW \qquad (11.1)$$

At this stage it is always a good idea to check that we would obtain roughly the same answer for the power by considering the work done per second at the load. The force (F) on the load is 9810 N, and the velocity (v) is 0.5 m/s, so the power (F_v) is 4.9 kW. This is 20% less than we obtained above, because here we have ignored the power lost in the gearbox.

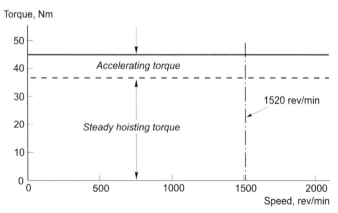

Figure 11.8 Torque requirements for motor in hoist application (Figure 11.7).

So far we have established that we need a motor capable of continuously delivering 5.9 kW at 1520 rev/min in order to lift the heaviest load at the maximum required speed. However, we have not yet addressed the question of how the load is accelerated from rest and brought up to the maximum speed. During the acceleration phase the motor must produce a torque greater than the load torque, or else the load will descend as soon as the brake is lifted. The greater the difference between the motor torque and the load torque, the higher the acceleration. Suppose we want the heaviest load to reach full speed from rest in say 1 second, and suppose we decide that the acceleration is to be constant. We can calculate the required accelerating torque from the equation of motion, i.e.

$$\text{Torque (Nm)} = \text{Inertia} \left(\text{kg m}^2\right) \times \text{Angular acceleration} \left(\frac{\text{rad}}{\text{sec}^2}\right) \quad (11.2)$$

We usually find it best to work in terms of the variables as seen by the motor, and therefore we first need to find the effective total inertia as seen at the motor shaft, then calculate the motor acceleration, and finally use equation (11.2) to obtain the accelerating torque.

The effective inertia consists of the inertia of the motor itself, the referred inertia of the drum and gearbox, and the referred inertia of the load on the hook. The term 'referred inertia' means the apparent inertia, viewed from the motor side of the gearbox. If the gearbox has a ratio of n:1 (where n is greater than 1), an inertia of J on the low-speed side appears to be an inertia of J/n^2 at the high-speed side. (This formula is the same as that for finding the referred impedance as seen at the primary of an ideal transformer, as discussed in Appendix 2.)

In this example the load actually moves in a straight line, so we need to ask what the effective inertia of the load is, as 'seen' at the drum. The geometry here is simple, and it is not difficult to see that as far as the inertia seen by the drum is concerned the load appears to be fixed to the surface of the drum. The load inertia at the drum is then obtained by using the formula for the inertia of a mass m located at radius r, i.e. $J = mr^2$, yielding the effective load inertia at the drum as $1000 \text{ kg} \times (0.25 \text{ m})^2 = 62.5 \text{ kg m}^2$.

The effective inertia of the load as seen by the motor is $1/(80)^2 \times 62.5 \approx 0.01 \text{ kg m}^2$. To this must be added first the motor inertia which we can obtain by consulting the manufacturer's catalogue for a 5.9 kW, 1520 rev/min motor. This will be straightforward for a d.c. motor, but a.c. motor catalogues tend to give ratings at utility frequencies only, and here a motor with the right torque needs to be selected, and the possible torque–speed curve for the type of drive considered. For simplicity let us assume we have found a motor of precisely the required rating which has a rotor inertia of 0.02 kg m^2. The referred inertia of the drum and gearbox must be added and this again we have to calculate or look up. Suppose this yields a further 0.02 kg m^2. The total effective inertia is thus 0.05 kg m^2, of which 40% is due to the motor itself.

The acceleration is straightforward to obtain, since we know the motor speed is required to rise from zero to 1520 rev/min in 1 second. The angular acceleration is given by the increase in speed divided by the time taken, i.e.

$$\left(1520 \times \frac{2\pi}{60}\right) \div 1 \; = \; 160 \, \text{rad/sec}^2$$

We can now calculate the accelerating torque from equation (11.2) as

$$T \; = \; 0.05 \times 160 \; = \; 8 \, \text{Nm}$$

Hence in order to meet both the steady-state and dynamic torque requirements, a drive capable of delivering a torque of 45 Nm ($=37 + 8$) at all speeds up to 1520 rev/min is required, as indicated in Figure 11.8.

In the case of a hoist, the anticipated pattern of operation may not be known, but it is likely that the motor will spend most of its time hoisting rather than accelerating. Hence, although the peak torque of 45 Nm must be available at all speeds, this will not be a continuous demand, and will probably be within the short-term overload capability of a drive which is continuously rated at 5.9 kW.

We should also consider what happens if it is necessary to lower the fully loaded hook. We allowed for friction of 20% of the load torque (31 Nm), so during descent we can expect the friction to exert a braking torque equivalent to 6.2 Nm. But in order to prevent the hook from running away, we will need a total torque of 31 Nm, so to restrain the load, the motor will have to produce a torque of 24.8 Nm. We would naturally refer to this as a braking torque because it is necessary to prevent the load on the hook from running away, but in fact the torque remains in the same direction as when hoisting. The speed is, however, negative, and in terms of a 'four-quadrant' diagram (e.g. Figure 3.12) we have moved from quadrant 1 into quadrant 4, and thus the power flow is reversed and the motor is regenerating, the loss of potential energy of the descending load being converted back into electrical form (and losses). Hence if we wish to cater for this situation we must go for a drive that is capable of continuous regeneration: such a drive would also have the facility for operating in quadrant 3 to produce negative torque to drive down the empty hook if its weight was insufficient to lower itself.

In this example the torque is dominated by the steady-state requirement, and the inertia-dependent accelerating torque is comparatively modest. Of course if we had specified that the load was to be accelerated in one-fifth of a second rather than 1 second, we would require an accelerating torque of 40 Nm rather than 8 Nm, and as far as torque requirements are concerned the acceleration torque would be more or less the same as the steady-state running torque. In this case it would be necessary to consult the drive manufacturer to determine the drive rating, which would depend on the frequency of the start/stop sequence.

The question of how to rate the motor when the loading is intermittent is explored more fully later, but it is worth noting that if the inertia is appreciable the stored rotational kinetic energy ($\frac{1}{2} J \omega^2$) may become very significant, especially when the drive is required to bring the load to rest. Any stored energy has either to be dissipated in the motor and drive itself, or returned to the supply. All motors are inherently capable of regenerating, so the arrangement whereby the kinetic energy is recovered and dumped as heat in a resistor within the drive enclosure is the cheaper option, but is only practicable when the energy to be absorbed is modest. If the stored kinetic energy is large, the drive must be capable of returning energy to the supply, and this inevitably pushes up the cost of the converter.

In the case of our hoist, the stored kinetic energy is only

$$\frac{1}{2} \times 0.05 \left(1520 \times \frac{2\pi}{60} \right)^2 = 633 \text{ joules}$$

or about 1% of the energy needed to heat up a mug of water for a cup of coffee. Such modest energies could easily be absorbed by a resistor, but given that in this instance we are providing a regenerative drive, this energy would also be returned to the supply.

4.2 Inertia matching

There are some applications where the inertia dominates the torque requirement, and the question of selecting the right gearbox ratio has to be addressed. In this context the term 'inertia matching' often causes confusion, so it is worth explaining what it means.

Suppose we have a motor with a given torque, and we want to drive an inertial load via a gearbox. As discussed previously, the gear ratio determines the effective inertia as 'seen' by the motor: a high step-down ratio (i.e. load speed much less than motor speed) leads to a very low referred inertia, and vice versa.

If the specification calls for the acceleration of the load to be maximized, it turns out that the optimum gear ratio is that which causes the referred inertia of the load to be equal to the inertia of the motor. Applications in which load acceleration is important include all types of positioning drives, e.g. in machine tools and phototypesetting. This explains why brushless permanent magnet synchronous motors, discussed in Chapter 9, are available from a wide range of manufacturers with different inertias. (There is an electrical parallel here – to get the most power into a load from a source with internal resistance R, the load resistance must be made equal to R.)

It is important to note, however, that inertia matching only maximizes the *acceleration* of the load. Frequently it turns out that some other aspect of the specification (e.g. the maximum required load speed) cannot be met if the gearing is chosen to satisfy the inertia matching criterion, and it then becomes necessary to accept reduced acceleration of the load in favor of higher speed.

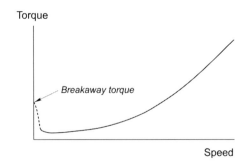

Figure 11.9 Torque–speed characteristics for fan-type load.

4.3 Fan and pump loads

Fans and centrifugal pumps have steady-state torque–speed characteristics which generally have the shape shown in Figure 11.9.

These characteristics are often approximately represented by assuming that the torque required is proportional to the square of the speed, and thus power is proportional to the cube of the speed. We should note, however, that the approximation is seldom valid at low speeds because most real fans or pumps have a significant static friction or breakaway torque (as shown in Figure 11.9) which must be overcome when starting.

When we consider the power–speed relationships the striking difference between the constant-torque and fan-type load is underlined. If the motor is rated for continuous operation at the full speed, it will be very lightly loaded (typically around 12% rated power) at half speed, whereas with the constant-torque load the power rating will be 50% at half speed. Fan-type loads which require speed control can therefore be handled by drives such as the inverter-fed cage induction motor, which can run happily at reduced speed and very much reduced torque without the need for additional cooling. If we assume that the rate of acceleration required is modest, the motor will require a torque–speed characteristic which is just a little greater than the load torque at all speeds. This defines the operating region in the torque–speed plane, from which the drive can be selected.

Many fans, which are required to operate almost continuously at rated speed, do not require speed control of course, and are well served by utility-frequency induction motors.

5. GENERAL APPLICATION CONSIDERATIONS

5.1 Regenerative operation and braking

All motors are inherently capable of regenerative operation, but in drives the basic power converter as used for the 'bottom of the range' version will not normally be capable of continuous regenerative operation. The cost of providing for fully regenerative operation is usually considerable, and users should always ask the question 'do I really need it?'

In most cases it is not the recovery of energy for its own sake which is of prime concern, but rather the need to achieve a specified dynamic performance. Where rapid reversal is called for, for example if kinetic energy has to be removed quickly, and, as discussed in the previous section, this implies that the energy is either returned to the supply (regenerative operation) or dissipated (usually in a braking resistor). An important point to bear in mind is that a non-regenerative drive will have an asymmetrical transient speed response, so that when a higher speed is demanded, the extra kinetic energy can be provided quickly, but if a lower speed is demanded, the drive can do no better than reduce the torque to zero and allow the speed to coast down.

5.2 Duty cycle and rating

This is a complex matter, which in essence reflects the fact that whereas all motors are governed by a thermal (temperature rise) limitation, there are different patterns of operation which can lead to the same ultimate temperature rise.

Broadly speaking the procedure is to choose the motor on the basis of the root mean square of the power cycle, on the assumption that the losses (and therefore the temperature rise) vary with the square of the load. This is a reasonable approximation for most motors, especially if the variation in power is due to variations in load torque at an essentially constant speed, as is often the case, and the thermal time-constant of the motor is long compared with the period of the loading cycle. (The thermal time-constant has the same significance as it does in relation to any first-order linear system, e.g. an R/C circuit. If the motor is started from ambient temperature and run at a constant load, it takes typically four or five time-constants to reach its steady operating temperature.) Thermal time-constants vary from more than an hour for the largest motors (e.g. in a steel mill) through tens of minutes for medium-power machines down to minutes for fractional-horsepower motors and seconds for small stepping motors.

To illustrate the estimation of rating when the load varies periodically, suppose a constant-frequency cage induction motor is required to run at a power of 4 kW for 2 minutes, followed by 2 minutes running light, then 2 minutes at 2 kW, then 2 minutes running light, this 8-minute pattern being repeated continuously. To choose an appropriate power rating we need to find the r.m.s. power, which means exactly what it says, i.e. it is the square root of the mean (average) of the square of the power. The variation of power is shown in the upper part of Figure 11.10, which has been drawn on the basis that when running light the power is negligible. The 'power squared' is shown in the lower part of the figure.

The average power is 1.5 kW, the average of the power squared is 5 kW^2, and the r.m.s. power is therefore $\sqrt{5}$ kW, i.e. 2.24 kW. A motor that is continuously

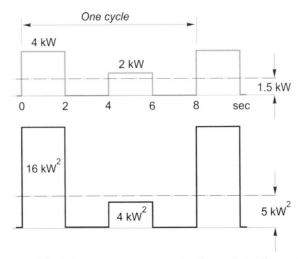

Figure 11.10 Calculation of r.m.s. power rating for periodically varying load.

rated at 2.24 kW would therefore be suitable for this application, provided of course that it is capable of meeting the overload torque associated with the 4 kW period. The motor must therefore be able to deliver a torque that is greater than the continuous rated torque by a factor of 4/2.25, i.e. 178%: this would be within the capability of most general-purpose induction motors.

Motor suppliers are accustomed to recommending the best type of motor for a given pattern of operation, and they will typically classify the duty type in one of eight standard categories which cover the most commonly encountered modes of operation. As far as rating is concerned the most common classifications are maximum continuous rating, where the motor is capable of operating for an unlimited period, and short time rating, where the motor can only be operated for a limited time (typically 10, 30 or 60 minutes) starting from ambient temperature. NEMA, in the USA, refer to motor duty cycles in terms of continuous, intermittent or special duty. IEC 60034-1 defines duty cycles in eight ratings S1…S8.

It is important to add that in controlled-speed applications, the duty cycle also impacts the power electronic converter and here the thermal time constants are *very* much shorter than in the motor. The duty cycle can be very important in some applications, and should be clearly stated in a torque and speed–time profile to ensure that there is no ambiguity that could lead to subsequent disappointment.

5.3 Enclosures and cooling

There is clearly a world of difference between the harsh environment faced by a winch motor on the deck of an ocean-going ship and the comparative comfort enjoyed by a motor driving the drum of an office photocopier. The former must be protected against the ingress of rain and sea water, while the latter can rely on a dry and largely dust-free atmosphere.

Classifying the extremely diverse range of environments poses a potential problem, but fortunately this is one area where international standards have been agreed and are widely used. The International Electrotechnical Committee (IEC) standards for motor enclosures are now almost universal and take the form of a classification number prefixed by the letters IP, and followed by two digits. The first digit indicates the protection level against ingress of solid particles ranging from 1 (solid bodies greater than 50 mm diameter) to 5 (dust), while the second relates to the level of protection against ingress of water ranging from 1 (dripping water) through 5 (jets of water) to 8 (submersible). A zero in either the first or second digit indicates no protection.

Methods of motor cooling have also been classified and the more common arrangements are indicated by the letters IC followed by two digits, the first of which indicates the cooling arrangement (e.g. 4 indicates cooling through the surface of the frame of the motor) while the second shows how the cooling circuit power is provided (e.g. 1 indicates motor-driven fan).

The most common motor enclosures for standard industrial electrical machines are IP21 and 23 for d.c. motors; IP44 and 54 for induction motors; and IP65 and 66 for brushless permanent magnet synchronous motors. These classifications are known in the United States, but it is common practice for manufacturers there to adopt less formal designations, for example the following:

Open drip proof (ODP): Motors with ventilating openings, which permit passage of external air over and around the windings. (Similar to IP23)

Totally enclosed fan cooled (TEFC): A fan is attached to the shaft to push air over the frame during operation to enhance the cooling process. (Similar to IP54)

Washdown duty (W): An enclosure designed for use in the food processing industry and other applications routinely exposed to washdown, chemicals, humidity and other severe environments. (Similar to IP65)

5.4 Dimensional standards

Standardization is improving in this area, though it remains far from universal. Such matters as shaft diameter, center height, mounting arrangements, terminal box position, and overall dimensions are fairly closely defined for the mainstream motors (induction, d.c.) over a wide size range, but standardization is relatively poor at the low-power end because so many motors are tailor-made for specific applications.

Standardization is also poor in relation to brushless permanent magnet motors. Paradoxically, the lack of standardization, particularly in terms of frame size, mounting arrangements and shaft dimensions, has been cited as a significant driver of innovation in these machines, whereas the induction motor has been handicapped by its universal success and resulting standardization!

5.5 Supply interaction and harmonics

As we have seen in section 5 of Chapter 8, most converter-fed drives cause distortion of the mains voltage which can upset other sensitive equipment, particularly in the immediate vicinity of the installation. We have also discussed some of the options available to mitigate such effects. This is, however, a complex subject area and one which, in most cases, requires a systems overview rather than consideration of a single drive or even group of drives. It requires an intimate knowledge of the supply system, particularly its impedances.

With more and larger drives being installed, the problem of mains distortion is increasing, and supply authorities therefore react by imposing increasingly stringent statutory limits governing what is allowable.

The usual pattern is for the supply authority to specify the maximum amplitude and spectrum of the harmonic currents at the point of supply to a particular customer. If the proposed installation exceeds these limits, appropriate filter circuits must be connected in parallel with the installation. These can be costly, and their design is far from simple because the electrical characteristics of the supply system need to be known in advance in order to avoid unwanted resonance phenomena. Users need to be alert to the potential problem and, where appropriate, ensure that the drive supplier takes responsibility for handling it.

Measurement of harmonic currents is not straightforward and here the standards help, with IEC 61000-4-7:2002 providing valuable guidance.

Introduction to Closed-loop Control

A1.1. OUTLINE OF APPROACH

The aim of this Appendix is to help readers who are not familiar with closed-loop control and feedback to feel confident when they meet such ideas in the drives context. In line with the remainder of the book, the treatment avoids mathematics where possible, and in particular it makes only passing reference to transform techniques. This approach is chosen deliberately, despite the limitations it imposes, in the belief that it is more useful for the reader to obtain a sound grasp of what really matters in a control system, rather than to become adept at detailed analysis or design.

Traditionally, control ideas are introduced at the same time as the Laplace transform technique. The Laplace approach to the analysis and design of linear systems (i.e. those that obey the principle of superposition) is unchallenged for the very good reason that it works wonderfully well. It allows the differential equations that describe the dynamic behavior of the systems we wish to control to be recast into algebraic form. Instead of having to solve differential equations in the time domain, we are able to transform the equations and emerge in a parallel universe in which algebraic equations in the complex frequency or 's' domain replace differential equations in time. For modeling control systems this approach is ideal because we can draw simple block diagrams containing algebraic 's-domain transfer functions' that fully represent both the steady-state and dynamic (transient) relationships of elements such as motors, amplifiers, filters, and transducers. We can assess whether the performance of a proposed system will be satisfactory by doing simple analysis in the s-domain, and decide how to improve matters if necessary. When we have completed our studies in the s-domain, we have a set of inverse transforms to allow us to 'beam back' to our real time domain.

Despite the undoubted power of the transform method, the fact that the newcomer has first to learn what amounts to a new language presents a real challenge. And in the process of concentrating on developing transfer functions, understanding how system order influences transient response, and other supporting matters, some of the key ideas that are central to successful control systems can easily be overlooked.

In addition to having to come to grips with new mathematical techniques, the newcomer has to learn many new terms (e.g. bandwidth, integrator, damping factor, dB/octave), and is expected to divine the precise meaning of other casually used terms (e.g. gain, time–constant) by reference to the particular context in which they are used.

Our aim here is to avoid these potential pitfalls by restricting our analytical exploration to steady-state performance only. In the context of a speed-controlled drive, for example, this means that although our discussion will reveal the factors that determine how well the actual steady-state speed corresponds with the target speed, we will have to accept that it will not tell us how the speed gets from one value to another when the speed reference is changed (i.e. the transient response). By accepting this limitation we can avoid the need to use transform techniques, and concentrate instead on understanding what makes the system tick.

A1.2. CLOSED-LOOP (FEEDBACK) SYSTEMS

A closed-loop system is one where there is feedback from the output, feedback being defined by the *Oxford English Dictionary* as 'the modification of a process by its results or effects, especially the difference between the desired and actual result'. We will begin by looking at some examples of feedback systems and identify their key features.

A1.2.1 Error-activated feedback systems

An everyday example of a feedback system is the traditional lavatory cistern (Figure A1.1), where the aim is to keep the water level in the tank at the full mark. The valve through which water is admitted to the tank is controlled by the position of the arm carrying a ball that floats in the water. The steady-state condition is shown in diagram (a), the inlet valve being closed when the arm is horizontal.

When the WC is flushed (represented by the bottom valve being opened – see diagram (b)), the water level falls rapidly. As soon as the water level falls the ball drops, thereby opening the inlet valve so that fresh water enters the tank in an effort to maintain the water level. Given that the purpose of the system is to maintain the water level at the target level, it should be clear from diagram (b) why, in a control

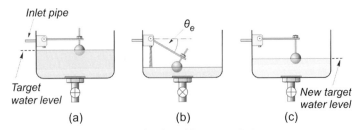

Figure A1.1 Simple closed-loop regulating system.

systems context, the angle θ_e is referred to as the 'error angle': when the water is at the desired level, the error is zero.

This is an example of an 'error-activated' system, because any error automatically initiates corrective action. The fact that the water level has fallen is communicated or fed back to the valve, which responds by admitting more water to combat the 'error'. In traditional cisterns the inlet water rate is proportional to the angle θ_e, so when the tank is empty it fills at a high rate and the water level rises rapidly. As the level rises the valve begins to close and the rate at which the tank refills reduces until finally when the target level is reached the valve is fully closed and the water level is restored to its target value (i.e. full). In this particular example the 'feedback' of the ball position takes the form of a direct mechanical connection between the water level detector (the floating ball) and the inlet valve. Most of the control systems that we meet in a drives context are also error-activated, though in the majority of cases the feedback is less direct than in this example.

Alert readers will have spotted that although the aim of the system is to maintain a constant water level, the level must change every time corrective action is required. The only time that the system can have no error is when it reaches the steady state, with no water going out and the inlet valve fully closed. If there were a leak in the tank, causing water to drain away, the water level would stabilize at a lower level, at which the inlet valve was just open sufficiently wide to admit water at the same rate as it was leaking away. Clearly, an inlet valve that yields a high flow rate for a very small angle will be better able to maintain the correct level than one that needs a large angle error to yield the same flow rate. In control terms, this is equivalent to saying that the higher the 'gain' (expressed as *flow rate/error angle*), the better.

Another important observation is that although it is essential that there is a supply of water available at the inlet valve, the water pressure itself is not important as far as the target level is concerned: if the pressure is high the tank will fill up more quickly than if the pressure is low, but the final level will be the same. This is because the amount of water admitted depends on the integral with respect to time of the inlet flow rate, so the target level will ultimately be reached regardless of the inlet pressure. In fact this example illustrates the principal advantage of 'integral control', in that the steady-state error ultimately becomes zero. We will return to this later.

Finally, it is worth pointing out that this is an example of a regulator, i.e. a closed-loop system where the target level does not change, whereas in most of the systems we meet in drives the target or reference signal will vary. However, if it was ever necessary to alter the target level in the cistern, it could be accomplished by adjusting the distance between ball and arm, as in diagram (c).

A1.2.2 Closed-loop systems

To illustrate the origin and meaning of the term 'closed loop' we will consider another familiar activity, that of driving a car, and in particular we will imagine that we are required to drive at a speed of exactly 50 km/h, the speed to be verified by an auditor from the bureau of standards.

The first essential is an accurate speedometer, because we must measure the output of the 'process' if we are to control it accurately. Our eyes convey the 'actual speed' signal (from the speedometer dial) to our brain, which calculates the difference or error between the target speed and the actual speed. If the actual speed was 30 km/h, the brain would conclude that we were 20 km/h too slow and instruct our foot to press much harder on the accelerator. As we see the speed rising towards the target, the brain will decide that we can ease off the accelerator a little so that we don't overshoot the desired speed. If a headwind springs up, or we reach a hill, any drop in speed will be seen and we will press harder on the gas pedal; conversely, if we find speed is rising due to a following wind or a falling gradient, we apply less accelerator, or even apply the brake. This set of interactions can be represented diagrammatically as in Figure A1.2.

Diagrams of this sort – known as block diagrams – are widely used in control system studies because they provide a convenient pictorial representation of the inter-relationships between the various elements of the system. The significance of the lines joining the blocks depends on the context: for example, the line entering the speedometer represents the fact that there is a hardware input from the wheels (perhaps a mechanical one, or more likely a train of electronic pulses whose frequency corresponds to speed), while the line leaving the speedometer represents the transmission of information via reflected light to the eyes of the driver. In many instances (e.g. an electronic amplifier) the output from a block is directly proportional to the input, but in others (for example, the 'vehicle dynamics' block in Figure A1.2) the output depends on the integral of the input with respect to time.

The lower part of the block diagram represents the speed feedback path; the upper part (known as the forward path) represents the process itself. The fact that all the blocks are connected together so that any change in the output of one affects all the others, gives rise to the name *closed-loop system*.

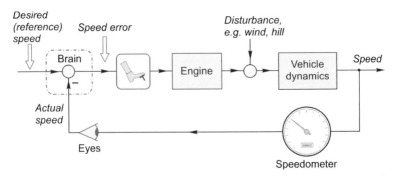

Figure A1.2 Block diagram representation of driving a car at a target speed.

The circles represent summing junctions, i.e. points where signals are added or subtracted. The output signal from the summing junction is the sum of the input signals, or the difference if one is negative. The most important summing junction is the one on the left (inside the driver's brain!) where the actual speed (via the eyes) is subtracted from the reference speed (stored in the brain) to obtain the speed error. The negative sign on the speed feedback signal indicates that a subtraction is required, and it is this that gives rise to the term *negative feedback*.

Factors that clearly have a bearing on whether we will be able to achieve the desired speed are the ability of the car to achieve and maintain the speed; the extent to which the driver's eyes can detect movements in the speedometer pointer; the need for the brain to subtract the actual speed from the desired speed; and the requirement for the brain to instruct the muscles of the leg and foot, and their ability to alter the pressure on the accelerator pedal. But all of these will be of no consequence if the speedometer is not accurate, because even if the driver thinks he is doing 50 km/h, the man from the standards bureau will not be impressed. A VW beetle with an accurate speedometer is potentially better than a Rolls-Royce with an inaccurate one – a fact that underlines the paramount importance of the feedback path, and, in particular, highlights the wisdom of investing in a good transducer to measure the quantity that we wish to control. This lesson will emerge again when we look at what constitutes a good closed-loop system.

Returning to how the speed-control system operates, we can identify the driver as the controller, i.e. the part of the system where the error is generated, and corrective action initiated, and it should be clear that an alert driver with keen eyesight will be better at keeping the speed constant than one whose eyesight is poor and whose actions are somewhat sluggish. We will see that in most automatic control systems expressions such as 'alert' and 'keen eyesight' translate into the 'gain' of the controller: a controller with a high gain gives a large output in response to a small change at its input. Finally, we should also acknowledge that the block diagram only accounts for some of the factors that influence the human driver: when the car approaches a hill, for example, the driver knows from experience that he will need more power and acts on this basis without waiting for the speedometer to signal a drop in speed. Although this is a fascinating aspect of control it is beyond our current remit.

A1.3. STEADY-STATE ANALYSIS OF CLOSED-LOOP SYSTEMS

A typical closed-loop system is shown in block diagram form in Figure A1.3. As explained in the introduction, our aim is to keep the mathematical treatment as simple as possible, so each of the blocks contains a single symbol that represents the ratio of output to input under steady-state conditions, i.e. when any transients have died out and all the variables have settled.

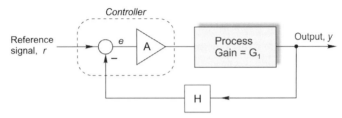

Figure A1.3 Typical arrangement of negative-feedback closed-loop control system.

A useful question to pose whenever a new block diagram is encountered is 'What is this control system for?' The answer is that the variable that is sensed and fed back (i.e. the output of the block labeled 'process' in Figure A1.3) is the quantity we wish to control.

For example, the 'process' block might represent say an unloaded d.c. motor, the input being the armature voltage and the output being the speed. We would then deduce that the system was intended to provide closed-loop control of the motor speed. The gain G_1 is simply the ratio of the steady-state speed to the armature voltage, and it would typically be expressed in rev/min/volt. The triangular symbol usually represents an analogue electronic amplifier, in which case it has a (dimensionless) gain of A, i.e. the input and output are both voltages. We conclude from this that the reference input must also be a voltage, because the inputs and output from a summing junction must necessarily be of the same type. It also follows that the signal fed back into the summing junction must also be a voltage, so that the block labeled H (the feedback transducer) must represent the conversion of the output (speed, in rev/min) to voltage. In this case we know that the speed feedback will indeed be obtained from a tachogenerator, with, in this case, an e.m.f. constant of H volts/rev/min.

In electronic systems where the summing junction is usually an integral part of the amplifier, the combination forms the controller, as shown by the dashed line in Figure A1.3.

At this point we should pause and ask what we want of a good control system. As has already been mentioned, the output of the summing junction is known as the error signal: its pejorative name helps us to remember that we want the error to be as small as possible. Because the error is the difference between the reference signal and the fed-back signal, we deduce that in an ideal (error-free) control system the fed-back signal should equal the reference signal.

The sharp-witted reader will again wonder how the system could possibly operate if the error was zero, because then there would be no input to the forward path and unless the gain of the controller was infinite there would be nothing going into the process to produce the output that we are trying to control. So for the present let us assume that the gain of the controller is very high, and make the justifiable assumption that we have a good system and that therefore the feedback signal is almost, but not quite, equal to the reference signal.

In terms of the symbols in Figure A1.3, the almost-zero-error condition is represented by approximating the reference signal (r) to the fed-back signal (Hy), i.e.

$$r = Hy, \text{ or } y = \frac{1}{H}r \tag{A1.1}$$

This equation is extremely important because it indicates that in a good control system, the output is proportional to the input, with the constant of proportionality or 'gain' of the closed-loop system depending only on the feedback factor (H). We came to the same conclusion when we looked at the example of driving a car at a given speed, when we concluded the most important aspect was that we had a reliable speedometer to provide accurate feedback.

Of course we must not run away with the idea that the forward path is of no consequence, so next we will examine matters in a little more detail to see what conditions have to be satisfied in order for the steady-state performance of a closed-loop system to be considered good. For this analysis we will assume that all the blocks in the forward path are combined together as a single block, as shown in Figure A1.4, where the gain G represents the product of all the gains in the forward path. (For example, if we want to reduce Figure A1.3 to the form shown in Figure A1.4, we put $G = AG_1$.)

The signals in Figure A1.4 are related as follows:

$$y = Ge \text{ and } e = r - Hy \tag{A1.2}$$

To establish the relationship between the output and input we eliminate e, to yield

$$y = \left\{\frac{G}{1 + GH}\right\}r \tag{A1.3}$$

This shows us that the output is proportional to the input, with the closed-loop gain (output y over input r) being given by $G/(1 + GH)$. Note that so far we have made no approximations, so we should not be surprised to see that the gain of the closed-loop system does in fact depend not only on the feedback, but also on the forward path. This result is summarized in Figure A1.4, where the single block on the right behaves exactly the same as the complete closed-loop system on the left.

Figure A1.4 Closed-loop system (left) and its representation by a single equivalent block.

It is worth mentioning that if we had been using the Laplace transform method to express relationships in the s-domain, the forward and feedback paths would be written $G(s)$ and $H(s)$, where $G(s)$ and $H(s)$ represent algebraic functions describing – in the s-domain – both steady-state and dynamic properties of the forward and feedback paths, respectively. These functions are known as 'transfer functions' and although the term strictly relates to the s-domain, there is no reason why we cannot use the same terminology here, provided that we remember that we have chosen to consider only steady-state conditions, and therefore what we mean by transfer function is simply the gain of the block in the steady state. In the context of Figure A1.4, we note that the transfer function (i.e. gain) of the single block on the right is the same as the gain of the closed-loop system on the left.

Returning to equation (A1.3), it is easy to see that if the product GH (which must always be dimensionless) is very much greater than 1, we can ignore the 1 in the denominator and then equation (A1.3) reduces to equation (A1.1), derived earlier by asking what we wanted from a good control system. This very important result is shown in Figure A1.5, where the single block on the right approximates the behavior of that on the left, which itself precisely represents the complete system shown in Figure A1.4.

So now we can be precise about the conditions we seek to obtain a good system. First, we need the product GH to be much greater than 1, so that overall the performance is not dependent on the forward path (i.e. the process), but instead depends on the gain of the feedback path.

The product GH is called the 'loop gain' of the system, because it is the gain that is incurred by a signal passing once round the complete loop. In electronic control systems there is seldom any difficulty in achieving enough loop gain because the controller will be based on an amplifier whose gain can be very high. In drive systems (e.g. speed control) it would be unusual for the overall loop gain (error amplifier, power stage, motor, tacho feedback) to be less than 10, but unlikely to be over a thousand. We will see later, however, that if we use integral control, the steady-state loop gain will be infinite.

Secondly, having ensured, by means of a high loop gain, that the overall gain of the closed-loop system is given by $1/H$, it is clearly necessary to ensure that the feedback gain (H) can be precisely specified and guaranteed. If the value of H is not correct for the closed-loop gain we are seeking, or if it is different on a hot day from a cold day, we will not have a good closed-loop system. This underlines again the importance of spending money on a good feedback transducer.

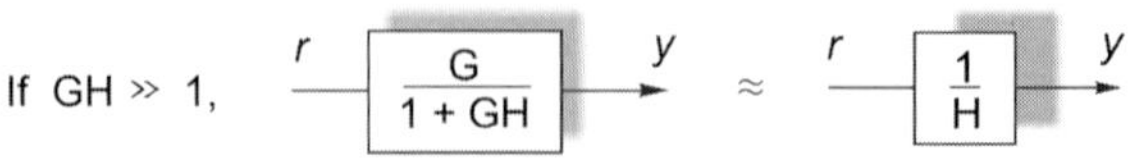

If $GH \gg 1$, $\quad \dfrac{G}{1 + GH} \quad \approx \quad \dfrac{1}{H}$

Figure A1.5 Simplification of closed-loop system model when loop gain (GH) is much greater than 1.

A1.3.1 Importance of loop gain – example

To illustrate the significance of the forward and feedback paths with regard to the overall gain of a closed-loop system we will use as an example an operational amplifier of the type frequently used in traditional analogue control. In their basic form these amplifiers typically have very high gains, but the gain can vary considerably even among devices from a single batch. We will see that as normally employed (with negative feedback) the unpredictability of the open-loop gain is not a problem.

The op-amp itself is represented by the triangle in Figure A1.6(a). Voltages applied to the non-inverting input are amplified without sign change, while voltages applied to the inverting input are amplified and the sign is changed. If the open-loop gain of the amplifier is A, and the input voltages on the positive and negative inputs are v_1 and v_2, respectively, the output voltage will be $A(v_1 - v_2)$. The two inputs therefore perform the differencing function that we have seen is required in a negative feedback system.

In Figure A1.6(a), we can see that a potential divider network of two high-precision resistors has been connected across the output voltage, with resistors chosen so that a fraction (0.2) of the output voltage can be fed back to the inverting input terminal. In block diagram terms the circuit can therefore be represented as shown in Figure A1.6(b), where r is the input or reference voltage, y is the output voltage, and e represents the error signal.

We will assume that because of variations between amplifiers, the gain of the forward path (G in control systems terms) could vary between 100 and 10,000 (though for an op-amp the lower limit is unlikely to be as low as 100), and the feedback gain (H) is 0.2.

If the feedback system is good, equation (A1.1) will apply and we would expect the overall (closed-loop) gain to be given by $1/0.2$, i.e. 5. Table A1.1 shows that the actual closed-loop gain is close to 5, the discrepancy reducing as the loop gain increases. When the loop gain is 2000 (corresponding to a forward gain of 10,000), the closed loop gain of 4.9975 is within 0.05% of the ideal, while even a loop gain as low as 20 still gives a closed-loop gain that is only 4.8% below the ideal.

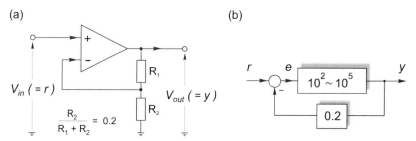

Figure A1.6 Operational amplifier with negative feedback: (a) circuit diagram, (b) equivalent block diagram.

Table A1.1 Closed-loop gain is relatively insensitive to forward-path gain provided that the loop gain is much greater than 1.

Forward gain, G	Feedback factor, H	Loop gain, GH	$1 + GH$	Closed-loop gain $G/(1 + GH)$
100	0.2	20	21	4.7619
1000	0.2	200	201	4.9751
10,000	0.2	2000	2001	4.9975

The important point to note is that the closed–loop gain is relatively insensitive to the forward path gain, provided that the loop gain remains large compared with 1, so if our op-amp happens to need replacing and we change it for one with a different gain, there will be very little effect on the overall gain of the system. On the other hand, it is very important for the gain of the feedback path to be stable, as any change is directly reflected in the closed-loop gain; for example, if H were to increase from 0.20 to 0.22 (i.e. a 10% increase), the closed-loop gain would fall by almost 10%. This explains why high–precision resistors with excellent temperature stability are used to form the potential divider.

We should conclude this example by acknowledging that we have had to sacrifice the high inherent open-loop gain for a much lower overall closed-loop gain in the interests of achieving an overall gain that does not depend on the particular op-amp we are employing. But since such circuits are very cheap, we can afford to cascade them if we require more gain.

A1.3.2 Steady-state error – integral control

An important criterion for any closed-loop system is its steady-state error, which ideally should be zero. We can return to the op-amp example studied above to examine how the error varies with loop gain, the error being the difference between the reference signal and the feedback signal. If we make the reference signal unity for the sake of simplicity, the magnitudes of the output and error signals will be as shown in Figure A1.7, for the three values of forward gain listed in Table A1.1. At the highest loop gain the error is only 0.05%, rising to only 4.8% at the lowest loop gain. These figures are impressive, but what if even a very small error cannot be tolerated? It would seem that we would need infinite loop gain, which at first sight seems unlikely. So how do we eliminate steady-state error?

The clue to the answer lies in the observation above, that in order to have zero error we need infinite loop gain, i.e. at some point in the forward path we need a block that gives a steady output even when its input is zero. This is far less fanciful than it sounds, because all we need is for the forward path to contain a block that represents an element or process whose output depends not simply on its input at that time, but on the integrated effect of the input up to the time in question.

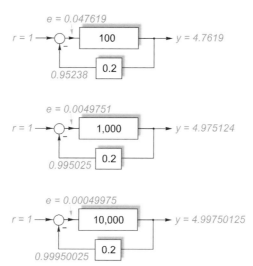

Figure A1.7 Diagram showing how the magnitude of the error signal (e) reduces as the gain of the loop increases.

An example that is frequently encountered in drives is a block whose input is torque and whose output represents angular velocity. The torque (T) determines the rate of change of the angular velocity, the equation of motion being

$$\frac{d\omega}{dt} = \frac{T}{J}$$

where J is the inertia. So the angular velocity is given by

$$\omega = \frac{1}{J}\int T dt,$$

i.e. the speed is determined by the integral with respect to time of the torque, not by the value of torque at any particular instant. The block diagram for this is shown in Figure A1.8: the presence of the integral sign alerts us to the fact that at any instant, the value of the angular speed depends on the past history of the torque to which it has been subjected.

To illustrate the variation in the 'effective gain' of a block containing an integrator, we can study a simple example of an inertia of $0.5\ \text{kgm}^2$, initially at rest, that

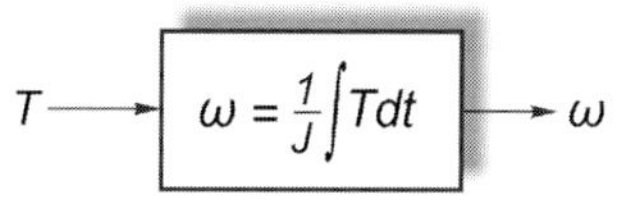

$$T \longrightarrow \boxed{\omega = \frac{1}{J}\int T dt} \longrightarrow \omega$$

Figure A1.8 Block diagram representing the integration of torque to obtain angular velocity.

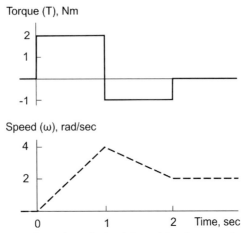

Figure A1.9 Variation of angular velocity (ω) resulting from application of torque (T).

is subjected to a torque of 2 Nm for 1 second, followed by a torque of -1 Nm for a further second, and zero torque thereafter. Plots of the input torque and the corresponding output angular speed with time are shown in Figure A1.9.

For the first second the acceleration is 4 rad/s/s, so the speed is 4 rad/s after 1 second. For the next second the torque is negative and the deceleration is 2 rad/s/s, so the speed after 2 seconds has fallen to 2 rad/s. Subsequently the torque and acceleration are zero so the speed remains constant.

If we examine the graphs, we see that the ratio of output (speed) to input (torque), i.e. the quantity that hitherto we have referred to as gain, varies. For example, just before $t = 1$, the gain is $4/2 = 2$ rad/s/Nm, whereas just after $t = 1$ it is $4/-1 = -4$ rad/s/Nm. Much more importantly, however, we see that from $t = 2$ onwards, the output is constant at 2 rad/s, but the input is zero, i.e. the gain is infinite when the steady state is reached.

The fact that the gain changes until the steady state is reached means that we can only use the simple 'gain block' approach to represent an integrator when conditions have settled into the steady state, i.e. all the variables are constant. The output of an integrator can only remain constant if its input is zero: if the input is positive, the output will be increasing, and if the input is negative the output will be decreasing. So in the steady state, an integrator has infinite gain, which is just what we need to eliminate steady-state error.

The example discussed above includes an integration that reflects physical properties, i.e. the fact that the angular acceleration is proportional to the torque, and hence the angular velocity is proportional to the integral of torque. Another 'natural' integration applies to the water tank we looked at earlier, where the level of water is proportional to the integral of the flow rate with respect to time.

In many systems there is no natural integration in the forward-path process, so if it is important to achieve zero steady-state error, an integrating term is included in the controller. This is discussed next.

A1.3.3 PID controller

We saw earlier that the simplest form of controller is an amplifier, the output of which is proportional to the error signal. A control system that operates with this sort of controller is said to have 'proportional' or 'P' control. An important feature of proportional control is that as soon as there is any change in the error, proportionate action is initiated immediately.

We have also seen that in order to completely eliminate steady-state error we need to have an integrating element in the forward path, so we may be tempted to replace the proportional controller in Figure A1.3 by a controller whose output is the integral of the error signal with respect to time. This is easily done in the case of an electronic amplifier, yielding an 'integral' or 'I' controller.

However, unlike the proportional controller where the output responds instantaneously to changes in the error, the output of an integrating controller takes time to respond. For example, we see from Figure A1.9 that the output builds progressively when there is a step input. If we were to employ integral control alone, the lag between output and input signals may well cause the overall (closed-loop) response to be unacceptably sluggish.

To obtain the best of both worlds (i.e. a fast response to changes and elimination of steady-state error), it is common to have both proportional and integral terms in the controller, which is then referred to as a PI controller. The output of the controller ($y(t)$) is then given by the expression

$$y(t) = Ae(t) + k \int e(t)\mathrm{d}t$$

where $e(t)$ is the error signal, A is the proportional gain and k is a parameter that allows the rate at which the integrator ramps up to be varied. The latter adjustment is also often – and rather confusingly – referred to as *integrator gain*.

We can easily see that because raising the proportional gain causes a larger output of the controller for a given error, we generally get a faster transient response. On the other hand, the time-lag associated with an integral term generally makes the transient response more sluggish, and increases the likelihood that the output will overshoot its target before settling.

Some controllers also provide an output term that depends on the rate of change or differential of the error. This has the effect of increasing the damping of the transient response, an effect similar to that demonstrated in Figure 4.16. PI controllers that also have this differential (or D) facility are known as PID controllers.

A1.3.4 Stability

So far we have highlighted the benefits of closed-loop systems, but not surprisingly there is a potential negative side that we need to be aware of. This is that if the d.c. loop gain is too high, some closed-loop systems exhibit self-sustaining oscillations, i.e. the output 'rings' – generally at a high frequency – even when there is no input to the system. When a system behaves in this way it is said to be unstable, and clearly the consequences can be extremely serious, particularly if large mechanical elements are involved. (It is worth mentioning that whereas in control systems instability is always undesirable, if we were in the business of designing electronic oscillators we would have an entirely different perspective, and we would deliberately arrange our feedback and loop gain so as to promote oscillation at the desired frequency.)

A familiar example of spontaneous oscillation is a public address system (perhaps in the village hall) that emits an unpleasantly loud whistle if the volume (gain) is turned too high. In this case the closed loop formed by sound from the loudspeaker feeding back to the microphone is an unwanted but unavoidable phenomenon, and it may be possible to prevent the instability by shielding the microphone from the loudspeakers (i.e. lowering the loop gain by reducing the feedback), or by reducing the volume control.

Unstable behavior is characteristic of linear systems of third or higher order and is well understood, though the theory is beyond our scope. Whenever the closed-loop system has an inherently oscillatory transient response, increasing the proportional gain and/or introducing integral control generally makes matters worse. The amplitude and frequency of the 'ringing' of the output response may become larger, and the settling time may increase. If the gain increases further the system becomes unstable and oscillation grows until limited by saturation of one or more elements in the loop.

As far as we are concerned it is sufficient to accept that there is a potential drawback to raising the gain in order to reduce error, and to take comfort from the fact that there are well-established design criteria that are used to check that a system will not be unstable before the loop is closed. Stability of the closed-loop system can be checked by examining the frequency response of the open-loop system, the gain being adjusted to ensure (by means of design criteria known as gain and phase margins) that, when the loop is closed for the first time, there is no danger of instability.

A1.3.5 Disturbance rejection – example using d.c. machine

We will conclude our brief look at the benefits of feedback by considering an example that illustrates how a closed-loop system combats the influence of inputs (or disturbances) that threaten to force the output of the system from its target value. We already referred to the matter qualitatively when we looked at how we would drive a car at a constant speed despite variations in wind or gradients.

Throughout this book the self-regulating properties of electric motors have been mentioned frequently. All electric motors are inherently closed-loop systems, so it is fitting that we finish by revisiting one of our first topics (the d.c. machine), but this time we take a control-systems perspective to offer a fresh insight as to why the performance of the machine is so good.

The block diagram of a separately excited d.c. machine (with armature inductance neglected) is shown in Figure A1.10.

This diagram is a pictorial representation of the steady-state equations developed in Chapter 3, together with the dynamic equation relating resultant torque, inertia and angular acceleration. This set of equations is repeated here for convenience.

$$\text{Motor torque, } T_m = kI$$

$$\text{Motional e.m.f.}, E = k\omega$$

$$\text{Armature circuit, } V = E + IR$$

$$\text{Equation of motion, } T_\mathrm{m} - T_\mathrm{L} = T_\mathrm{res} = J\frac{\mathrm{d}\omega}{\mathrm{d}t} \text{ or } \omega = \frac{1}{J}\int T_\mathrm{res}\mathrm{d}t$$

Each section of the diagram corresponds to one of the equations: for example, the left-hand summing junction and the block labeled $1/R$ yield the armature current, by implementing the rearranged armature circuit equation, i.e.

$$I = \frac{V - E}{R}$$

The summing junction in the middle of the forward path allows us to represent a 'disturbance' entering the system: in this case the disturbance is the load torque (including any friction), and is shown as being subtracted from the motor torque as this is usually what happens.

A control person who was unfamiliar with motors would look at Figure A1.10 and describe it as a closed-loop speed-control system designed to make the speed ω

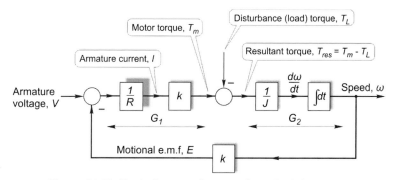

Figure A1.10 Block diagram of separately excited d.c. motor.

track the reference voltage V. Examination of the forward path (and ignoring the disturbance input for the moment) would reveal an integrating element, and this would signal that the loop gain was infinite under steady-state conditions. The control specialist would then say that when the loop gain is infinite, the closed-loop gain is $1/H$, where, in this case, $H = k$. Hence the speed at any voltage V is given by

$$\omega = \frac{1}{k}V$$

We obtained this result in Chapter 3 by arguing that if there was no friction or load torque, and the speed was steady (i.e. the acceleration was zero), the motor torque would be zero, and therefore the current would have to be zero, and hence the back e.m.f. must equal the applied voltage. We then equated $V = E = k\omega$ to obtain the result above.

Turning now to the effect of the load torque (the 'disturbance'), we will focus on the steady-state condition. When the speed is steady, the signal entering the integrator block must be zero, i.e. the resultant torque must be zero, or in other words the motor torque must be equal and opposite to the load torque. But from the diagram the motor torque is directly proportional to the error signal (i.e. $V - E$). So we deduce that as the load torque increases, the steady-state error increases in proportion, i.e. the speed (E) has to fall in order for the motor to develop torque. It follows that in order to reduce the drop in speed with load, the gain of section G_1 in Figure A1.10 must be as high as possible, which in turn underlines the desirability of having a high motor constant (k) and a low armature resistance (R). As expected, this is exactly what we found in Chapter 3.

We can see that the mechanism whereby the closed-loop system minimizes the effect of disturbances is via the feedback path: as the speed begins to fall following an increase in the disturbance input, the feedback reduces, the error increases, and more motor torque is produced to compensate for the load torque.

To quantify matters, we can develop a model that allows us to find the steady-state output due to the combined effect of the reference input (V) and the load torque (T_L), using the principle of superposition. We find the outputs when each input acts alone, and then sum them to find the output when both are acting simultaneously.

If we let G_1 and G_2 denote the steady-state gains of the two parts of the forward path shown in Figure A1.10, we can see that as far as the reference input (V) is concerned the gain of the forward path is $G_1 G_2$, and the gain of the feedback path is k. Hence, using equation (A1.3), the output is given by

$$\omega_V = \left\{ \frac{G_1 G_2}{1 + G_1 G_2 k} \right\} V$$

From the point of view of the load torque, the forward path consists only of G_2, with the feedback consisting of k in series with G_1. Hence, again using equation

(A1.3) and noting that the load torque will usually be negative, the effect of the load on the output is given by

$$\omega_{\mathrm{L}} = \left\{\frac{G_2}{1 + G_1 G_2 k}\right\}(-T_{\mathrm{L}}) = -\left\{\frac{G_2}{1 + G_1 G_2 k}\right\}T_{\mathrm{L}}$$

Hence the speed (ω) is given by

$$\omega = \left\{\frac{G_1 G_2}{1 + G_1 G_2 k}\right\}V - \left\{\frac{G_2}{1 + G_1 G_2 k}\right\}T_{\mathrm{L}}$$

The second term represents the influence of the disturbance – in this case the amount that the speed falls due to the load torque T_{L}. Looking at the block diagram, we see that in the absence of feedback, the effect on the output of an input T_L would simply be $G_2 T_{\mathrm{L}}$. Comparing this with the second term in the equation above we see that the feedback reduces the effect of the disturbance by a factor of

$$\frac{1}{1 + G_1 G_2 k}$$

so if the loop gain ($G_1 G_2 k$) is high compared with 1, the disturbance is attenuated by a factor of 1/loop gain.

We can simplify these expressions in the case of the d.c. motor example by noting that

$$G_1 = \frac{k}{R}$$

and, because of the integration, G_2 is infinite. Hence the steady-state speed is given by

$$\omega = \frac{V}{k} - \frac{R}{k^2}T_{\mathrm{L}}$$

which is the same result as we obtained in equation (3.10). The first term confirms that when the load torque is zero the speed is directly proportional to the armature voltage, while the second term is the drop in speed with load, and is minimized by aiming for a low armature resistance.

APPENDIX TWO

Induction Motor Equivalent Circuit

A2.1. INTRODUCTION

This appendix explores the steady-state equivalent circuit of the induction motor, and the reasons for including it are that it provides a simple method for quantitative performance predictions; the parameters that appear (e.g. leakage reactance, magnetizing reactance) are common currency in any discussion of induction motors, and understanding what they mean is necessary if we wish to engage effectively with those who use the language; and finally a knowledge of the behavior of the circuit with slip brings a new perspective to support the 'physical' basis followed in Chapters 5 and 6.

It has to be admitted that the superiority of digital simulation packages that model both transient and steady-state behavior has meant that the traditional equivalent circuit has fallen from favor. We think this is a pity, because it has a lot to offer. It provides relatively simple explanations for many of the most important aspects of induction motor behavior, not least of which is the influence of frequency, historically seen as constant, but now perhaps the most important variable.

A2.1.1 Outline of approach

Physically, the construction of the wound-rotor induction motor has striking similarities with that of the 3-phase transformer, with the stator and rotor windings corresponding to the primary and secondary windings of a transformer, so it is not surprising that the induction motor equivalent circuit can be derived from that of the transformer.

We begin with the 'ideal' transformer, for which the governing equations are delightfully simple. We then extend the equivalent circuit to include the modest imperfections of the real transformer. In the course of this discussion we will establish the meaning of the terms magnetizing reactance and leakage reactance, which also feature in the induction motor equivalent circuit. And perhaps even more importantly for what comes later, we will be in a position to appreciate the benefit of being able to assess what effect a load connected to the secondary winding has on the primary winding by making use of the concept of a 'referred' load impedance.

383

The emphasis throughout will be on how good a transformer is, and how for most purposes a very simple equivalent circuit is adequate. The approach taken differs from that taken in many textbooks, which begins with the all-singing, all-dancing circuit which not only looks frighteningly complicated, but also tends to give the erroneous impression that the transformer is riddled with serious imperfections.

When we reach the induction motor equivalent circuit, a really clever leap of the imagination is revealed. We saw in Chapter 5 that the magnitude and frequency of the rotor currents depend on the slip, and that they interact with the air-gap flux to produce torque. So how are we to represent what is going on in the rotor in a single equivalent circuit that must necessarily also include the stator variables, and in which all the voltages and currents are at supply frequency? We will discover that, despite the apparent complexity, all of the electromechanical interactions can be represented by means of a transformer equivalent circuit, with a hypothetical slip-dependent 'electromechanical resistance' connected where the secondary 'load' would normally be. The equivalent circuit must have created quite a stir when it was first developed, but is now taken for granted: this is a pity because it represents a major intellectual achievement.

A2.1.2 Similarity between induction motor and transformer

The development of a wound-rotor induction motor from a single-phase transformer is depicted in Figure A2.1.

In most transformers there will be many turns on both windings, but for the sake of simplicity only four coils are shown. The purpose of a transformer is to take in electrical power at one voltage and supply it at a different voltage. When an a.c. supply is connected to the primary winding, a pulsating magnetic flux (shown by the hashed lines in Figure A2.1) is set up. The pulsating flux links the secondary winding, inducing a voltage in each turn, so by choosing the number of turns in series the desired output voltage is obtained. Because no mechanical energy conversion is involved there is no need for an air-gap in the magnetic circuit, which therefore has an extremely low reluctance.

In Figure A2.1(b), we see a hypothetical set-up in which two small air-gaps have been introduced into the magnetic circuit to allow for the essential motion in the induction motor. Needless to say we would not do this deliberately in a transformer as it would cause an unnecessary increase the reluctance of the flux path (though the effect on performance would be much less than we might fear). The central core and winding space have also been enlarged somewhat (anticipating the need for two more phases), without materially altering the functioning of the transformer. Finally, in Figure A2.1(c) the two sets of coils are arranged as they would be in a 2-pole wound-rotor induction motor with full-pitched coils.

The most important points to note are first that the flux produced by the stator (primary) winding links the rotor (secondary) winding in much the same

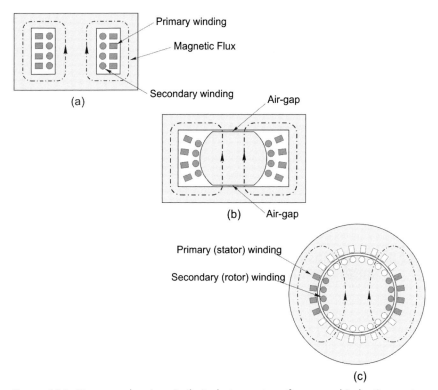

Figure A2.1 Diagrams showing similarity between transformer and induction motor.

way as it did in Figure A2.1(a), i.e. the two windings remain tightly coupled by the magnetic field. Secondly, only one-third of the slots are taken up because the remaining two-thirds will be occupied by the windings of the other two phases. Thirdly, the magnetic circuit has two air-gaps, and because of the slots that accommodate the coils, the flux threads its way down the teeth, so that it not only fully links the aligned winding, but also partially links the other two phase-windings.

If the two windings highlighted in Figure A2.1(c) were used as primary and secondary, this set-up would work perfectly well as a single-phase transformer.

A2.2. THE IDEAL TRANSFORMER

A quasi-circuit model is shown diagrammatically in Figure A2.2. This represents the most common application of the transformer, with the primary drawing power from an a.c. constant-voltage source (V_1) and supplying it to a load impedance (Z_2) at a different voltage (V_2).

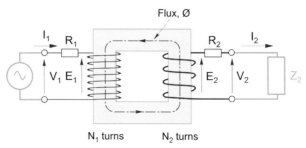

Figure A2.2 Single-phase transformer supplying secondary load Z_2.

The primary winding has N_1 turns with total resistance R_1, and the secondary winding has N_2 turns with total resistance R_2; they share a common magnetic circuit with no air-gap and therefore very low reluctance.

A2.2.1 Ideal transformer – no-load condition – flux and magnetizing current

We begin with the primary connected to the voltage source as shown in Figure A2.2, but the secondary is open-circuited – the no-load condition. The secondary has no current so it cannot influence matters. What we want to know is what determines the magnitude of the flux in the core, and how much current (known for obvious reasons as the magnetizing current, I_m) will be drawn from the voltage source to set up the flux.

We apply Kirchhoff's voltage law to relate the applied voltage (V_1) to the voltage induced in the primary winding by the pulsating flux (E_1) and to the volt-drop across the primary resistance, yielding

$$V_1 = E_1 + I_m R_1 \tag{A2.1}$$

We also apply Faraday's law to obtain the induced e.m.f. in terms of the flux, i.e.

$$E_1 = N_1 \frac{d\phi}{dt} \tag{A2.2}$$

We can simplify equation (A2.1) further in the case of the 'ideal' transformer, which not unexpectedly is assumed to have windings with zero resistance, i.e. $R_1 = 0$. So we deduce that for an ideal transformer, the induced e.m.f. is equal to the applied voltage, i.e. $E_1 = V_1$. This is very important and echoes the discussion in Chapter 5: but what we really want is to find out what the flux is doing, so we recast equation (A2.2) in the form

$$\frac{d\phi}{dt} = \frac{E_1}{N_1} = \frac{V_1}{N_1} \tag{A2.3}$$

Equation (A2.3) shows that the rate of change of flux at any instant is determined by the applied voltage, so if we want to know how the flux behaves in time we must specify the nature of the applied voltage. We will look at two cases: the first voltage waveform is good for illustrative purposes because it is easy to derive the flux waveform, while the second represents the commonplace situation, i.e. use on a sinusoidal a.c. supply.

First, we suppose that the applied voltage is a square wave. The rate of change of flux is then positive and constant while V_1 is positive, so the flux increases linearly with time; conversely, when the applied voltage is negative the flux ramps down, so overall the voltage and flux waves are as shown in the upper half of Figure A2.3.

From Figure A2.3 we see that the maximum flux in the core (ϕ_m) is determined not only by the magnitude of the applied voltage (which determines the slope of the flux/time plot) but also by the frequency (which determines for how long the positive slope continues). Normally we want to utilize the full capacity of the magnetic circuit, so we must adjust the voltage and frequency together in order to keep ϕ_m at its rated value. So both voltage and frequency have been doubled in (b) to keep ϕ_m the same as in (a).

Turning now to the everyday situation in which the transformer is connected to a sinusoidal voltage of peak value $\widehat{V}$ and frequency f, we can integrate equation (A2.3), to yield the maximum flux as

$$\phi_m = \frac{1}{2\pi N_1} \frac{\widehat{V}}{f} \qquad (A2.4)$$

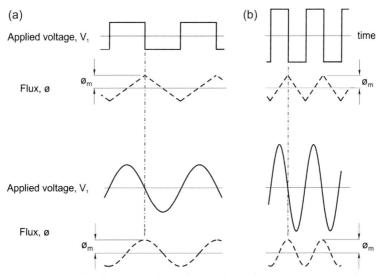

Figure A2.3 Flux and voltage waveforms for ideal transformer operating with square-wave (upper) and sinusoidal (lower) primary voltage. (The voltage and frequency in (b) are doubled compared with (a), but the peak flux remains the same.)

Typical primary voltage and flux waves are shown in the lower part of Figure A2.3. Equation (A2.4) shows that the amplitude of the flux wave is proportional to the applied voltage, and inversely proportional to the frequency. As mentioned above, we normally aim to keep the peak flux constant in order fully to utilize the magnetic circuit, and this means that changes to voltage or frequency must be done so that the ratio of voltage to frequency (V/f) is maintained. This is shown in Figure A2.3 where, to keep the peak flux (ϕ_m) in (a) the same when the frequency is doubled to that in (b), the voltage must also be doubled. We drew the same conclusion in relation to the induction motor, in Chapters 5 and 7.

Although equation (A2.4) was developed for an ideal transformer, it is also applicable with very little error to the real transformer, and is in fact a basic design equation.

The important message to take from this analysis is that under sinusoidal conditions at a fixed frequency, the flux in a given transformer is determined by the applied voltage. Interestingly, however, the only assumption we have to make to arrive at this result is that the resistance of the windings is zero: the argument so far is independent of the magnetic circuit, so we must now see how the reluctance of the transformer core makes its presence felt.

We know that although the amplitude of the flux waveform is determined by the applied voltage, frequency and turns, there will need to be an m.m.f. (i.e. a current in the primary winding) to drive the flux around the magnetic circuit: the 'magnetic Ohm's' law tells us that the m.m.f. required to drive flux ϕ around a magnetic circuit that has reluctance $\mathscr{R}$ is given by m.m.f. $= \mathscr{R}\phi$.

But in this section we are studying an ideal transformer, so we can assume that the magnetic circuit is made of infinitely permeable material, and therefore has zero reluctance. This means that no m.m.f. is required, so the current drawn from the supply (the 'magnetizing current' I_m) in the ideal transformer is zero.

To sum up, the flux in the ideal transformer is determined by the applied voltage, and the no-load current is zero. This hypothetical situation is never achieved in practice, but real transformers (especially large ones) come close to it. Viewed from the supply, the ideal transformer at no-load looks like an open circuit, as it draws no current. (A real transformer at no-load draws a small current, lagging the applied voltage by almost $90°$, and from the supply viewpoint it therefore has a high inductive reactance, known for obvious reasons as the 'magnetizing reactance'.) An ideal transformer is thus seen to have an infinite magnetizing reactance.

A2.2.2 Ideal transformer – no-load condition – voltage ratio

We now consider the secondary winding, but leave it disconnected from the load so that its current is zero, in which case it can clearly have no influence on the flux. Because the magnetic circuit is perfect, none of the flux set up by the primary leaks out, and all of it therefore links the secondary winding. We can therefore apply

Faraday's law and make use of equation (A2.2) to obtain the secondary induced e.m.f. as

$$E_2 = N_2 \frac{d\phi}{dt} = N_2 \frac{V_1}{N_1} = \frac{N_2}{N_1} V_1 \qquad (A2.5)$$

There is no secondary current, so there is no volt-drop across R_2 and therefore the secondary terminal voltage V_2 is equal to the induced e.m.f. E_2. Hence the voltage ratio is given by

$$\frac{V_1}{V_2} = \frac{N_1}{N_2} \qquad (A2.6)$$

This equation shows that any desired secondary voltage can be obtained simply by choosing the number of turns on the secondary winding. It is worth mentioning that equation (A2.6) applies regardless of the nature of the waveform, so if we applied a square wave voltage to the primary, the secondary voltage would also be a square wave with amplitude scaled according to the turns ratio.

We will see later that when the transformer supplies a load, the primary and secondary currents are inversely proportional to their respective voltages: so if the secondary voltage is lower than the primary there will be fewer secondary turns but the current will be higher and therefore the cross-sectional area of the wire used will be greater. The net result is that the total volumes of copper in primary and secondary are virtually the same, as is to be expected since they both handle the same power.

A2.2.3 Ideal transformer on load

We now consider what happens when we connect the secondary winding to a load impedance Z_2. We have already seen that the flux is determined solely by the applied primary voltage, so when current flows to the load it can have no effect on the flux, and hence because the secondary winding resistance is zero, the secondary voltage remains as it was at no-load, given by equation (A2.6). (If the voltage were to change when we connected the load we could be forgiven for beginning to doubt the validity of the description 'ideal' for such a device!)

The current drawn by the load will be given by $I_2 = V_2/V_2$ and the secondary winding will therefore produce an m.m.f. of $N_2 I_2$ acting around the magnetic circuit. If this m.m.f. went unchecked it would tend to reduce the flux in the core, but, as we have seen, the flux is determined by the applied voltage, and cannot change. We have also seen that because the core is made of ideal magnetic material it has no reluctance, and therefore the resultant m.m.f. (due to both the primary and secondary windings) is zero. These two conditions are met by the primary winding drawing a current such that the sum of the primary and secondary m.m.f.s is zero, i.e.

$$N_1 I_1 + N_2 I_2 = 0, \text{ or } \frac{I_1}{I_2} = -\frac{N_2}{N_1} \tag{A2.7}$$

In other words, as soon as the secondary current begins to flow, a primary current automatically springs up to neutralize the demagnetizing effect of the secondary m.m.f. The minus sign in equation (A.7) serves to remind us that primary and secondary m.m.f.s act in opposition. It has no real meaning until we define what we mean by the positive direction of winding the turns around the core, so since we are not concerned with transformer manufacture we can afford to ignore it from now on.

The current ratio in equation (2.7) is seen to be the inverse of the voltage ratio in equation (A.6). We could have obtained the current ratio by a different approach if we had argued from a power basis, by saying that in an ideal transformer, the instantaneous input power and the instantaneous output power must be equal. This would lead us to conclude that $V_1 I_1 = V_2 I_2$ and hence from equation (A.6) that

$$\frac{V_1}{V_2} = \frac{N_1}{N_2} = \frac{I_2}{I_1} \tag{A2.8}$$

To conclude our look at the ideal transformer, we should ask what the primary winding of an ideal transformer 'looks like' when the secondary is connected to a load impedance Z_2. As far as the primary supply is concerned, the apparent impedance looking into the primary of the transformer is simply the ratio V_1/I_1, which can be expressed in secondary terms as

$$\frac{V_1}{I_1} = \left(\frac{N_1}{N_2}\right)^2 \frac{V_2}{I_2} = \left(\frac{N_1}{N_2}\right)^2 Z_2 = Z_2' \tag{A2.9}$$

So when we connect an impedance Z_2 to the secondary, it appears from the primary side as if we have connected an impedance Z_2' across the primary terminals. This equivalence is summed up diagrammatically in Figure A2.4.

The ideal transformer effectively 'scales' the voltages by the turns ratio and the currents by the inverse turns ratio, and from the point of view of the input terminals, the ideal transformer and its secondary load (inside the shaded area in Figure A2.4(a)) is indistinguishable from the circuit in Figure A2.4(b), in which the impedance Z_2' (known as the 'referred' impedance) is connected across the supply. We should note that when we use referred impedances, the equivalent circuit of the ideal transformer simply reduces to a link between primary and referred secondary circuits: this point has been stressed by showing the primary and secondary terminals in Figure A2.4, even though there is nothing between them. When we model the real transformer we will find that there are circuit elements between the input and output terminals.

Finally we should note that although both windings of an ideal transformer have infinite inductance, there is not even a vestige of inductance in the equivalent circuit. This remarkable result is due to the perfect magnetic coupling between the

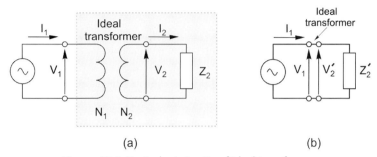

Figure A2.4 Equivalent circuits of ideal transformer.

windings. As we will see shortly, the real transformer can come very close to the ideal, but for reasons which will also become apparent, ultimate perfection is not usually what we seek.

A2.3. THE REAL TRANSFORMER

Real transformers behave much like ideal ones (except in very small sizes), and the approach is therefore to extend the model of the ideal transformer to allow for the imperfections of the real one.

We will establish the so-called 'exact' equivalent circuit on a step-by-step basis first; it can be used to predict all aspects of transformer performance, but we will find that it looks rather fearsome. Fortunately, we are not seeking great accuracy, so we can retreat to the much less daunting 'approximate' circuit which is perfectly adequate for our purposes.

A2.3.1 Real transformer – no-load condition – flux and magnetizing current

Here we take account of the finite resistances of the primary and secondary windings; the finite reluctance of the magnetic circuit; and the losses due to the pulsating flux in the iron core.

The winding resistances are included by adding resistances R_1 and R_2, respectively, in series with the primary and secondary of the ideal transformer, as shown in Figure A2.5.

The transformer in Figure A2.5 is shown without a secondary load; i.e. it is under no-load conditions. As previously explained, the component of the current drawn from the supply that sets up the flux in this condition is known as the magnetizing current (I_m). Because the ideal transformer (within the dotted line in Figure A2.5) is on open-circuit, the secondary current is zero and so the primary current must also be zero. So to allow for the fact that the real transformer has

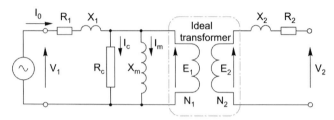

Figure A2.5 Development of equivalent circuit of unloaded transformer.

a no-load magnetizing current we add an inductive branch (known as the 'magnetizing reactance') in parallel with the primary of the ideal transformer, as shown in Figure A2.5. In most transformers the reluctance of the core is small, and as a result the magnetizing current is much smaller than the normal full-load (rated) primary current. The magnetizing reactance is therefore high, and we will see that it can be neglected for many purposes.

The next refinement of the no-load model takes account of the fact that the pulsating flux induces eddy current and hysteresis losses in the core. The core is laminated to reduce the eddy current losses, but the total loss is often significant in relation to overall efficiency and cooling, and we need to allow for it in our model. The iron losses depend on the square of the peak flux density, and they are therefore proportional to the square of the induced e.m.f. This means that we can model the losses simply by including a resistance (R_c – where the subscript denotes 'core loss') in parallel with the magnetizing reactance, as shown in Figure A2.5.

The no-load current (I_0) consists of the reactive magnetizing component (I_m) lagging the applied voltage by $90°$, and the core-loss (power) component (I_c) that is in phase with the induced voltage.

The magnetizing component is usually much greater than the core-loss component, so the real transformer looks predominantly reactive under open-circuit (no-load) conditions.

A2.3.2 Real transformer – leakage reactance

In the ideal transformer it was assumed that all the flux produced by the primary winding linked the secondary, but in practice some of the primary flux will exist outside the core (see Figure 1.8) and will not link with the secondary. This leakage flux, which is proportional to the primary current, will induce a voltage in the primary winding whenever the primary current changes, and it can therefore be represented by a 'primary leakage inductance' (l_1) in series with the primary winding of the ideal transformer. We will be using the equivalent circuit under steady-state operating conditions at a given frequency (f), in which case we are more likely to refer to the 'primary leakage reactance' ($X_1 = \omega l_1 = 2\pi f l_1$) as in Figure A2.5.

The induced e.m.f. in the secondary winding arising from the secondary leakage flux is modeled in a similar way, and this is represented by the 'secondary leakage reactance' (X_2) shown in Figure A2.5

It is important to point out that under no-load conditions when the primary current is small, the volt-drop across X_1 will be much less than V_1, so all our previous arguments relating to the relationship between flux, turns and frequency can be used in the design of the real transformer.

Because the leakage reactances are in series, their presence is felt when the currents are high, i.e. under normal full-load conditions, when the volt–drops across X_1 and X_2 may not be negligible, and especially under extreme conditions (e.g. when the secondary is short-circuited) when they provide a vitally important current-limiting function.

Looking at Figure A2.5 the conscientious reader could be forgiven for beginning to feel a bit downhearted to discover how complex matters appear to be getting, so it will come as good news to learn that we have reached the turning point. The circuit in Figure A2.5 represents all aspects of transformer behavior, but it is more elaborate than we need, so from now on things become simpler, particularly when we take the usual step and refer all secondary parameters to the primary.

A2.3.3 Real transformer on load – exact equivalent circuit

The equivalent circuit showing the transformer supplying a secondary load impedance Z_2 is shown in Figure A2.6(a). This diagram has been annotated to show

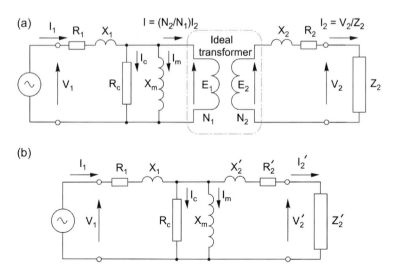

Figure A2.6 'Exact' equivalent circuit of real transformer supplying load impedance Z_2. (The circuit in (a) includes the actual secondary parameters, while in (b) the parameters have been 'referred' to the primary side.)

how the ideal transformer at the center imposes the relationships between primary and secondary currents. Provided that we know the values of the transformer parameters we can use this circuit to calculate all the voltages, currents and powers when either the primary or secondary voltage is specified.

However, we seldom use the circuit in this form, as it is usually much more convenient to work with the referred impedances, as discussed previously and shown in Figure A2.6(b). Note that we have not only referred the secondary load impedance Z_2 but also the secondary resistance and leakage reactance of the transformer itself: as before, the referred or effective load impedance, secondary resistance and leakage reactance as seen at the primary are denoted by Z_2', R_2' and X_2', respectively. When circuit calculations have been done using this circuit, the actual secondary voltage and current are obtained from their referred counterparts using the equations

$$V_2 = V_2' \times \frac{N_2}{N_1} \text{ and } I_2 = I_2' \times \frac{N_1}{N_2} \qquad \text{(A2.10)}$$

At this point we should recall the corresponding circuit of the ideal transformer, as shown in Figure A2.4(b). There we saw that the referred secondary voltage V_2' was equal to the primary voltage V_1. For the real transformer, however (Figure A2.6(b)), we see that the imperfections of the transformer are reflected in circuit elements lying between V_1 and V_2', i.e. between the input and output terminals.

In the ideal transformer the series elements (resistances and leakage reactances) are zero, while the parallel elements (magnetizing reactance and core-loss resistance) are infinite. The principal effect of the non-zero series elements of the real transformer is that, because of the volt-drops across them, the secondary voltage V_2' will be less than it would be if the transformer were ideal. And the principal effect of the parallel elements is that the real transformer draws a (magnetizing) current and consumes (a small amount of) power even when the secondary is open-circuited.

Although the circuit of Figure A2.6(b) represents a welcome simplification compared with that in Figure A2.6(a), calculations are still cumbersome, because for a given primary voltage, every volt-drop and current changes whenever the load on the secondary alters. Fortunately, for most practical purposes a further gain in terms of simplicity (at the expense of only a little loss of accuracy) is obtained by simplifying the 'exact' circuits to obtain the so-called approximate equivalent circuit.

A2.3.4 Real transformer – approximate equivalent circuit

The justification for the approximate equivalent circuit (Figure A2.7) rests on the fact that for all transformers except very small ones (i.e. for all transformers that would cause serious harm if they fell on one's foot) the series elements in Figure A2.6 are of low impedance and the parallel elements are of high impedance.

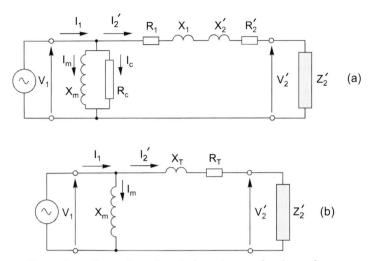

Figure A2.7 'Approximate' equivalent circuits of real transformer.

Actually, to talk of low or high impedances without qualification is nonsense: what the rather loose language in the paragraph above really means is that under normal conditions, the volt-drop across R_1 and X_1 will be a small fraction of the applied voltage V_1. Hence the voltage across the magnetizing branch is almost equal to V_1: so by moving the magnetizing branch to the left-hand side, the magnetizing current will be almost unchanged. Keen eyed readers will have spotted that the core-loss resistance has been omitted: the justification is that the losses in the iron are only of interest when we are calculating efficiency, which is not our concern here.

Moving the magnetizing branch brings massive simplification in terms of circuit calculations. First because the current and power in the magnetizing branch are independent of the load current, and secondly because primary and referred secondary impedances now carry the same current (I_2') so it is easy to calculate V_2' from V_1 or vice versa. Further simplification results if we combine the primary and secondary resistances and leakage reactances to yield total resistance (R_T) and total leakage reactance (X_T), as shown in Figure A2.7(b).

Given the difference in the numbers of turns and wire thickness between primary and secondary windings it is not surprising that the actual primary and secondary resistances are sometimes very different. But when referred to the primary, the secondary resistance is normally of similar value to the primary resistance, so both primary and secondary contribute equally to the total effective resistance. This is in line with common sense: since both windings contain the same quantities of copper and handle the same power, both would be expected to have similar influence on the overall performance.

When leakage reactances are quoted separately for primary and secondary they too may have very different ohmic values, but again when the secondary is referred its value will normally be the same as that of the primary. (In fact, the validity of regarding leakage reactance as something that can be uniquely apportioned between primary and secondary is questionable, as they cannot be measured separately, and for most purposes all that we need to know is the total leakage reactance X_T.)

A2.3.5 Measurement of parameters

Two simple tests are used to measure the transformer parameters – the open-circuit or no-load test and the short-circuit test. We will see later that very similar tests are used to measure induction motor parameters. In the open-circuit test, the secondary is left disconnected and with rated voltage applied to the primary winding, the input current and power are measured. From these, the values of X_m and R_c are calculated. In the short-circuit test the secondary terminals are shorted together and a low voltage is applied to the primary – just sufficient to cause rated current in both windings. The voltage, current and input power are measured, and from these the parameters R_T and X_T are calculated.

A great advantage of the open-circuit and short-circuit tests is that neither requires more than a few per cent of the rated power, so both can be performed from a modestly rated supply.

A2.3.6 Significance of equivalent circuit parameters

If it were possible to choose whatever values we wished for the parameters, we would probably look back at the equivalent circuit of the ideal transformer, and try to emulate it by making the parallel elements infinite, and the series elements zero. And if we only had to think about normal operation (from no-load up to rated load) this would be a sensible aim, as there would then be no magnetizing current, no iron losses, no copper losses, and no drop in secondary voltage when the load is applied.

Most transformers do come close to this ideal when viewed from an external perspective. At full-load the input power is only slightly greater than the output power (because the losses in the resistive elements are a small fraction of the rated power); and the secondary voltage falls only slightly from no-load to full-load conditions (because the volt-drop across Z_T at full-load is only a small fraction of the input voltage).

But in practice we have to cope with the abnormal as well as the normal, and in particular we need to be aware of the risk of a short-circuit at the secondary terminals. If the transformer primary is supplied from a constant-voltage source, the only thing that limits the secondary current if the secondary is inadvertently short-circuited is the series impedance Z_T. But we have already seen that the volt-drop across Z_T at rated current is only a small fraction of the rated voltage, which means that if rated voltage is applied across Z_T, the current will be many times rated value. So the lower the impedance of the transformer, the higher the short-circuit current.

From the transformer viewpoint, a current of many times rated value will soon cause the windings to melt, to say nothing of the damaging inter-turn electromagnetic forces produced. And from the supply viewpoint the fault current must be cleared quickly to avoid damaging the supply cables, so a circuit breaker will be required with the ability to interrupt the short-circuit current. The higher the fault current, the higher the cost of the protective equipment, so the supply authority will normally specify the maximum fault current that can be tolerated.

As a transformer designer we then face a dilemma: if we do what seems obvious and aim to reduce the series impedance to improve the steady-state performance, we find that the short-circuit current increases and there is the problem of exceeding the permissible levels dictated by the supply authority. Clearly the so-called ideal transformer is not what we want after all, and we are obliged to settle for a compromise. We deliberately engineer sufficient impedance (principally via the control of leakage reactance) to ensure that under abnormal (short-circuit) conditions the system does not self-destruct!

In Chapters 5 and 6 we mentioned something very similar in relation to the induction motor: we saw that when the induction motor is started direct-on-line it draws a heavy current, and limiting the current requires compromises in the motor design. We will see shortly that this behavior is exactly like that of a short-circuited transformer.

We have developed our ideas in relation to a single-phase transformer, but the equivalent circuit is also applicable on a per-phase basis to a 3-phase transformer operating under balanced conditions.

A2.4. DEVELOPMENT OF THE INDUCTION MOTOR EQUIVALENT CIRCUIT

A2.4.1 Stationary conditions

We saw earlier that on a per-phase basis, the stationary induction motor is very much like a transformer, so to model the induction motor at rest we can use any of the transformer equivalent circuits we have looked at so far.

We can represent the cage induction motor at rest (the so-called locked rotor condition) simply by setting $Z_2' = 0$ in Figure A2.7. Given the applied voltage we can calculate the current and power that will be drawn from the supply, and if we know the effective turns ratio we can also calculate the rotor current and the power being supplied to the rotor. But although our induction motor resembles a transformer, its purpose in life is very different because it is designed to convert electrical power to mechanical power, which of necessity involves movement. Our locked-rotor calculations will therefore be of limited use unless we can calculate the starting torque developed. Far more importantly, we need to be able to represent the electromechanical processes that take place over the whole speed range, so that we can predict the input current, power and developed torque at any speed. The remarkable simple and effective way that this can be achieved is discussed next.

A2.4.2 Modeling the electromechanical energy conversion process

A very important observation in relation to what we are now seeking to do is that we recognized earlier that although the rotor currents were at slip frequency, their effect (i.e. their m.m.f.) was always reflected back at the stator windings at the supply frequency. This suggests that it must be possible to represent what takes place at slip frequency on the rotor by referring the action to the primary (fixed-frequency) side, using a transformer-type model; and it turns out that we can indeed model the entire energy-conversion process in a very simple way. All that is required is to replace the referred rotor resistance (R_2') with a fictitious slip-dependent resistance $((R_2')/s)$ in the short-circuited secondary of our transformer equivalent circuit.

Hence if we build from the exact transformer circuit in Figure A2.6, we obtain the induction motor equivalent circuit shown in Figure A2.8.

At any given slip, the power delivered to this fictitious 'load' resistance represents the power crossing the air-gap from rotor to stator. We can see straight away that because $(R_2')/s$ is inversely proportional to slip, it reduces as the slip increases, thereby causing the power across the air-gap to increase and resulting in more current and power being drawn in from the supply. This behavior is of course in line with what we already know about the induction motor. We can check the extremes by putting $s = 1$ to represent locked rotor, as discussed above, in which case the current becomes very large, as expected; and $s \to 0$ to represent no-load running, in which case the rotor current and power are very small, again as expected.

We will see how to use the equivalent circuit to predict and illuminate motor behavior in the next section, but first there are two points worth making. First, given the complexity of the spatial and temporal interactions in the induction motor it is extraordinary that everything can be properly represented by such a simple

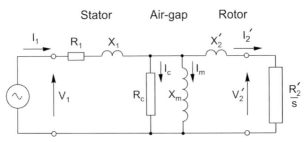

Figure A2.8 'Exact' per-phase equivalent circuit of induction motor. The secondary (rotor) parameters have been referred to the primary (stator) side.

equivalent circuit, and it has always seemed a pity to the authors that something so elegant receives little by way of commendation in the majority of textbooks. Secondly, the following brief discussion is offered for the benefit of readers who are seeking at least some justification for introducing the fictitious resistance $(R_2')/s$, though it has to be admitted that full treatment is beyond our scope.

The key to developing the representation lies in ensuring that the magnitude and phase of the referred rotor current (at supply frequency) in the transformer model is in agreement with the actual current (at slip frequency) in the rotor. We argued in Chapter 5 that the induced e.m.f. in the rotor at slip s would be sE_2 at frequency sf, where E_2 is the e.m.f. induced under locked-rotor ($s = 1$) conditions, when the rotor frequency is the same as the supply frequency, i.e. f. This e.m.f. acts on the series combination of the rotor resistance R_2 and the rotor leakage reactance, which at frequency sf is given by sX_2, where X_2 is the rotor leakage reactance at supply frequency. Hence the magnitude of the rotor current is given by

$$I_2 = \frac{sE_2}{\sqrt{R_2^2 + s^2 X_2^2}}$$ (A2.11)

In the supply-frequency equivalent circuit, e.g. Figure A2.6, the secondary e.m.f. is E_2, rather than sE_2, so in order to obtain the same current in this model as given by equation (A2.11), we require the slip-frequency rotor resistance and reactance to be divided by s, in which case the secondary current would be correctly given by

$$I_2 = \frac{E_2}{\sqrt{\left(\frac{R_2}{s}\right)^2 + X^2}} = \frac{sE_2}{\sqrt{R_2^2 + s^2 X_2^2}}$$ (A2.12)

A2.5. PROPERTIES OF INDUCTION MOTORS

We have started with the exact circuit in Figure A2.8 because the air-gap in the induction motor causes its magnetizing reactance to be lower than for a transformer of similar rating, while its leakage reactance will be higher. We therefore have to be a bit more cautious before we make major simplifications, though we will find later that for many purposes the approximate circuit (with the magnetizing branch on the left) is adequate. In this section we concentrate on what can be learned from a study of the rotor section, which is the same in exact and approximate circuits, so our conclusions from this section are completely general.

Of the power that is fed across the air-gap into the rotor, some is lost as heat in the rotor resistance, and the remainder (hopefully the majority!) is converted to useful mechanical output power. To represent this in the referred circuit we split the ficti-tious resistance $(R_2')/s$ into two parts, R_2' and $R_2'((1 - s)/s)$, as shown in Figure A2.9.

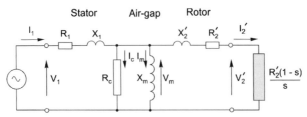

Figure A2.9 Exact equivalent circuit with effective rotor resistance (R_2'/s) split into R_2' and $R_2'(1 - s)/s$. The power dissipated in R_2' represents the rotor copper loss per phase, while the power in the shaded resistance $R_2'(1 - s)/s$ corresponds to the mechanical output power per phase, when the slip is s.

The rotor copper loss is represented by the power in R_2', and the useful mechanical output power is represented by the power in the 'electromechanical' resistance $R_2'((1 - s)/s)$, shown shaded in Figure A2.9. To further emphasize the intended function of the motor – the production of mechanical power – the electromechanical element is shown in Figure A2.9 as the secondary 'load' would be in a transformer. For the motor to be a good electromechanical energy converter, most of the power entering the circuit on the left must appear in the electromechanical load resistance. This is equivalent to saying that for good performance, the output voltage V_2' must be as near as possible equal to the input voltage V_1. And if we ignore the current in the center magnetizing branch, the 'good' condition simply requires that the load resistance is large compared with the other series elements. This desirable condition is met under normal running condition, when the slip is small and hence $R_2'((1 - s)/s)$ is large.

Our next step is to establish some important general formulae, and draw some broad conclusions.

A2.5.1 Power balance

From Figure A2.8, the total power entering the rotor (the 'air-gap' power P_2) is given by

$$P_2 = \left(I_2'^{\,2} \right) \frac{R_2'}{s} \tag{A2.13}$$

And from Figure A2.9, the power lost as heat in the rotor conductors is given by

$$P_{\text{Loss}} = \left(I_2'^{\,2} \right) R_2' = sP_2 \tag{A2.14}$$

while the mechanical power output is given by

$$P_{\text{Mech}} = \left(I_2'^{\,2} \right) R_2' \left(\frac{1 - s}{s} \right) = (1 - s)P_2 \tag{A2.15}$$

The rotor efficiency is therefore given by

$$\frac{P_{\text{Mech}}}{P_2} = (1 - s) \times 100\%$$

These relationships were mentioned in Chapters 5 and 6, and they are of fundamental importance and universal applicability. They show that of the power delivered across the air-gap, a fraction s is inevitably lost as heat, leaving the fraction $(1 - s)$ as useful mechanical output. Hence an induction motor can only operate efficiently at low values of slip.

Throughout this book we have emphasized that all electrical machines can act as motors or generators, so it is appropriate to mention that the induction machine equivalent circuit is valid in the generating mode, i.e. when the slip s is negative. The effective rotor resistance simply becomes negative, which in circuit terms indicates that it is a source of power, and all the other expressions presented here are equally applicable with negative slip.

A2.5.2 Torque

Returning to the motoring mode, we can now obtain the relationship between the power entering the rotor and the torque developed. Mechanical power is torque times speed, and when the slip is s the speed is $(1 - s)\omega_s$, where ω_s is the synchronous speed. Hence from the power equations above we obtain

$$\text{Torque} = \frac{\text{Mechanical power}}{\text{Speed}} = \frac{(1 - s)P_2}{(1 - s)\omega_s} = \frac{P_2}{\omega_s} \qquad (A2.16)$$

Again this is of fundamental importance, showing that the torque developed is proportional to the power entering the rotor.

All of the relationships derived in this section are universally applicable and do not involve any approximations. Further useful deductions can be made when we simplify the equivalent circuit, but first we will look at an example of performance prediction based on the exact circuit.

A2.6. PERFORMANCE PREDICTION – EXAMPLE

The per-phase equivalent circuit parameters (referred to the stator) of a 4-pole, 60 Hz, 440 V, 3-phase, delta-connected induction motor are as follows:

Stator resistance, $R_1 = 0.2\,\Omega$; Stator and rotor leakage reactances,

$X_1 = X_2 = j1.0\,\Omega$; Rotor resistance, $R_2' = 0.3\,\Omega$;

Magnetizing reactance, $X_m = j40\,\Omega$; Iron loss resistance, $R_c = 250\,\Omega$

The mechanical frictional losses at normal speed amount to 2.5 kW.

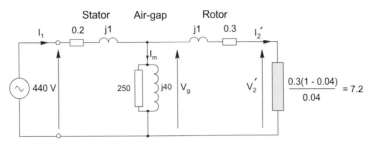

Figure A2.10 Exact equivalent circuit for exemplar induction motor, under full-load conditions (i.e. slip $s = 0.04$). All of the impedance values are in ohms.

We will calculate the input line current, the output power and the efficiency at the full-load speed of 1728 rev/min.

The per-phase equivalent circuit, with all values in ohms, is shown in Figure A2.10. The synchronous speed is 1800 rev/min and the slip speed is thus $1800 - 1728 = 72$ rev/min, giving a slip of $72/1800 = 0.04$. The electromechanical resistance of $7.2\,\Omega$ is calculated from the expression $R_2'((1 - s)/s)$, with $R_2' = 0.3\,\Omega$. Given that the power in this resistor represents the mechanical output power, we will see that, as expected, it dominates our calculations.

Although the various calculations in this example are straightforward, they are not for the faint-hearted as they inevitably involve frequent manipulation of complex numbers, so most of the routine calculations will be omitted and we will concentrate on the results and their interpretation.

A2.6.1 Line current

To find the line current we must find the effective impedance looking in from the supply, so we first find the impedances of the rotor branch and the magnetizing (air-gap) branches. Expressed in 'real and imaginary' and also 'modulus and argument' form these are:

$$\text{Rotor branch impedance} = 7.5 + j1\,\Omega = 7.57\angle 7.6°\,\Omega$$

$$\text{Magnetizing branch impedance} = 6.25 + j39.0\,\Omega = 39.5\angle 80.9°\,\Omega$$

The parallel combination of these two branches has an effective impedance of $6.74 + j2.12\,\Omega$ or $7.07\angle 17.5\,\Omega$. At roughly $7\,\Omega$, and predominantly resistive, this is in line with what we would expect: the magnetizing branch impedance is so much higher than for the rotor branch that the latter remains dominant. We must now add the stator impedance of $0.2 + j1\,\Omega$ to yield the total motor impedance per phase as $6.94 + j3.12\,\Omega$, or $7.61\angle 24.2°\,\Omega$.

The motor is mesh connected so the phase voltage is equal to the line voltage, i.e. 440 V. We will use the phase voltage as the reference for angles, so the phase current is given by

$$I_1 = \frac{440 \angle 0°}{7.61 \angle 24.2°} = 57.82 \angle -24.2 \text{ A}$$

The line current is $\sqrt{3}$ times the phase current, i.e. 100 A.

A2.6.2 Output power

The output power is the power converted to mechanical form, i.e. the power in the fictitious load resistance of 7.2 Ω, so we must first find the current through it.

The input current I_1 divides between the rotor (I_2') and magnetizing branches (I_m); and straightforward calculation yields $I_2' = 53.99 \angle -14.38° A$, and $I_m = 10.35 \angle -87.7° A$.

Again we see that these currents line up with our expectations. Most of the input current flows in the rotor branch, which is predominantly resistive and the phase lag is small; conversely relatively little current flows in the magnetizing branch and its large phase lag is to be expected from its predominantly inductive nature.

The power in the load resistor is given by $P_{Mech} = (53.99)^2 \times 7.2 = 20{,}987 \text{ W/phase}$; there are three phases so the total mechanical power developed is 62.96 kW.

The generated torque follows by dividing the power by the speed in rad/s, yielding torque as 348 Nm, or 116 Nm/phase.

We are told that the windage and friction power is 2.5 kW, so the useful output power is $62.96 - 2.5 = 60.46$ kW.

A2.6.3 Efficiency

We can proceed in two alternative ways, so we will do both to act as a check. First, we will find the total loss per phase by summing the powers in the winding resistances and the iron-loss resistor in Figure A2.10. The losses in the two winding resistances are easy to find because we already know the currents through both of them: the losses are $I_1^2 R_1 = (57.82)^2 \times 0.2 = 668 \text{ W/phase}$ and $I_2' R_2' = (53.99)^2 \times 0.3 = 874.5 \text{ W/phase}$, for the stator and rotor, respectively.

The iron loss is a bit more tricky, as first we need to find the voltage across the magnetizing branch (sometimes called the gap voltage V_g, as in Figure A2.10). We can get this from the product of the rotor current and the rotor impedance, which yields

$$V_g = 53.99 \angle -14.38° \times 7.57 \angle 7.6° = 408.5 \angle -6.8° V$$

The power in the 250 Ω iron loss resistor is now given by $(408.5)^2/250 = 667$ W/phase.

The total loss (including the windage and friction) is then given by

$$\text{Total loss} = 3(668 + 874.5 + 667) + 2500 = 9.13 \text{ kW}$$

The efficiency is given by

$$\text{Efficiency} = \frac{P_{\text{Mech}}}{P_{\text{in}}} = \frac{P_{\text{Mech}}}{P_{\text{Mech}} + \text{losses}} = \frac{60.46}{60.46 + 9.13} = \frac{60.46}{69.59}$$
$$= 0.869 \text{ or } 86.9\%$$

Alternatively we could have found the input power directly from the known input voltage and current, using $P_{\text{in}} = 3(V_1 I_1 \cos\phi_1) = 3 \times 440 \times 57.82 \times \cos 24.3° = 69.59\,\text{kW}$, which agrees with the result above.

A2.6.4 Phasor diagram

It is instructive to finish by looking at the phasor diagrams showing the principal voltages and currents under full-load conditions, as shown in Figure A2.11, which is drawn to scale.

From the voltage phasor diagram, we note that in this motor the volt-drop due to the stator leakage reactance and resistance is significant, in that the input voltage of 440 V is reduced to 408.5 V, i.e. a reduction of just over 7%.

The voltage across the magnetizing branch determines the magnetizing current and hence the air-gap flux density, so in this motor the air-gap flux density will fall by about 7% between no-load and full-load. (At no-load the slip is very small and the impedance of the rotor branch is high, so that there is negligible rotor current

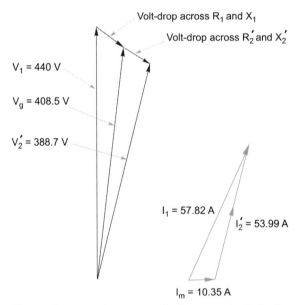

Figure A2.11 Phasor diagrams, drawn to scale, for exemplar induction motor under full-load conditions.

and very little volt-drop due to the stator impedance, leaving almost the full 440 V across the magnetizing branch.)

A similar volt-drop occurs across the rotor impedance, leaving just under 390 V applied at the effective load. Note, however, that because the stator and rotor impedances are predominantly reactive, the appreciable volt-drops are not responsible for corresponding power losses.

These significant volt-drops reflect the fact that this motor is suitable for direct-on-line starting from a comparatively weak supply, where it is important to limit the starting current. The designer has therefore deliberately made the leakage reactances higher than they would otherwise be in order to limit the current when the motor is switched on.

To estimate the starting current we ignore the magnetizing branch and put the slip $s = 1$, in which case the rotor branch impedance becomes $0.3 + j1$, so that the total impedance is $0.5 + j2$ or $2.06\,\Omega$. When the full voltage is applied the phase current will be $440/2.06 = 214$ A. The full-load current is 57.8 A, so the starting current is 3.7 times the full-load current. This relatively modest ratio is required where the supply system is weak: a stiff supply might be happy with a ratio of 5 or 6.

We have already seen that the torque is directly proportional to the power into the rotor, so it should be clear that when we limit the starting current to ease the burden on the supply, the starting torque is unavoidably reduced. This is discussed further in the following section.

The current phasor diagram underlines the fact that when the motor is at full-load, the magnetizing current is only a small fraction of the total. The total current lags the supply voltage by $24.3°$, so the full-load power-factor is $\cos 24.3° = 0.91$, i.e. very satisfactory.

A2.7. APPROXIMATE EQUIVALENT CIRCUITS

This section is devoted to what can be learned from the equivalent circuit in simplified form, beginning with the circuit shown in Figure A2.12, in which the magnetizing branch has been moved to the left-hand side. This makes calculations very much easier because the current and power in the magnetizing branch are

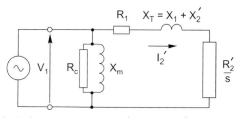

Figure A2.12 Approximate equivalent circuit for induction motor.

independent of the load branch. The approximations involved in doing this are greater than in the case of a transformer because for a motor the ratio of magnetizing reactance to leakage reactance is lower, but algebraic analysis is much simpler and the results can be illuminating.

A cursory examination of electrical machines textbooks reveals a wide variety in the approaches taken to squeeze value from the study of the approximate equivalent circuit, but there are often so many formulae that the reader becomes over-whelmed. So here we will focus on two simple messages. First, we will develop an expression that neatly encapsulates the fundamental behavior of the induction motor, and illustrates the trade-offs involved in design; and secondly, we will examine how the relative values of rotor resistance and reactance influence the shape of the torque–speed curve.

A2.7.1 Starting and full-load relationships

Straightforward circuit analysis of the circuit in Figure A2.12, together with equation (A2.16) for the torque, yields the following expressions for the load component of current and for the torque per phase:

$$I_2' = \frac{V_1}{\sqrt{\left(R_1 + \dfrac{R_2'}{s}\right)^2 + X_T^2}} \quad \text{and} \quad T = \frac{1}{\omega_s}\left\{\frac{V_1^2}{\left(R_1 + \dfrac{R_2'}{s}\right)^2 + X_T^2}\right\}\frac{R_2'}{s} \quad (A2.17)$$

The second expression (with the square of voltage in the numerator) reminds us of the sensitivity of torque to voltage variation, in that a 5% reduction in voltage gives a little over 10% reduction in torque.

If we substitute $s = 1$ and $s = s_{fl}$ (full-load slip) in these equations we obtain expressions for the starting current, the full-load current, the starting torque and the full-load torque. Each of these quantities is important in its own right, and they all depend on the rotor resistance and reactance. But by combining the four expressions we obtain the very far-reaching result given by equation (A2.18):

$$\frac{T_{st}}{T_{fl}} = \left(\frac{I_{st}}{I_{fl}}\right)^2 s_{fl} \quad (A2.18)$$

The remarkable thing about equation (A2.18) is that it does not contain the rotor or stator resistances, or the leakage reactances. This underlines the fact that this result, like those given in equations (A2.15) and (A2.16), reflects fundamental properties that are applicable to any induction motor.

The left-hand side of equation (A2.18) is the ratio of starting torque to full-load torque, an important parameter for any (direct-on-line) application as it is clearly no good having a motor that can drive a load once up to speed, but either has insufficient torque to start the load from rest, or (perhaps less likely) more starting torque than is necessary leading to a too-rapid acceleration.

On the right-hand side of equation (A2.18), the importance of the ratio of starting to full-load current has already been emphasized: in general it is desirable to minimize this ratio in order to prevent voltage regulation at the supply terminals during a direct-on-line start. The other term is the full-load slip, which, as we have already seen, should always be as low as possible in order to maximize the efficiency of the motor.

The inescapable design trade-off faced by the designer of a constant-frequency machine is revealed by equation (A2.18). We usually want to keep the full-load slip as small as possible (to maximize efficiency), and for direct starting the smaller the starting current, the better. But if both terms on the right side are small, we will be left with a low starting torque, which is generally not desirable, and we must therefore seek a compromise between the starting and full-load performances, as was explained in Chapter 6.

(It has to be acknowledged that the current ratio in equation (A2.18) relates the load (rotor branch) currents, the magnetizing current having been ignored, so there is inevitably a degree of approximation. But for the majority of machines (i.e. 2-pole and 4-pole) equation (A2.18) holds good, and in view of its simplicity and value it is surprising that it does not figure in many 'machines' textbooks.)

A2.7.2 Torque vs slip and pull-out torque

The aim here is to quantify the dependence of the maximum or pull-out torque on the rotor parameters, for which we make use of the simplest possible (but still very useful) model shown in Figure A2.13. The magnetizing branch and the stator resistance are both ignored, so that there is only one current, the referred rotor current I_2' being the same as the stator current I_1.

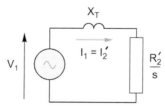

Figure A2.13 Equivalent circuit for induction motor with magnetizing branch neglected.

To obtain the torque/slip relationship we will make use of equation (A2.16), which shows that the torque is directly proportional to the total rotor power. The total rotor power per phase is given by

$$P_2 = P_{rotor} = \left(I_2'\right)^2 \frac{R_2'}{s} = V_1^2 \left\{ \frac{sR_2'}{R_2'^2 + s^2 X_T^2} \right\} \qquad (A2.19)$$

The bracketed expression on the right-hand side of equation (A2.19) indicates how the torque varies with slip.

At low values of slip (i.e. in the normal range of continuous operation) where sX_T is much less than the rotor resistance R_2', the torque becomes proportional to the slip and inversely proportional to the rotor resistance. This explains why the torque–speed curves we have seen in Chapters 5 and 6 are linear in the normal operating region, and why the curves become steeper as the rotor resistance is reduced.

At the other extreme, when the slip is 1 (i.e. at standstill), we usually find that the reactance is larger than the resistance, in which case the right-hand bracketed term in equation (A2.19) reduces to $(R_2')/(X_T^2)$. The starting torque is then proportional to the rotor resistance, a result also discussed in Chapters 5 and 6.

By differentiating this bracketed expression in equation (A2.19) with respect to the slip, and equating the result to zero, we can find the slip at which the maximum torque occurs. The slip for maximum torque turns out to be given by

$$s = \frac{R_2'}{X_T} \tag{A2.20}$$

(A circuit theorist could have written down this expression by inspection of Figure A2.13, provided that he knew that the condition for maximum torque was that the power in the rotor was at its peak.)

Substituting the slip for maximum torque we obtain an expression for the maximum torque per phase as:

$$T_{max} = \frac{V_1^2}{\omega_s} \frac{1}{2X_T} \tag{A2.21}$$

From these two equations we see that the slip at which maximum torque is developed is directly proportional to the rotor resistance, but that the peak torque itself is independent of the rotor resistance, and depends inversely on the leakage reactance.

A2.8. MEASUREMENT OF INDUCTION MOTOR PARAMETERS

The tests used to obtain the equivalent circuit parameters of a cage induction motor are essentially the same as those described for the transformer. To simulate the 'open-circuit' test we would need to run the motor with a slip of zero, so that the secondary (rotor) referred resistance $((R_2')/s)$ became infinite and the rotor current was absolutely zero, but because the motor torque is zero at synchronous speed, the only way we could achieve zero slip would be to drive the rotor from another source, e.g. a synchronous motor. This is hardly ever necessary because when the shaft is unloaded the no-load torque is usually very small and the slip is sufficiently close to zero that the difference does not matter. From readings of input voltage, current and power (per phase), the parameters of the magnetizing branch are derived as described earlier. The no-load friction and windage losses will be combined with the iron losses and represented in the parallel resistor – a satisfactory practice as long as the voltage and/or frequency remain(s) constant.

The short-circuit test of the transformer becomes the locked-rotor test for the induction motor. By clamping the rotor so that the speed is zero and the slip is one, the equivalent circuit becomes the same as the short-circuited transformer. The total resistance and leakage reactance parameters are derived from voltage, current and power measurements as already described. With a cage motor the rotor resistance clearly cannot be measured directly, but the stator resistance can be obtained from a d.c. test and hence the referred rotor resistance can be obtained.

A2.9. EQUIVALENT CIRCUIT UNDER VARIABLE-FREQUENCY CONDITIONS

More and more induction motors now operate from variable-frequency inverters, the frequency (and voltage) being varied not only to control steady-state speed, but also to profile torque during acceleration and deceleration. Two questions that we might therefore ask are (a) does the equivalent circuit remain valid for other than 50 or 60 Hz operation; and (b) if so, do the approximations that have been developed remain useful?

The answer to question (a) is that the form of the equivalent circuit is independent of the supply frequency. This is to be expected because we are simply representing in circuit form the linking of the electric and magnetic circuits that together constitute the motor, and these physical properties do not depend on the excitation frequency.

The answer to the question of the validity of approximations is less straight-forward, but broadly speaking all that has so far been said is applicable except at low frequencies, say below about 10 Hz for 50 or 60 Hz motors. As the frequency approaches zero, the volt-drop due to stator resistance becomes important, as the example below demonstrates.

Consider the example studied previously, and suppose that we reduce the supply frequency from 60 to 6 Hz. All the equivalent circuit reactances reduce by a factor of 10, as shown in Figure A2.14, which assumes operation at the same (rated) torque for both cases. We can see immediately that whereas at 60 Hz the stator and rotor leakage reactances are large compared with their respective resistances, the reverse is true at 6 Hz, so we might immediately expect significant differences in the circuit behavior. (It should be pointed out that the calculations

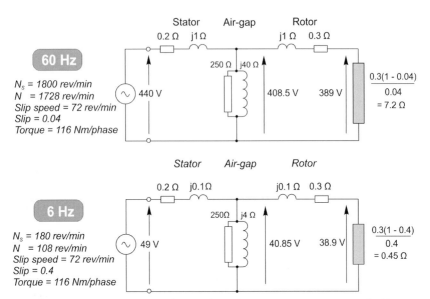

Figure A2.14 Comparison of equivalent circuit parameters at 60 and 6 Hz.

required to derive the voltages at 6 Hz are not trivial, as the reader may like to verify.)

We saw in Chapter 7 that to get the best out of the motor we need the flux to be constant and the slip speed (in rev/min) to be the same when the motor is required to develop full-load torque.

To keep the air-gap flux constant we need to ensure that the magnetizing current remains the same when we reduce the frequency. We know that at 6 Hz the magnetizing reactance is one-tenth of its value at 60 Hz, so the voltage V_m must be reduced by a factor of 10, from 408.5 to 40.85 V, as shown in Figure A2.14. (We previously established that the condition for constant flux in an ideal transformer was that the V/f ratio must remain constant: here we see that the so-called air-gap voltage (V_m) is the one that matters.)

Previously we considered operation at 60 Hz, for which the synchronous speed was 1800 rev/min and the rated speed 1728 rev/min. This gave a slip speed of 72 rev/min and a slip of 0.04. At 6 Hz the synchronous speed is 180 rev/min, so for the same slip speed of 72 rev/min the new slip is 0.4. This explains why the 'load' resistance of 7.2 Ω at 60 Hz becomes only 0.45 Ω at 6 Hz, while the total referred rotor resistance reduces from 7.5 Ω at 60 Hz to 0.75 Ω at 6 Hz.

The 60 Hz voltages shown in the upper half of Figure A2.14 were calculated previously. To obtain the same torque at 6 Hz as at 60 Hz, we can see from equation (A2.16) that the rotor power must reduce in proportion to the frequency, i.e. by a factor of 10. The rotor resistance has reduced by a factor of 10 so we would expect the rotor voltage to also reduce by a factor of 10. This is confirmed in Figure A2.14, where the rotor voltage at 6 Hz is 38.9 V. (This means the rotor current (51.9 A) at 6 Hz is the same as it was at 60 Hz, which is what we would expect given that for the same torque we would expect the same active current). Working backwards (from V_m) we can derive the input voltage required, i.e. 49 V.

Had we used the approximate circuit, in which the magnetizing branch is moved to the left-hand side, we would have assumed that to keep the amplitude of the flux wave constant, we would have to change the voltage in proportion to the frequency, in which case we would have decided that the input voltage at 6 Hz should be 44 V, not the 49 V really needed. And if we had supplied 44 V rather than 49 V, the torque at the target speed would be almost 20% below our expectation, which is clearly significant and underlines the danger of making unjustified assumptions when operating at low frequencies.

In this example the volt-drop of 10 V across the stator resistance is much more significant at 6 Hz than at 60 Hz, for two reasons. First, at 6 Hz the useful (load) voltage is 38.9 V whereas at 60 Hz the load voltage is 389 V: so a fixed drop of 10 V that might be considered negligible compared with 389 V is certainly not negligible in comparison with 38.9 V. And secondly, when the load is predominantly resistive, as here, the reduction in the magnitude of the supply voltage due to a given series impedance is much greater if the impedance is resistive than if it is reactive. This matter was discussed in section 1 of Chapter 6.

FURTHER READING

Acarnley PP. Stepping Motors: A Guide to Modern Theory and Practice. 4th ed. London: IET; 2002.
A comprehensive treatment at a level which will suit both students and users.

Beaty HW, Kirtley JL. Electric Motor Handbook. New York: McGraw-Hill; 1998.
Comprehensive analytical treatment including chapters on motor noise and servo controls.

Bose BK. Power Electronics and Variable Frequency Drives: Technology and Applications. Wiley-Blackwell; 1996.
Authorative book covering many key areas of motor drive analysis, performance and design. Contains many references to key books and technical papers on the subject.

Chaisson J. Modeling and High-Performance Control of Electrical Machines. John Wiley & Sons, IEEE Press series on power engineering; 2005.
A good book for anyone wanting to model an electrical machine as part of a simulation package or in a control system. Good treatment of a number of practical issues including noise on feedback signals.

Drury W. The Control Techniques Drives and Controls Handbook. 2nd ed. London: IET; 2009.
A practical guide to the technology underlying drives and motors and consideration of many of the design issues faced with their use in many applications and environments.

Hendershot JR, Miller TJE. Design of Brushless Permanent-Magnet Motors. LLC: Motor Design Books; 2010.
A comprehensive guide to the design and performance of brushless permanent magnet motors.

Hindmarsh J, Renfrew A. Electrical Machines and Drives: Worked Examples. 3rd ed. Butterworth-Heinemann; 1996.
Worked examples, which put theory in context. Good discussion of engineering implications. Numerous problems are also provided, with answers supplied!

Jordan HE. Energy-Efficient Motors and their Applications. 2nd ed. New York: Springer; 1994.
Clearly written specialist text.

Kenjo T. Electric Motors and their Controls. New York: Oxford Science Publications; 1991.
A general (basic) introduction with beautiful illustrations, and covering many small and special purpose motor types.

Mohan N, Undeland TM, Robbins WP. Power Electronics: Converters, Applications, and Design. 3rd ed. John Wiley & Sons; 2002.
Cohesive presentation of power electronics fundamentals for applications and design. Lots of worked examples.

Neale M, Needham P, Horrell R. Couplings and Shaft Alignment. Professional Engineering Publishing Limited; 1991.
A practical guide to coupling selection and problems of shaft alignment.

Vas P. Electrical Machines and Drives. Oxford University Press; 1992.
A definitive, highly analytical, book on space vector theory of electrical machines.

Vas P. Sensorless Vector and Direct Torque Control. Oxford University Press; 1998.
Comprehensive analytical treatment of field orientation and direct torque control of a.c. machines.

Wu B. High Power Converters and AC Drives. John Wiley & Sons; 2006.
Good practical design book on high-power a.c. drive topologies.

INDEX

Note: Page numbers with 'f' denote figures.

Color Plates

Plate 2.1 Emerson – Control Techniques Unidrive M700 industrial a.c. drives. These three are from the lower end of the range of seven that extends to 315 kW. The right hand unit is 4 kW/5 HP (380 mm high), while the left hand unit is 45 kW/60 HP. *(Courtesy of Emerson – Control Techniques)*

Plate 3.1 Cutaway view of 4-pole d.c. motor. The skewed armature windings on the rotor are connected at the right-hand-end via risers to the commutator segments. There are four sets of brushes, and each brush arm holds four brush-boxes. The sectioned coils surrounding two of the four poles are visible close to the left-hand armature end-windings. *(Courtesy of ABB)*

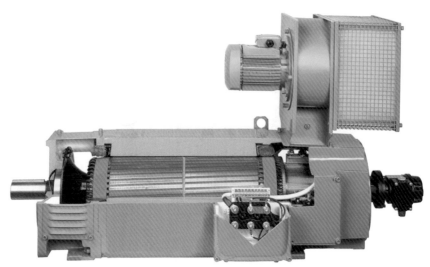

Plate 4.1 High performance force-ventilated d.c. motor. The motor is of all-laminated construction and designed for use with a thyristor converter. The small blower motor is an induction machine that runs continuously, thereby allowing the main motor to maintain full torque at low speed without overheating. *(Courtesy of Emerson – Leroy Somer)*

Plate 5.1 Stator of three-phase induction motor. The semi-closed slots of the stator core obscure the active sides of the stator coils, but the ends of the coils are just visible beneath the binding tape. *(Courtesy of Brook Crompton)*

Plate 5.2 Cage rotor for induction motor. The rotor conductor bars and end rings are cast in aluminium, and the blades attached to the end rings serve as a fan for circulating internal air. An external fan will be mounted on the non-drive end to cool the finned stator casing (as shown in Figure 8.4). *(Courtesy of Brook Crompton)*

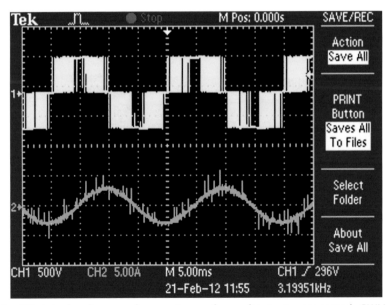

Plate 8.2 Actual voltage and current waveforms for a star-connected, PWM-fed induction motor. Upper trace – voltage across U and V terminals; lower trace – U phase motor current.

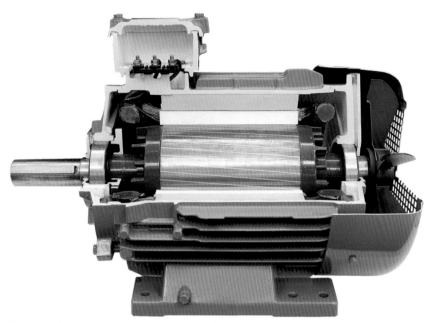

Plate 8.4 Typical shaft-mounted external cooling fan on an a.c. induction motor. *(Courtesy of Emerson – Leroy Somer)*

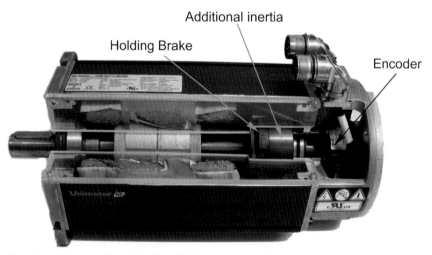

Additional inertia

Holding Brake

Encoder

Plate 9.1 Unimotor fm – 18.4 Nm, 2000 rev/min permanent magnet servo motor. The holding brake is only used when the windings are not energized. *(Courtesy of Emerson – Control Techniques Dynamics)*

Plate 9.12 Permanent magnet servo motors. *(Courtesy of Emerson – Control Techniques Dynamics.)*

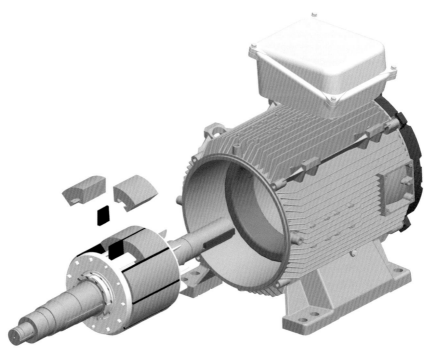

Plate 9.14 Permanent magnet industrial motor in standard IEC frame. *(Courtesy of Emerson Leroy Somer)*

Plate 10.1 Hybrid 1.8° stepping motors, of sizes 34 (3.4 inch diameter), 23, 17 and 11. *(Courtesy of Astrosyn)*

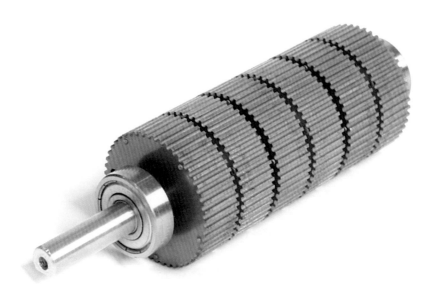

Plate 10.2 Rotor of size 34 (3.4 inch or 8 cm diameter) 3-stack hybrid 1.8° stepping motor. The dimensions of the rotor end-caps and the associated axially-magnetized permanent magnet are optimized for the single-stack version. Extra torque is obtained by adding a second or third stack, the stator simply being stretched to accommodate the longer rotor. *(Courtesy of Astrosyn)*

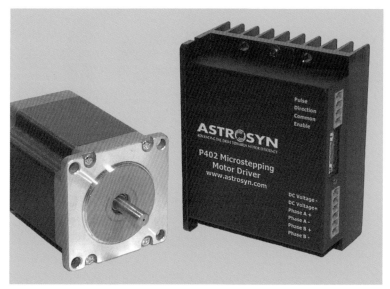

Plate 10.3 This bipolar chopper mini-stepping drive with integral heatsink provides a range of step divisions up to 25,600 steps/rev and allows drive current to be set from 0.25 A to 2.0 A with supply voltage from 14 V to 40 V. *(Courtesy of Astrosyn)*

Plate 10.4 These injection molded steppers offer improved efficiency and reduced noise, and find application in security cameras, stage lighting, medical equipment, semiconductor manufacture and office automation products such as scanners and printers. *(Courtesy of Astrosyn)*

Plate 10.5 The die-cast aluminum construction of these self-contained mini-stepping bipolar drivers provides efficient heat dissipation, and they are designed for retrofitting to existing size 17 motors. *(Courtesy of Astrosyn)*

Plate 10.6 Switched Reluctance Motor. The toothed rotor does not require windings or magnets, and is therefore exceptionally robust. *(Courtesy of Nidec S R Drives Ltd.)*